Palgrave Studies in Environmental Sustainability

Series Editor

Robert Brinkmann, Hofstra University, Hempstead, NY, USA

The series will take advantage of the growing interest in a number of environmental issues in sustainability, particularly in those that focus on interdisciplinary approaches to solving real-world problems. Unfortunately, many of the themes of sustainability address short- and long-term survivability of our planet and there is a pressing need for access to information to those who seek to solve sustainability problems. Thus, the scope of the book series seeks to be comprehensive within an environmental sustainability framework. The topics covered in the series range from climate change to public land management. While no series can ever be truly comprehensive, the series will provide one of the most definitive surveys of knowledge in the area of environmental sustainability.

Claudia Baldwin · Séverine van Bommel
Editors

# Rural Development for Sustainable Social-ecological Systems

## Putting Communities First

palgrave
macmillan

*Editors*
Claudia Baldwin
School of Law and Society
University of the Sunshine Coast
Peregian Beach, QLD, Australia

Séverine van Bommel
School of Agriculture
and Food Science
University of Queensland
Brisbane, QLD, Australia

Palgrave Studies in Environmental Sustainability
ISBN 978-3-031-34224-0    ISBN 978-3-031-34225-7 (eBook)
https://doi.org/10.1007/978-3-031-34225-7



This Palgrave Macmillan imprint is published by the registered company Springer Nature Switzerland AG
The registered company address is: Gewerbestrasse 11, 6330 Cham, Switzerland

# Foreword

I am honoured to be invited by the editors, Claudia Baldwin and Severine von Bommel, to write this Foreword. Published in honour of Professor Helen Ross on the occasion of her retirement, this book is a *Liber Amicorum*, a Latin phrase meaning "book of friends". It may also be called a *Festschrift*, a book honouring a respected person, usually in the form of an edited volume (as this one is) containing contributions from the honouree's colleagues and former students.

Albert Einstein once suggested that "We cannot solve our problems with the same thinking we used when we created them". We do have a major problem with human-environment relationships, and it seems we need to change our thinking! In fact, we have been changing our thinking, although our practices have not quite caught up with our thinking. As well, there are some major barriers, political and economic, before we could transform to more sustainable practices. The big question is, what might be the elements of a new approach to community-environment interactions?

*Putting Communities First* is a remarkable book that captures some of the revolutionary changes in several areas related to community-environment interactions. The Introduction chapter written by the editors traces in some detail the context of major historical shifts in theory and practice that have been occurring in the overlapping areas of rural community development, community-based natural resource management, and social aspects of resilience. As an educator, researcher, and writer, I have been fascinated by the idea of major conceptual changes, or paradigm shifts, especially in the area of community-environment interactions, in which Prof. Helen Ross and I have a shared interest.

Here I use "paradigm shift" in the original sense of Thomas Kuhn's *The Structure of Scientific Revolutions* (1962), as a fundamental change in the basic concepts and experimental practices of a scientific discipline—a fundamental change that happens when the usual way of thinking about or doing something is replaced by a new and different way. As a philosopher and physicist, Kuhn was thinking about natural sciences, but the concept surely applies to social sciences as well, including transdisciplinary social sciences in which Prof. Helen Ross has excelled in so many ways.

I see the area of community-environment interactions to be in the midst of three paradigm changes captured by Putting Communities First. They are: (1) the paradigm change of participation in rural development; (2) the paradigm change in environmental governance; and (3) the adoption of a systems view. These changes are not specific to community-environment interactions but to environmental management and community development in general. They provide the context and benchmarks in many areas of applied social science that deal with people-and-environment issues. This takes us to the concept of social-ecological systems that is in the subtitle of the book—the integrated concept of coupled, interdependent, and co-evolutionary relationships between humans and their lands and resources.

**Paradigm change of participation in rural development.** The emergence of participatory development and resource management is a remarkable story. Back in the 1950s and 1960s, central governments were generally seen as the main vehicle to implement the development agenda and to manage the environment. Both resource management and development were considered to be technical matters, and central governments were the repository of the technical expertise to carry out the various tasks required. By the 1980s, however, there was a general disillusionment in the ability of central governments to carry out these tasks. They often seemed to be too remote from the people they were supposed to be serving and not responsible to them. Scholars and practitioners were beginning to question the wisdom of policies that assumed central governments were solely in charge.

In both development and resource management areas, there was increasing interest in community-based approaches, and the ability of local institutions to manage local resources and to help set development goals. The subtitle of Robert Chambers' 1983 book, *Putting the Last First*, is probably part of the inspiration for the title of the present book. Chambers' book was also a milestone in the paradigm change for participation in rural and agricultural development, stakeholder participation, community engagement, and decentralization, the idea being to change the locus of decisions from the central government down to the local government.

The logic of this new thinking was compelling: bringing governance closer to those governed, and enabling people whose livelihood and well-being would be affected by policy decisions to have a say in those decisions. Widespread recognition of the notion of user participation and problem-solving at the lowest feasible level of organization resulted even in the incorporation of this principle (the *subsidiarity principle*) into the Maastricht Treaty of 1992 establishing the European Community. (However, the implementation of this principle in the EC is another story.) Thus, by the 1990s, governance focus in development had shifted to the local level. Instead of top-down planning and management, "grassroots" and bottom-up planning became entrenched in various areas of environment, resource management, planning, and development. Some major shifts in global geopolitics were also part of this change: the fiscal crises of the 1980s and the collapse of socialist economies from 1989 onwards.

Chapters 2 to 9 (Part 1) of the book deal with the remarkable radiation and flowering of these ideas in Australia (rural community development, rural economy, extension, women's economic empowerment, stakeholder participation, and Landcare), as well as rural community involvement in Chinese agriculture and community development for international aid.

**Paradigm change in environmental governance.** Participation revolution in the two areas, rural development and environmental management, has common roots and shows parallel evolution. However, there are additional lessons from a focus on resource management and environmental governance. The book wisely makes a separate section for the latter area, with focus on local institutions, partnerships, and co-management. Here, the role of institutions becomes important. (The book could have used a couple of more examples in this area.) Elinor Ostrom won the 2009 Nobel Prize in Economic Sciences with her brilliant and readable 1990 book, *Governing the Commons*, which is about the involvement of local institutions for collective action in management. Ostrom argued not so much for the devolution of governance to local institutions, but rather the recognition of local institutions that already govern the commons (shared resources).

Community-based management is important of course not only for commons but also in the wider area of partnerships involving local resource users and governments in a variety of settings, for example, conservation and tourism. Co-management, or the sharing of management power and responsibility, has been a major challenge for governments, a revolution and perhaps a paradigm shift in its own right. Imagine, governments not only sharing decision-making power but also the responsibility of carrying out the provisions of management!

To bring this change into sharper focus, consider Indigenous co-management in conservation. The early colonists in Australia thought Indigenous peoples were backward and irresponsible, even setting fire to the bush! (Indigenous fire management in Australia is now understood to be an ecologically sophisticated system.) So transforming thinking from the early colonizer views of Indigenous-environment relationship, to present Indigenous community-based conservation is nothing short of a revolutionary change! And a successful change at that. Indigenous lands constitute a major part of Australia's protected areas network, and the Indigenous community-based conservation system is arguably the best in the world.

Chapters 10 to 15 (Part 2) of the book deal with a selection of topics regarding this paradigm change and its implications: waterways partnerships, collaborative models to engage Indigenous peoples in conservation (both Australia), followed by three chapters on Southeast Asia: women's empowerment in community-based tourism, coastal resource co-management, and coastal community perceptions of plastics pollution.

**Adoption of a systems view.** This shift has been long in coming in ecology and allied fields. Most ecologists, like most natural scientists, still pursue reductionism—it is believed that this is how science advances. However, those in environmental management are beginning to understand the importance of a holistic approach. The idea of holism is technically known as complex adaptive systems, wherein adaptive refers to positive and negative feedbacks in the system, and the system itself is thought to be a set of interacting (or interrelated) components that form a unified whole. In contrast to natural sciences, many social science disciplines such as anthropology, human ecology, and planning naturally take a holistic view. Most social scientists probably do embrace holism, in the sense of looking at the larger picture of connections, but prefer not to use the mechanistic-sounding language of systems theory and complexity.

I would argue, however, that an explicit systems view is necessary when it comes to bringing humans back into the environment. Are there any ecosystems in the world with no humans or human influences? Many natural scientists and social scientists have embraced the complexity of dealing with humans and their biophysical environment together. This calls for treating people as an integral part of the environment, the social-ecological systems approach. This approach, under different names, developed in the 1930s in anthropology and human ecology. But the social-ecological systems terminology and concepts developed in the late 1990s as an explicit recognition of the integral role of humans in all ecosystems (Berkes and Folke, *Linking Social and Ecological Systems*, 1998). It developed as a part of resilience thinking. Up until then, resilience was an ecological idea (with counterparts in psychology and engineering) applied to ecosystems in which human interaction was implicit but not explicit. The social-ecological systems approach made the argument that the social and ecological parts of the system were equal and intertwined, and that they could not be teased apart.

Resilience is only a small part of complex adaptive systems thinking. But it is a key application of the systems view to human-environment relationships. Prof. Helen Ross' major contribution to resilience has been in the area of community resilience using the social-ecological systems approach, and the chapters in Part 3 reflect this, at least partially. Chapters 16 to 22 deal with a selection of topics regarding the adoption of a systems view and its implications: integrated approaches to community resilience, attributes of social resilience, resilience in the mining sector, mental models and values, agri-food supply chains in China, and boundary organizations in governance.

In *The Structure of Scientific Revolutions*, Kuhn argued that science does not progress by the gradual accumulation of facts and theories, but rather develops episodically in lurches. Gradually developing knowledge and understanding in a given field proceeds until cracks appear in the then-current theory. The discovery of anomalies leads to the overthrow of the old theory and to a paradigm shift, a scientific revolution. The new paradigm then asks novel questions, leading to new directions ad insights. This book illustrates how the students and colleagues of Prof. Helen Ross have followed her insights, built on them, and contributed to paradigm change in community-environment relationships.

Winnipeg, Canada
May 2022

Fikret Berkes
Natural Resources Institute
University of Manitoba Winnipeg
Manitoba, Canada
Fikret.Berkes@umanitoba.ca

**Fikret Berkes** Ph.D. (McGill), FRSC, is Distinguished Professor Emeritus, Natural Resources Institute, University of Manitoba, Winnipeg, MB R3T 2N2 Canada. He is former Tier I Canada Research Chair in Community-based Resource Management. His work combines social and ecological perspectives for the study of relations between societies and their environments. He is interested in community-based management, commons, traditional ecological knowledge, and social-ecological resilience. His honours include the IUCN International Union for Conservation of Nature CEESP Inaugural Award for Meritorious Research; IASC Elinor Ostrom Award for Senior Scholar; and the ESA Ecological Society of America Sustainability Science Award for the book *Sacred Ecology*.

# Acknowledging and Welcoming Emeritus Professor Helen Ross Contributions and Care for Country

I am a *trawlwulwuy* woman from *tebrakunna* country in north-east Tasmania, Australia, and our island home is host to extraordinary First Peoples and other communities that comprise our shared societies. When the ancestors of the night sky country created our first person, the *palawa* being that is both human and kangaroo and which landed into *trouwunnan*/Tasmania, our peoples have carried forward for over 40,000 years the rights and actions to care for our precious and sacred country. Even though the rupture of genocide, when the British strangers in their boats came to our *trouwunnan* shores in 1803 and smashed our lives and societies, country has been a regenerative presence. While country is crying with sorrow and longing for its peoples and the practices that have nurtured it, country is also capable of welcome and accommodation to others.

I first knew of Emeritus Professor Helen Ross as a Fellow Tasmanian. A doctoral supervisor introduced me to her work and shared connection of geography, from there I was able to engineer a first meeting at the VIth World Parks Congress in Sydney 2014. Following on, I held in comfort that I had found a Ph.D. examiner that I wanted to test myself against, to be in the company of the best and brightest conservation minds. Helen and I found a collective belonging from being connected to *trouwunnan* lands and seas; country grows and sustains good people and communities, regardless of origins, from those who respect Aboriginal culture and heritage.

There are many elders, peoples, and communities across Australia that have welcomed Helen to country; Tasmania is not alone in hosting her and many of her family. Indeed, Helen has a welcome from diverse Indigenous communities across the world. The love of people, place, and healthy lifehoods is what drives Helen and she has contributed to many spectacular achievements for Indigenous Peoples and Local Communities to gain rights to our territories of life. In Australia, she is known for her commitment to securing joint management and Indigenous Protected Areas in partnership with our peoples and expanding out the definitions and aims of mainstream conservation to be more inclusive.

These essays in honour of Helen's work speak to the depths of her commitment to making greater societies on country, usually as a mentor to at least two generations of researchers and practitioners. For Helen, it is not enough that we have good methodologies that are centred on community participation to account for our actions and behaviours, but that we also build knowledge to create resilience in all parts of our lives and uncertain futures. At the heart of Helen's scholarship is the focus on relationships with the communities and countries that have welcomed her. Her ethical research and benefit sharing, which underpins her community development, is what differentiates Helen's lifelong obligations to producing positive social impact.

Helen is a wisdom-maker who has created a suite of rights for peoples like me, Indigenous custodians and stewards, to contribute to global conservation and community development arenas. She is a researcher with 'many eyes and ears' that take in so much more of the concepts of country to give texture and richness to our voices and practices. Helen is an ally and kin who works *with* her communities, rather than do things in our name. The authors and their work are a snapshot of the lifetime of connections that Helen has nurtured through her promise to care and the benefits we all receive when working with her.

To read this volume of essays is to know Helen: endearing, thoughtful, and generous essays that reflect on journeys, stories, and action in the company of others. Healthy country leads people to do marvellous things for stable futures and Helen's work powers the lights that contribute to caring for country well. Of those who have known or been influenced by Helen, we are connected together as much as she is to the territories and communities that have welcomed her. May country keep you and yours safe.

May 2022

Dr. Emma Lee
Centre for Social Impact
Swinburne University
of Technology
Melbourne, VIC, Australia
ejlee@swin.edu.au

**Emma Lee** Ph.D., is Aboriginal and Torres Strait Research Fellow, Centre for Social Impact, Swinburne University of Technology, Australia. She is a *trawl-wulwuy* woman of *tebrakunna* country, north-east Tasmania, Australia. Her research fields over the last 25 years have focused on Indigenous affairs, land and sea management, policy, and governance of Australian regulatory environments. In 2021, she became the first Indigenous Australian editor of a Best Practice Guideline for the International Union for the Conservation of Nature (IUCN). She has been the recipient of various prestigious awards, and she currently has a role on the National Co-Design Group, Indigenous Voice.

# Contents

# Notes on Contributors

**Roswita M. Aboe** is Lecturer in the Faculty of Teachers Training and Education, Universitas Khairun (Unkhair), Ternate, Indonesia. Since 2018 she has actively participated in supervising students from Unkhair during their regular community service in the village of Cengkeh Afo where the Cengkeh Afo and Gamalama Spices (CAGS) community-based tourism project is located.

**Claudia Baldwin** B.A., M.A. (UW, Canada), Ph.D. (UQ), Sustainability Research Centre, University of the Sunshine Coast, Australia. She uses participatory and visual methods to research social-environmental change re water, coastal and land use planning, climate change adaptation as well as age and ability-friendly communities. She taught Regional and Urban Planning at UNiSC for 16 years, as well as water planning courses for IWC Brisbane and IWRM Adelaide. Books include *Integrated Water Planning: Achieving Sustainable Outcomes* (2014, translated into Chinese) and *Natural Hazards and Disaster Justice: Challenges for Australia and its Neighbours* (2020). Prior to academia, she had over 20 years experience with governments and international agencies.

**Harriot Beazley** is Associate Professor (Human Geography) and Program Coordinator for the Master of International Development at the University of the Sunshine Coast, Australia. She is a social and cultural geographer and community development practitioner. She also facilitates modules on *Gender and Social Inclusion in Tourism* for the DFAT funded

Australia Awards Sustainable Tourism programmes to participants from Indonesia, led by Prof Bill Carter and prof Noel Scott. The *Cengkeh Afo Gamalama Spice* (CAGS) community-based tourism project is the outcome of one of these programmes, created by Australia Award Fellow Kris Syamsudin.

**Jennifer Bellamy** is Honorary Senior Research Fellow, School of Agriculture and Food Sciences, the University of Queensland, Australia (2012–2021), and prior to that Principal Research Fellow (2006–2011), in the School of Agriculture and Food Sciences (SAFS) at the University of Queensland, Australia. As a geographer and interdisciplinary research scientist, she previously worked for over 25 years in CSIRO Australia. She has published widely on governance of social-ecological systems; collaboration as a solution to wicked natural resource policy problems; Integrated approaches to natural resource management and planning; linking science, policy, and community for sustainable development; and women in agriculture and rural development.

**Katrina Brown** Ph.D., is Professor Emerita of Social Science, College of Life and Environmental Sciences, University of Exeter, Exeter, UK. She is an environmental social scientist, specializing in environmental change, development, vulnerability, and resilience. Her research focuses on how individuals and societies understand and respond to change, and their different capacities for adaptation and transformation. She is committed to interdisciplinary research on sustainability and has led international research teams to examine environmental change and poverty alleviation in developing countries. Her book, Resilience, Development and Global Change, presents a re-visioning of resilience for development.

**Valerie A. Brown** Ph.D., is Professorial Fellow, Australian National University, Canberra, Australia. She is an internationally recognized leader in the field of collective thinking, known for *Tackling Wicked Problems* (2011). During her career, she has published 13 books and more than 110 refereed articles in the following areas: collective thinking, learning and action, transformational change, and environmental management. She has supervised nearly 100 PhD theses. She specializes in supervising students working on collective responses to complex socio-environmental issues in public health, environmental management, Indigenous and community-based natural resources management, and urban planning.

**R. W. Carter** Ph.D., was Professor Heritage Resource Management, Sustainability Research Centre, School of Law and Society, University of the Sunshine Coast, Australia. He has led numerous research projects relating to heritage resource management. He has been advising on and researching community-based tourism in Indonesia, Sri Lanka, and Cambodia for the last 15 years. He has published over 100 scholarly articles in this area and supervised over 35 postgraduate research students, from honours to Ph.D. level. He has also held senior positions within the parks management sector.

**Jim Cavaye** Ph.D., is Pro Vice Chancellor, Cradle Coast Region, University of Tasmania, Burnie TAS, Australia. He is an accomplished practitioner, educator, and researcher in community development with 38 years' experience working with rural and regional communities. He has assisted over 140 local communities across Australia and internationally with community development, engagement processes, and economic development strategies. He was formerly a well-known Consultant and a Principal Rural Development Officer with the Queensland Government.

**Tracy Cheung** is a Postdoctoral Research Fellow at the Centre for Natural Gas, the University of Queensland in Australia. She holds a joint-Ph.D. degree in Geography awarded by the University of Hamburg, Germany, and Macquarie University, Australia. Her research focuses on the interdisciplinary field of energy transition, climate change, and urban and regional governance and policy. In addition, she has extensive experience in researching multiple stakeholders and communities, considering their interests and expectations towards low-carbon energy development in Australia, Germany, and Hong Kong.

**Jeff Coutts** Ph.D., has a Ph.D. in Agricultural and Environmental Science and has extensive experience in extension and its evaluation in Australia and internationally. He is a life member of the Australasian Pacific Extension Network (APEN), a member of Australasian Evaluation Society (AES), a member of the Editorial Committee of the Journal of Agricultural Education and Extension, and an Adjunct Professor with the Centre for Applied Climate Sciences at UniSQ. He was the Director of the then Rural Extension Centre at the University of Queensland (1996–2001) and now works as a consultant in his company, Coutts J&R.

**Inne Dwiastuti** graduated with Bachelor of Economics, University of Padjadjaran, Indonesia; Master of Industrial Engineering, University of

Indonesia; and Master of Public Policy, ANU, Australia. She is completing her Ph.D. study on the role of Indonesia's small to medium enterprises in rural empowerment. She is a member of The Economic Research Center BRIN, the National Research and Innovation Agency of Indonesia, where she specializes in public policy, the development of Small Medium Enterprises (SMEs), and poverty eradication.

**Ursula Harman** Ph.D., is Sociologist specializing in social inclusion in science, technology, and innovation (STI) processes. Currently, she is Lecturer in the field of science and technology studies, Researcher, and International Senior Consultant. She is a member of the Pro-Woman Committee since 2021 on Science, Technology, and Innovation at the National Council of Science, Technology, and Innovation (CONCYTEC Peru), Scaling Advisor for Canada's International Development Research Centre (IDRC), 2019 Latin America and Caribbean Eisenhower Fellow, and District Councilwoman of San Bartolo for the period 2019–2022. She is also an Australia Alumni Change Agent Awards 2022 awardee.

**Brian Head** Ph.D., is Professor of Public Policy and Governance at the University of Queensland. He is a Fellow of the Academy of Social Sciences Australia, Vice-President of the International Association for Public Policy, and a member of the advisory board of several international journals. His applied research interests include programme evaluation, evidence-informed policymaking, managing controversial and wicked problems, improving collaboration, and strengthening expertise in the era of post-truth politics. He has edited or written more than a dozen books, including *Wicked problems in public policy* (2022), together with more than 150 research articles, chapters, and research reports.

**Nga Ho** Ph.D., is Academic Staff, School of Agriculture and Food Sciences, the University of Queensland, St Lucia, Qld, Australia. She had over 10 years working for international organizations in community development and natural resources management. Since 2009 she has been teaching undergraduate and postgraduate courses in Vietnam and at the University of Queensland. She has a particular interest in studying people-environment relationships for sustainable environmental and rural development with a focus on community-based natural resource management and system approaches in natural resources management.

**Arturo Izurieta Valery** has extensive experience in research and management of protected areas and local communities in islands, emphasizing the relationship between environmental health and human well-being. He has worked for more than 30 years on strategic projects and facilitating relations between government authorities, local actors, academia, and science from Ecuador, the Galapagos Islands, Australia, Malaysia, and Central America. He was Director of the Galapagos National Park and Marine Reserve from and Executive Director of the Charles Darwin Foundation for the Galapagos Islands from. He coordinates the Oceans, Coasts, and Islands Group of the Ecuadorian NGO, CEDENMA.

**John James** Ph.D., is a recognized national and international thought leader in the use of contemporary extension models and approaches. He specializes in using online engagement to enable change and innovation. After many years of government service, he now runs his own consulting business, Enablers of Change, based in Hobart. He is an enthusiastic supporter of APEN and is both a Past President and a Life Member. He enjoys sharing his extension knowledge and experience, doing so through his role as Adjunct Associate Professor at the University of Queensland, and Casual Lecturer at the University of Tasmania.

**Natalie A. Jones** Ph.D., is Research Fellow and Lecturer, School of Agriculture & Food Sciences, the University of Queensland. She is an applied anthropologist that is dedicated to advancing the role of social science in natural resource management and agriculture within interdisciplinary teams. Her research interests involve understanding how people perceive and interact with environmental systems. She has a particular interest in studying mental models and values to understand how people make sense of and relate to their environment. She has a track record in implementing participatory processes to support decision-making.

**Ken Keith** M.Sc., lectured and researched in Rural Extension, the University of Queensland. Three significant strands of his work in soil conservation were community education, social research into the adoption of conservation measures, and the early establishment of Landcare groups in Queensland. He led a team developing citizen science projects for environmental monitoring and co-authored a soils education package *Understanding Soil Ecosystem Relationships* (1992). He was also coordinator of early steps to support community groups taking action on local land management issues (Landcare groups). He co-edited with Dr Chamala the book *Participative Approaches for Landcare: perceptions, policies, programs* (1995).

**Chengdong Liao** B.Sc., M.Sc., and Ph.D. (UQ), is Research Fellow at Institute for Science, Technology, and Society of South China Normal University (SCNU) since 2020. He teaches undergraduate and graduate level courses in science ethics, systems science, and research methodology, and is leading science grants in agricultural extension and systems sciences. His PhD thesis studied interactions between the extension system that is top-down and smallholder rice farmers that show adaptiveness and agency in China. His research currently is on revisiting agricultural extension and innovation systems using systems-thinking theories.

**Yiyu Liu** B.A., M.A. and Ph.D. (BNU, China), is Associate Professor and Assistant Dean of Institute for Science, Technology and Society, Project Director of Center for Systems Science and Systems Management, South China Normal University (SCNU), Director of Chinese Society for Philosophy of Complexity and Systems Science and Chinese Society for Dialectics of Nature, Visiting Academic at the University of Queensland (2020). His educational discipline is philosophy of science. His areas of specialization include systems methodology, systems thinking and systems management, philosophy of social science, process thinking, and sustainable development.

**Li Liu** B.A., M.A., Ph.D.(UQ), is an Assistant Professor at Fudan Development Institute, Fudan University, Shanghai, China. From 2019 to 2021, she has worked as Postdoctoral Research Fellow at the Institute for Advanced Study in Social Sciences at Fudan University. Prior to academia she worked in local government in China, business enterprises in Australia, which has expanded her visions for her applied research. Currently, she has focused on the social impacts of big data and digital transformation on communities, theoretical examination, and case studies of E-agri-food supply chains driven rural community development.

**Price Livingstone** graduated with a Bachelor of Arts in 2018 and a Master of International Development from the University of the Sunshine Coast in 2020. In 2022 he graduated with a Master of Public Policy and Governance from the University of Queensland. He has a passion for sustainability and human geography. He has volunteered in Cambodia building and installing bio-sand water filters, and for an UNFAO project focused on climate change and food security. He joined the USC team to the Cengkeh Afo and Gamalama Spices (CAGS) community-based tourism project in Ternate, North Maluku, funded by a DFAT Endeavour grant.

**Kirsten Maclean** Ph.D. (ANU, 2007), is Senior Research Scientist, Human Geography, Sustainable Pathways Program, CSIRO, Brisbane, Qld, Australia. She uses co-research practice and participatory methodologies to investigate the role of diverse people, their knowledge and values in relation to their natural and cultural resource management and planning interests and aspirations in regional and rural Australia.

**Ruth Nettle** is Professor, Rural Innovation Research Group at the University of Melbourne, Australia. She has had a career in agricultural extension, agricultural education, and engaged research. As a social scientist working in rural innovation systems, her research interests are in understanding and improving rural innovation processes with a concern for the significant issues faced by farmers and farming systems as well as rural communities. Her research has focused on agricultural extension design and policy, the social organization and workforce relations in farming, the resilience of farming systems, and change management.

**Melissa Nursey-Bray** is Professor, Department of Geography, Environment, and Population, at the University of Adelaide. She is Director of the research group ACE (Adaptation, Community, Environment). She is also Deputy Director of the Fay Gale Centre for Gender Research. She has over 100 publications and has led multiple grants in fisheries, Indigenous resource management, climate adaptation, and placemaking. All of her work explores how knowledge, power, values, conflict, and culture all affect how communities are impacted by, and become involved in, environmental decision-making.

**Gomathy Palaniappan** is a Senior Research Fellow in the School of Agriculture and Food Sciences at the University of Queensland, specializing in socio-economic research and development. She brings over 20 years of experience working with rural communities in agriculture at the grassroots level. Her focus is on engaging smallholder farmers to access markets and improve livelihoods. She brings experience from working in Australia, India, Papua New Guinea, Philippines, Pakistan, Laos, and Cambodia.

**Anastasia Papadopoulos** has a Master in Environmental Studies, a community-based environmental social scientist with a background in protected areas governance planning, environmental stewardship, and ecological monitoring. She has coordinated and assisted on projects focused on elevating Indigenous and local voices in conservation, protected areas governance and planning, community-led climate change adaptation as well as youth engagement in environmental stewardship. She takes a complex systems-thinking approach to her work and thrives and values working in interdisciplinary settings that prioritize community needs, aspirations, and capacities through meaningful engagement while simultaneously benefitting the environment and nurturing resilient communities.

**Anya Phelan** is a Lecturer within the Strategy & Entrepreneurship discipline, at the University of Queensland Business School, and Entrepreneur in Residence with CSIRO Plastics Innovation Hub. Her current research includes plastic pollution, social entrepreneurship, responsible supply chains, circular economy, and sustainable livelihoods. She co-leads Cape York Recycling Project which aims to support the advancement of people living in Far North Queensland through innovative recycling solutions and was the Principal Investigator for the Global Change Institute Flagship Project—Small Island Initiative for a Plastic Free Ocean—investigating factors contributing to ocean plastic pollution in Indonesia.

**Michelle Prasad** is a Researcher, Project Manager, Author, and a descendent of the Global South. She graduated with a Master of International Development from the University of the Sunshine Coast, investigating the ability of the *Cengkeh Afo and Gamalama Spices* (CAGS) community-based tourism enterprise in North Maluku, Indonesia, to empower women, protect indigenous food culture, and conserve natural resources. Funded by a DFAT Endeavour grant, she worked with Indonesian community members in Ternate and Australia Award Fellow Kris Syamsudin to document life stories and recipes using participatory research methodologies. She and Kris Syamsudin are the co-authors of the recently published (2022) book, *The Heart of The Spice Forest*.

**Purwanto Purwanto** (Bachelor in Economics, Unsoed, Indonesia) (M.Econ.St., Ph.D., UQ, Australia) has conducted research on rural development issues, including food security and poverty reduction. He was awarded Ph.D. in 2020 from the School of Agriculture and Food Sciences, the University of Queensland. He continues to conduct research on rural development as the Research Coordinator of National Research

Priority grant in 2020. He is also the editor of the book, *Actualizing Concept and Model for Community Based Rural Development* (2021, forthcoming, Airlangga University Press).

**William Ramsay** obtained his Bachelor of Environmental Science (2019) and a Master of International Development (2021), from the University of the Sunshine Coast (USC). He has worked with a range of development-orientated organizations, notably Reforest Africa and The Australian Centre for Pacific Island Research at USC. He joined the USC team to the Cengkeh Afo and Gamalama Spices (CAGS) community-based tourism project in Ternate, North Maluku, funded by a DFAT Endeavour grant.

**Helen Ross** honouree of this publication, is Professor Emeritus in Rural Development at the University of Queensland, and former Leader of the Rural Development Group. Her research fields evolved from Indigenous housing and urban systems, to social impact assessment, Indigenous natural resource management, to mental models, values, and resilience. She has pioneered the use of systems and participatory methods. She has edited the *Australasian Journal of Environmental Management* for over 18 years and is an Honorary Life Member, Fellow, and winner of the national prize of the Environment Institute of Australia and New Zealand.

**Novie A. Setianto** Ph.D. University of Queensland, is Faculty of Animal Science, Jenderal Soedirman University, Indonesia. Since 2015, he has been involved in two international collaboration research projects, *Capturing Coral Reef & Related Ecosystem Services (CCRES)* in 2015–2017 and *Small Island Initiative for a Plastic Free Ocean* in 2018–2020. From 2018 to 2020 he was a co-researcher for the Global Change Institute Flagship Project—Small Island Initiative for a Plastic Free Ocean. From 2016 to 2018, he undertook research focused on developing strategies for beef farming in Central Java.

**Jonathan Simpson** is an early career Researcher from the University of Queensland. His qualification background spans environmental management, communications, and international relations, and as such he was interested to delve into the topic of resilience and appreciate the way Prof. Helen Ross's paper has impacted globally. His research interests include environmental and industrial impacts on communities and how they might mitigate or adapt to these.

**Kris Syamsudin** is a Recipient of an Australia Awards Fellowship and the Founder of the community-based tourism project Cengkeh Afo and Gamalama Spices (CAGS) in Ternate, North Maluku. Based on a hillside clove plantation home to Afo, the world's oldest clove tree, the initiative shares local agriculture, history, and heritage through spice production, supporting local livelihoods and generating income, especially for women. The initiative has attracted international attention, and together with Michelle Prasad Kris has written his first book about the CAGS project titled *The Heart of The Spice Forest*.

**Sérgio Rustichelli Teixeira** Ph.D., is trained as a Zootechnician and managed dairy farms prior to joining Brazil's national scientific organization, Embrapa, as a scientific researcher from 1985 to 2019. He undertook a master's degree in production engineering at the Federal University of Santa Catarina, awarded in 1993, and a Ph.D. in agricultural extension awarded in 2004 at the University of Queensland. That study compared Australian and Brazilian dairy farmers' needs and priorities for information, and how science and extension could become more farmer-focused in each country. He has led numerous research projects in Brazil and Africa.

**Ian Thomson** Ph.D., Principal of Shinglespit Consultants Inc., previously, he was a founding member and principal of On Common Ground Consultants Inc. He has held management positions with Orvana Minerals Corp and Placer Development Ltd and has more than 40 years of experience in the mining industry, working for the last two decades to advance and refine the management of social issues in resource development projects. The book, 'The Social License: The story of the San Cristobal mine', product of his collaboration with Robert Boutilier, was published by Routledge in September 2018.

**Séverine van Bommel** Ph.D., is Senior lecturer, School of Agriculture and Food Sustainability, the University of Queensland, Gatton Qld, Australia. She is an interdisciplinary social scientist working on rural community development and extension in agriculture and natural resource management. She holds a Ph.D. from Wageningen University (NL), where she worked as Associate Professor with the Communication Studies Group until 2018. Taking an qualitative, interpretive approach, she studies the construction of meaning in interaction with a focus on co-design, gender, and development. She is passionate about making the voices of marginalized populations heard in participatory research and community-based approaches.

**Bradd Witt** Ph.D. (UQ), is Lecturer, School of Earth and Environmental Sciences, The University of Queensland, St Lucia Qld, Australia. He is a multidisciplinary researcher who is passionate about rural and regional issues, in particular the management of rural and regional landscapes for diverse socio-ecological values, environmental, and sustainability performance of agriculture especially how these industries negotiate dynamic internal industry and external community and consumer expectations, stakeholder engagement in environmental management, and land use change and decade to century scale environmental change in rangelands.

**Katherine Witt** Ph.D., is Senior Research Fellow in sociology within the Engineering Faculty at the University of Queensland. She studies the social and economic effects of energy transition and the contribution of new energy projects and technologies to sustainable regional development. She uses 'socio-ecological systems' approaches in her research on the social dimensions of complex and contentious land and water use issues, including mining, agriculture, onshore gas, large-scale renewables, hydrogen, carbon capture, use and storage, and radioactive waste management. She is currently co-chair of the Social Impact Assessment section of the *International Association for Impact Assessment.*

**Melanie Zurba** Ph.D., is Associate Professor with the School for Resource and Environmental Studies (SRES) at Dalhousie University, Canada. Her lab is the Community-Engaged CoLab and her research is focused on environmental governance and the social aspects of the human-with-nature relationship. Her current research focuses on community engagement, collaboration and learning for improving governance systems, and arts-based approaches for understanding thoughts, emotions, values, and aspirations relating to social-ecological systems. She is also Chair of the International Union for the Conservation of Nature (IUCN) Theme for Governance, Equity, and Rights (TGER) and is a member of the IUCN Commission for Environmental, Economic, and Social Policy (CEESP) Steering Committee.

# List of Figures

# List of Tables

CHAPTER 1

# State of the Art, Trends and Progress in Work on the Social Aspects of Social-Ecological Systems in Research and Practice

*Séverine van Bommel and Claudia Baldwin*

## Introduction

This book celebrates trends and progress in research and practice on the role of communities in the sustainable management of social-ecological systems, and the roles of Professor Helen Ross in progressing relevant

S. van Bommel (✉)
School of Agriculture and Food Sciences, University of Queensland, Brisbane, Australia
e-mail: s.vanbommel@uq.edu.au

C. Baldwin
Sustainability Research Centre, University of the Sunshine Coast, Sunshine Coast, Australia
e-mail: CBaldwin@usc.edu.au; claudiab@westnet.com.au



C. Baldwin and S. van Bommel (eds.), *Rural Development for Sustainable Social-ecological Systems*, Palgrave Studies in Environmental Sustainability,
https://doi.org/10.1007/978-3-031-34225-7_1

theoretical concepts and their application. It reflects Helen's underlying commitment to ensuring people and communities can contribute to matters that affect their lives and interests, and to improving such processes through research, teaching, coaching, and facilitating.

This is a 'friends' book'—a so-called Liber Amicorum, typically published on the occasion of the honouree's retirement. It is a European tradition that we are introducing to Australia. This book has been published in honour of Professor Helen Ross on the occasion of her official retirement from the Rural Development Group at the School of Agriculture and Food Sciences at the University of Queensland, Australia.

It contains original contributions by the honoured academic's close colleagues, including former doctoral students. It is a testament to Helen's collaborative and nurturing (or people-centred) approach in working with others that her colleagues initiated this book. In a chat about the purpose of this book, Helen also acknowledges the contribution of her many academic colleagues and students over the years to the development of her own thinking.

Underlying Professor Ross's scholarship is a desire to understand what individuals and communities think and do, and how to improve management processes and practice. Over her career she has moved from examining people and built places, such as Aboriginal housing, to the interaction of people with their natural environment.

While the authors are from a variety of geographical locations and disciplines, they have in common, a history of working with Professor Helen Ross, either within Australia, which is her home base, or elsewhere in the world. Consequently in developing the book themes, we refer to global trends as well as provide Australian perspectives. The book is structured along three themes that have emerged in the domain of community-environment interactions, mainly since the 1980s, and interwoven with Helen's academic career: (1) community development and rural extension; (2) natural resource management, conservation and co-management; and (3) social aspects of resilience.

## Rural Community Development: Facilitating Change

Works in the first section of this book are inspired by Helen's early research and collaborations with students, focussed on actor-oriented approaches to building agency and structural elements in rural development. It acknowledges the mutual influence of people and environment on each other.

Agricultural research and extension in Australia was prompted by challenges of crop and animal production in an environment unlike Europe in the 1880s, resulting in government departments of agriculture with professionals advising farmers, based on research (Hunt et al., 2012). Over time it led to a series of research farms and agricultural colleges, and eventually universities (similar to the USA Land Grant system) as well as research institutes such as the current CSIRO (the Commonwealth Scientific and Industrial Research Organisation in 1949), with ongoing tension about a balance between research and extension efforts as part of 'RD&E'. From the 1960s to 1990s was a shift towards academic research and post-graduate courses in extension (not just agriculture), such as at University of Queensland (offering a position to Professor Ross in Rural Community Development in 2000). Around this time, the focus broadened beyond production to natural ecosystems, business-planning and sociological perspectives, recognising the need to embrace more participatory models (Hunt et al., 2012).

These Australian trends were in tandem with and reflected global trends. Thus the theme of the first section of the book draws on community development-oriented approaches to land management, dating back to the late 1980s. The publication of the World Commission on Environment and Development (the Our Common Future report) created awareness among governments and donors for the need to address environmental problems (Scoones et al., 2007). However, the solutions proposed and implemented largely followed a top-down blueprint approach not tailored to the realities of local communities and smallholder farmers (Chambers, 1987). In 1987, a workshop on 'Farmers and Agricultural Research' was organised which brought together social scientists and natural scientists from around the world at the Institute of Development Studies (IDS) in the UK to discuss and address this problem. This workshop became later known as the 'Farmers First' workshop (Scoones & Thompson, 2009). Participants challenged the Transfer of Technology values and practices which dominated business-as-usual agriculture extension practices. Drawing on earlier experiences by the integrative global Farming Systems Research movement, the idea was that extension staff needed to take the perceptions and lived experiences of local farmers and local communities into account. Experimentation by smallholder farmers and local knowledge was emphasised. It was argued that local people should not just be treated as recipients of technology and information (like a bucket to be filled) but instead they should be treated as actors that needed to be facilitated in determining their own development direction and creating their own plan of how to get there (Jennings et al., 2011; Scoones & Thompson, 2009).

In the late 1980s and early 1990s these events led to the emergence of participatory approaches focussed on community development as a new approach to agriculture as part of natural resource management that aimed to empower the community, in particular encouraging marginalised groups such as women, smallholder farmers or children, to take responsibility for their own development process (Chambers, 1994; Pretty, 1995; Röling & Wagemakers, 1998). Participation was thought to give legitimisation to the intervention process by providing a broad social basis of public support (Castro & Nielson, 2003). In addition to this, it was thought to enhance efficiency by 'linking up with local dynamics'; local knowledge in addition to technical knowledge was thought to create a better plan in the sense that it would be acceptable to a broader group of people (Hounkonnou, 2004). Last but not least, it was thought to enhance empowerment by means of democratisation (Castro & Nielson, 2003).

In Australia, Landcare came into its element (see the next section). Discussions focussed on different types of participation and on various methods of organising. A prime example of this is Ross, Abel & Manning's, 1996 work offering guidance to communicating about ecological processes on farm land. By the end of the 1990s, critical studies started to emerge that questioned whether the aims of the participatory projects—such as efficiency, legitimacy and empowerment—were indeed met in practice (Turnhout et al., 2010). Some studies concluded that the participatory processes had not been designed or facilitated appropriately (Chambers, 1994, 2002; Pretty, 1995; Turnhout et al., 2010). Other studies—often taking a post-structuralist approach—claimed that participatory processes were blind to power dynamics and therefore often ended up reproducing power inequalities instead of changing them

(Cooke & Kothari, 2001; Mosse, 2004; Turnhout et al., 2010). This debate, including the need for structural societal change, continues today.

In the meantime, in the mid and late 2000s attention shifted from participation at the farm level or community level in what Scoones and Thompson (2009) refer to as 'people-centered learning and innovation at a system level'. It had become clear that many resource problems at the farm level—such as water management or other common pool resources—were part of wider resource dilemmas[1] (Röling, 1994). This gave rise to the need to manage resources at a level or scale—such as a group of landowners or a water catchment—at which the perceived dilemma could be effectively addressed (Woodhill & Röling, 1998). It led to a wide variety of approaches that aimed at creating platforms that would allow multiple stakeholders to interact, negotiate, deliberate and learn their way out of these problems as long as the pre-condition of perceived interdependence was met. Parallel to the 'people-centered learning and innovation at a system level' approach, also an 'agricultural innovation systems' (AIS) approach was developed (Klerkx et al., 2012). Instead of focussing on natural resources at the regional scale, this approach focuses on the entire agricultural system and investigates how the different parts of the agricultural system—such as market, policy, society and knowledge institutes, all interact in agricultural innovation at the country level, sector level or in relation to a particular technology. The increase in the level of systemic thinking, allows AIS to work with problems—such as climate change or biodiversity loss—that span multiple sectors (Pigford et al., 2018).

In Australia, it was exemplified by industry-based Rural Development Corporations (RDCs) with increasing industry and private sector funding coinciding with less State and Commonwealth government funding. Government extension services were reduced according to the declining relative importance of agriculture in the economy and economic rationalism (Botha & Coutts, 2006). Many agricultural colleges were closed as did the UQ Rural Extension Centre. Rural extension and community development were based on interaction among people, information sharing, social learning, dialogue, action and evaluation (Hunt et al., 2012).

[1] A resource dilemma is a situation in which stakeholders with different interests use the same resources to different ends and in which use by one stakeholder reduces the access or value by others.

Chapters in the first section refer to these continuing contemporary issues. The rural agricultural community is experiencing additional challenges of globalisation and climate change-induced hazards of flood and drought, changing production conditions and risks, all of which still require the communication and interaction so valued in more recent extension services.

## Community-Based Natural Resource Management, Conservation and Co-Management

Parallel to the changes in rural community development and extension, approaches to nature conservation and natural resource management were also evolving towards more collaborative and participatory approaches. The second section of the book on 'Community-based management, Collaboration, and Co-management' captures ways of producing value from multi-stakeholder partnerships, exemplified by Professor Ross's work in this area.

While rural extension services became more participatory and holistic, expanding to address broader property and business issues, in Australia, community engagement in natural resource management (NRM), featuring land and water, also grew in prominence from the 1980s. A 'social experiment' in community-based natural resource management emerged from the grassroots Landcare and catchment management movements which had embraced local scale participation in building social capital and engaged in social learning with access to change agents and expertise (Curtis et al., 2014). A partnership between the Australian Conservation Foundation and the National Farmers Federation prompted Australian government funding into the Decade of Landcare program (1990–2000). By the early 1990s an estimated 30% of the farming community were volunteers in over 4000 Landcare type groups in which members worked collaboratively on local/district issues such as pest plant and animal control and riparian revegetation (Curtis & De Lacy, 1996). At this time, motivation for farmers to be involved in Landcare activities was assessed as being economic, to improve land productivity, with less emphasis on biodiversity. Hence the need for Prof Val Brown's edited book (1996) about different Landcare Languages to which (Ross et al., 1996) contributed. By the end of the Decade, additional ongoing funds to tackle Australia-wide NRM issues were funnelled through newly established Regional NRM bodies most of which emerged from catchment groups. A typical challenge was fostering a deliberative inclusive process that included social learning (Baldwin & Ross, 2012; Schusler et al., 2003), was seen to be legitimate, accountable, equitable, and having sufficient flexibility for adaptive governance (Lockwood et al., 2010). It was clear that community-based NRM was about managing people-environment relationships, corresponding to parallel developments in the theory of social-ecological systems (SES) (Curtis et al., 2014).

Over these same decades, concern emerged globally about the state of diminishing water resources and long-term sustainability leading to the promotion of integrated water resource management (IWRM) concepts by UN agencies and globally. The EU *Water Framework Directive* required member nations to involve stakeholders in developing River Basin Plans to achieve sustainable use across Europe. In the USA, basin plans were developed to resolve contentious issues of river use. In Australia, the National Water Initiative in 2004 formalised intergovernmental water reform agenda arrangements started a decade before, incentivising State governments to develop plans to reduce overallocation and overuse of surface and groundwater, provide for environmental and other public benefit outcomes, and ensure resource security of the consumptive pool. Water resource planning thus necessitated an understanding of human use and values about water. Prof. Helen Ross contributed to community engagement approaches in the water sector (Dean et al., 2016), as well as to guidelines for social impact assessment of environmental flows (Hassall & Associates, 2003). Initiatives around the world featured collaborative and participatory processes but it became clear that to achieve sustainable outcomes, sophisticated processes were needed to make trade-offs and resolve conflict. Helen was at the forefront of promoting negotiation, not just consultation, in the case of Indigenous people and development in the East Kimberley (Coombs & Ross, 1989), mediation as part of social impact assessment (Ross, 2003), and consensus-building approaches (Baldwin & Ross, 2012) in relation to water management.

Until then, conventional protected area approaches had included a 'hands off' approach to nature conservation (Arts et al., 2012, p. 16), often steering visitors to 'sacrificial' areas, with the remaining of protected land being off limits. In developing countries, communities were displaced and re-located if they were living in National Parks (Funder & Ladekjær Gravesen, 2021). This preservation approach was critiqued for not taking into account the needs and rights of local people and the cultural significance of land for their local identity. In response to this, in 1975 the IUCN Kinshasa resolution recognised the 'vulnerability of indigenous people and the great significance they attach to land ownership' (WCC resolution. Protection of Traditional Ways of Life, RES/005, 1975, p. 1). Soon after, the publication of the *World Conservation Strategy* in 1980 called for an integration of conservation and development (Corson et al., 2020). Integrated Conservation and Development Projects (or ICDPs) soon became the new approach to conservation and natural resource management, especially in developing countries (Alpert, 1996). This was based on the idea that if communities could benefit more from conservation economically, and that benefit was more than the value of their own resource practices, then they would support management and conservation of natural resources. Donors and international NGOs supported schemes that aimed at generating and sharing of conservation revenues with communities (Nelson et al., 2021; Pimbert & Pretty, 1997).

Community-based conservation was often used in an instrumental way in the sense that it aimed at achieving a particular conservation outcome. As such it often focussed on awareness raising, practice change, and development of alternative livelihoods (Funder & Ladekjær Gravesen, 2021). Major community-based conservation initiatives included community-based wildlife management in Africa, joint forest management in Asia and community-based marine protected areas in SE Asia and the Pacific (Western et al., 1994). The 2000s is characterised by a shift from ICDP to co-management, driven by an increased recognition of the role of Indigenous communities in biodiversity conservation. The IUCN World Parks Congress in Durban in 2003 raised awareness of the need for full incorporation of the rights of indigenous people and communities in the conservation and management of protected areas. This led to a greater focus on the institutional arrangements in community-based conservation to enable local communities to partnership in stewardship programs (Corson et al., 2020). Helen's past experience with Indigenous groups saw her well-poised to contribute to developing cooperative management arrangements for Australia's Great Barrier Reef (George et al., 2004; Robinson et al., 2006; Ross et al., 2004) and guiding chapter authors Izurieta (PhD student 2003–2006) on Indigenous co-management of the GBR World Heritage area and Nursey Bray (PhD student 2000–2006) on Indigenous co-management of hunting.

Over the last decade, the 2010 *Nagoya Protocol of the Convention on Biological Diversity* as well as the 2016 IUCN Natural Resources Governance Framework continued to pay explicit attention to the role of local communities and Indigenous groups in the conservation and management of natural resources.

Following Funder and Ladekjær Gravesen (2021), some new trends and developments that have influenced the discussions on community-based conservation, include—but are not limited to—Neo-liberal market-based approaches such as PES schemes and REDD schemes (Kareiva et al., 2011; Marvier & Kareiva, 2014); Convivial conservation (Buscher & Fletcher, 2020) and bio-cultural conservation (Gavin et al., 2015, 2018; Maffi & Woodley, 2012).

## Social Aspects of Resilience

The third theme addresses the social aspects of resilience of social-ecological systems. The concept of resilience was originally introduced by Holling in 1973 but he was only able to develop it further in the early 1990s through his work with the Beijer Institute and the University of Florida. It became further institutionalised in the form of the Resilience Alliance—an interdisciplinary consortium of people that all study the dynamics of social-ecological systems—with its associated journal *Ecology and Society*. For quite some time the work on resilience of Complex Adaptive Systems was mainly progressed by ecologists. It was only in the mid-1990s that the work on resilience was picked up by social scientists interested in social learning and adaptive management for facilitating sustainable resource use. Until then, the conventional tools for social scientists working in the field of human–environmental relations had been impact assessment and risk assessment. Helen contributed to the field of social impact assessment by looking through the eyes of those affected to improve outcomes, for example in regards to lead contamination (Ross & McGee, 2006), the impact of Coronation Hill mining on Indigenous communities (Ross, 2001), and as early as 1990 in East Kimberley indigenous communities (Ross, 1992).

But with increased awareness of the complexity, uncertainty and structural coupling of social-ecological systems, social scientists increasingly felt that the traditional impact assessment and risk assessment, with their reductionist scientific methods, were not able to adequately represent the interactions between people and their environment, let alone plan for more sustainable outcomes (Jiggins & Röling, 2000). Initially social scientists turned to the notions of adaptive management and social learning to address the management of complex social-ecological systems. It was only in the 2000s that the human side of social-ecological systems became more integrated into the concepts of social and community resilience. Work by Prof. Helen Ross and colleagues on mental models (Ross & Abel, 2000), was expanded by her PhD students Jones (2012) regarding natural resources systems and Baldwin (2008) on farmers' values about water resources.

In response to the literature on sustainability and sustainable management, scholars working on social resilience argued that we need to go beyond sustainability (with a static focus on maintaining the status quo) and actively build the capacity "to learn and to adjust its responses to the impacts of external drivers and internal change" (Berkes, 2017, p. 5). This includes the capacity to absorb shock, adapt to change or transform into a new system. Most of the work on social aspects of resilience has focussed on agency, learning, framing and mental models and self-organisation. During the last decade, community resilience has received more attention in literature mostly through the work of Professor Helen Ross together with Prof. Fikret Berkes. Bringing together insights from individual resilience and ecological resilience, Berkes and Ross have started to work towards an integrated approach for community resilience.

## Helen Ross's Intellectual Journey

Originally from Tasmania, Helen did her undergraduate degree in psychology, with a focus on social-psychology, becoming interested in anthropology during her honours year. She worked in public policy for Commonwealth Department of Aboriginal Affairs during the Whitlam era, before getting a scholarship to study at London School of Economics. Her PhD, completed in 1983, examined perceptions of Australian Aboriginal housing and living environments through a cultural lens, contributing to an emerging research area. After another two years in government in land rights and as a policy adviser to the Special Minister of State, she built credentials at Australia National University over 14 years in social impact assessment and Indigenous land management, new and flourishing applied research areas.

Arriving at the University of the Queensland (UQ) as a Professor in 2000, she became known for social aspects of land, water and marine collaborative management, co-management with First Nations, and rural women, where theory and practice are mutually informing. A restructure in 2011, saw her move from Natural and Rural Systems Management at UQ to the School of Agriculture coinciding with a structural demise in rural extension services.

About the trajectory of her career, Helen recently said that 'when one door closed, another opened' as she adapted to new settings, new teams, and funding opportunities. As natural resource issues emerged in prominence, national policies changed, and funding became available over time, she applied her expertise to a range of geographical locations—tropical savannahs, Murray-Darling Basin, Great Barrier Reef and other marine parks—and various governance systems. With the end of the Decade of Landcare and Integrated Catchment Management, the need for her collaborative governance skills was ever apparent with new Regional Natural Resource Management bodies.

What motivates Helen? She told us that she likes bringing together people with different mindsets to focus on a problem, whether as leader of a research team, teacher of post-grad students, or thesis supervisor.

From her academic home base in Australia, to the various countries in Asia and Latin America where she carried out her field work, Helen Ross embodies the ideal of an engaged academic who always puts people first both in her research and teaching. Her collegiality and warm personality have led to many friendships with colleagues from all around the world. She has supervised 43 PhD and three Masters students to completion, of which 24 were international students, who she nurtured and saw as a vehicle for positive international development. Interest in people-environment relationships has naturally led to thinking about mental models, complexity, and social-ecological systems. More recent work in South-east Asia and on Indigenous bush foods has allowed her to pursue her research interests.

After 38 years in academia, Helen 'retired' at the end of 2020. Of course, we know that academics never really retire and her appointment as Professor Emeritus confirms the value that the University of Queensland places on her work, and provides a base for continuing theoretical extension and practical application, on a more flexible basis.

## Conclusion

The contributions in this book honour Professor Helen Ross's legacy, bringing together a high level of experience and engagement across academia and practice, with a balance of established and emerging scholars. The contributions include a range of disciplinary lenses, geographic contexts, and topics. The contributions not only provide an overview of the current thinking on the three themes that put communities first in the management of social-ecological systems, the contributions also identify a number of areas for critical reflection and future directions of research, expanded on in the concluding chapter.

## References

Alpert, P. (1996). Integrated conservation and development projects: Examples from Africa. *BioScience, 46*(11), 845.

Arts, B., van Bommel, S., Ros-Tonen, M., & Verschoor, G. (2012). Forest-people interfaces: from local creativity to global concerns. In B. Arts, S. van Bommel, M. Ros-Tonen, & G. Verschoor (Eds.), *Forest-people Interfaces: understanding community forestry and biocultural diversity*. Wageningen Academic Publishers.

Baldwin, C. (2008). *Integrating values and interests in water planning using a consensus-building approach.* (Unpublished doctoral dissertation). University of Queensland

Baldwin, C., & Ross, H. (2012). Bridging troubled waters: Applying consensus-building techniques to water planning. *Society and Natural Resources*, 25(3).

Berkes, F. (2017). Environmental governance for the anthropocene? social-ecological systems, resilience, and collaborative learning. *Sustainability (basel, Switzerland), 9*(7), 1232. https://doi.org/10.3390/su9071232

Botha, N., & Coutts, J. (2006). Extension and consultancy, current thinking. *Primary Industry Management*, 9(4), 10Á12.

Brown, V. A. (1996). *Landcare languages: Talking to each other about living with the land.* Department of Primary Industries and Energy.

Buscher, B., & Fletcher, R. (2020). *The conservation revolution: Radical ideas for saving nature beyond the Anthropocene.* Verso Books.

Castro, A. P., & Nielson, E. (2003). *Natural resource conflict management case studies: An analysis of power, participation and protected areas.*

Chambers, R. (2002). Relaxed and Participatory Appraisal: Notes on practical approaches and methods for participants in PRA/PLA-related familiarisation workshops. Participation Group, Institute of Development Studies, University of Sussex ….

Chambers, R. (1987). *Sustainable livelihoods, environment and development : Putting poor rural people first*. Institute of Development Studies, University of Sussex.

Chambers, R. (1994). Participatory rural appraisal (PRA): Challenges, potentials and paradigm. *World Development, 22*(10), 1437–1454. https://doi.org/10.1016/0305-750X(94)90030-2

Cooke, B., & Kothari, U. (2001). Participation: The new tyranny? Zed books.

Coombs, H. C., & Ross, H. (1989). From consultation to negotiation. In H. C. Coombs, H. M'cCann, H. Ross, & N. Williams (Eds.), *Land of promises: Aborigines and development in the East Kimberley* (pp. 129–137). Aboriginal Studies Press.

Corson, C., Flores-Ganley, I., Worcester, J., & Rogers, S. (2020). From paper to practice? Assembling a rights-based conservation approach. *Journal of Political Ecology, 27*(1), 1128–1147. https://doi.org/10.2458/v27i1.23621

Curtis, A., & De Lacy, T. (1996). Landcare in Australia: Does it make a difference? *Journal of Environmental Management, 46*(2), 119–137. https://doi.org/10.1006/jema.1996.0011

Curtis, A., Ross, H., Marshall, G. R., Baldwin, C., Cavaye, J., Freeman, C., et al. (2014). The great experiment with devolved NRM governance: Lessons from community engagement in Australia and New Zealand since the 1980s. *Australasian Journal of Environmental Management, 21*(2), 175–199. https://doi.org/10.1080/14486563.2014.935747

Dean, A., Fielding, K., Newton, F., & Ross, H. (2016). Community Engagement in the Water Sector: An outcome-focused review of different engagement approaches.

Funder, M., & Ladekjær Gravesen, M. (2021) 'The Evolution Of Community Based Conservation And Implications For Danish Development Cooperation'. DIIS Working Paper.

Gavin, M., McCarter, J., Berkes, F., Mead, A., Sterling, E., Tang, R., et al. (2018). Effective Biodiversity Conservation Requires Dynamic, Pluralistic, Partnership-Based Approaches. *Sustainability (basel, Switzerland), 10*(6), 1846. https://doi.org/10.3390/su10061846

Gavin, M. C., McCarter, J., Mead, A., Berkes, F., Stepp, J. R., Peterson, D., et al. (2015). Defining biocultural approaches to conservation. *Trends in Ecology & Evolution, 30*(3), 140–145. https://doi.org/10.1016/j.tree.2014.12.005

George, M., Innes, J., & Ross, H. (2004). Managing sea country together: key issues for developing co-operative management for the Great Barrier Reef World Heritage Area. CRC Reef Research Centre.

Hassall & Associates (2003). *Introduced marine pests: scoping the socio-economic impacts,* prepared for Fisheries & Aquaculture Branch, Agriculture, Fisheries and Forestry, Australia.

Hounkonnou, D. (2002). Linking up with local dynamics: Learning to listen. In Leeuwis, C. & Pyburn, R. (Eds.), *Wheelbarrows full of frogs: Social learning in rural resource management*, pp. 105–120.

Hunt, W., Birch, C., Coutts, J., & Vanclay, F. (2012). The Many Turnings of Agricultural Extension in Australia. *The Journal of Agricultural Education and Extension, 18*(1), 9–26. https://doi.org/10.1080/1389224X.2012.638780

Jennings, J., Packham, R. G., & Woodside, D. P. (2011). *Shaping change: Natural resource management, agriculture and the role of extension*. Australasia Pacific Extension Network.

Jiggins, J., & Röling, N. G. (2000). Adaptive management: Potential and limitations for ecological governance. *International Journal of Agricultural Resources, Governance and Ecology, 1*(1), 28–42.

Jones, N. A. (2012) *Eliciting mental models of natural resource systems: A procedural comparison (Unpublished doctoral dissertation)*. University of Queensland

Kareiva, P., Lalasz, R., & Marvier, M. (2011). Conservation in the Anthropocene: Beyond solitude and fragility. *Breakthrough Journal, 2*(Fall), 29–37.

Klerkx, L., van Mierlo, B., & Leeuwis, C. (2012). *Farming systems research into the*. (pp. 457–483). Dordrecht: Springer Netherlands. https://doi.org/10.1007/978-94-007-4503-2_20.

Lockwood, M., Davidson, J., Curtis, A., Stratford, E., & Griffith, R. (2010). Governance Principles for Natural Resource Management. *Society & Natural Resources, 23*(10), 986–1001. https://doi.org/10.1080/08941920802178214

Maffi, L., & Woodley, E. (2012). *Biocultural diversity conservation: A global sourcebook*. Routledge.

Marvier, M., & Kareiva, P. (2014). The evidence and values underlying 'new conservation.' *Trends in Ecology & Evolution, 29*(3), 131–132. https://doi.org/10.1016/j.tree.2014.01.005

Mosse, D. (2004). Is Good Policy Unimplementable? Reflections on the Ethnography of Aid Policy and Practice. *Development and Change, 35*(4), 639–671. https://doi.org/10.1111/j.0012-155X.2004.00374.x

Nelson, F., Muyamwa-Mupeta, P., Muyengwa, S., Sulle, E., & Kaelo, D. (2021). Progress or regression? Institutional evolutions of community-based conservation in eastern and southern Africa. *Conservation science and practice*, 3(1), n/a. doi:https://doi.org/10.1111/csp2.302.

Pigford, A.-A.E., Hickey, G. M., & Klerkx, L. (2018). Beyond agricultural innovation systems? Exploring an agricultural innovation ecosystems approach for niche design and development in sustainability transitions. *Agricultural Systems, 164*, 116–121. https://doi.org/10.1016/j.agsy.2018.04.007

Pimbert, M. P., & Pretty, J. N. (1997). Parks, people and professionals: Putting 'participation' into protected area management. *Social Change and Conservation, 16*, 297–330.

Pretty, J. N. (1995). Participatory learning for sustainable agriculture. *World Development, 23*(8), 1247–1263. https://doi.org/10.1016/0305-750X(95)00046-F

Robinson, C. J., Ross, H., & Hockings, M. (2006). *Development of co-operative management arrangements in the Great Barrier Reef: An adaptive management approach*. CRC Reef Research Centre.

Röling, N. G. (1994). Platforms for decision-making about eco-systems. In F. e. al. (Ed.), *Future of the land: mobilizing and integrating knowledge for land-use options*. Chichester: John Wiley & Sons.

Röling, N. G., & Wagemakers, M. A. E. (1998). *Facilitating sustainable agriculture: participatory learning and adaptive management in times of environmental uncertainty*. Cambridge University Press.

Ross, H., Abel, N., & Manning, M. (1996). Understanding and communicating about ecological processes on farm land. In V. A. Brown (Ed.), *Landcare languages: Talking to each other about living with the land. A Communication Manual for Landcare*. (pp. 271–280). Department of Primary Industries and Energy.

Ross, H. (2003). Environmental Mediation. In H. A. Becker, & F. Vanclay (Eds.), *The International handbook of social impact assessment: conceptual and methodological advances*: Edward Elgar.

Ross, H., Innes, J., George, M., & Gorman, K. (2004). Traditional owner Aspirations towards cooperative management of the Great Barrier Reef World Heritage Area: Community Case Studies. *CRC Reef Research Technical Report*, no. 56.

Ross, H. (1992). Opportunities for aboriginal participation in Australian social impact assessment. *Impact Assessment, 10*(1), 47–75. https://doi.org/10.1080/07349165.1992.9725731

Ross, H. (2001). Social impact assessment: Coronation Hill. In R. Baker, J. Davies, & E. Young (Eds.), *Working on Country: Contemporary Indigenous Management of Australia's Lands and Coastal Regions* (pp. 320–336). Oxford University Press.

Ross, H., & Abel, N. (2000). Eliciting mental models of landscape processes: The transect method. In G. T. Moore, J. Hunt, & L. Trevillion (Eds.), *Environment-behaviour research on the Pacific Rim* (pp. 295–310). Faculty of Architecture.

Ross, H., & McGee, T. K. (2006). Conceptual frameworks for social impact assessment revisited: A cumulative effects study on lead contamination and economic change. *Impact Assessment and Project Appraisal, 24*(2), 139–149.

Schusler, T. M., Decker, D. J., & Pfeffer, M. J. (2003). Social learning for collaborative natural resource management. *Society & Natural Resources, 16*(4), 309–326. https://doi.org/10.1080/08941920390178874

Scoones, I., Leach, M., Smith, A., Stagl, S., Stirling, A., & Thompson, J. 1 (2007) 'Dynamic systems and the challenge of sustainability' *STEPS Working Paper*. STEPS Centre.

Scoones, I., & Thompson, J. (2009). *Farmer first revisited*. Practical Action Publishing.

Turnhout, E., van Bommel, S., & Aarts, N. (2010). How participation creates citizens: Participatory governance as performative practice. *Ecology and Society, 15*(4), 26. https://doi.org/10.5751/ES-03701-150426

Western, D., Wright, R. M., & Strum, S. C. (1994). *Natural connections: Perspectives in community-based conservation*. Island Press.

Woodhill, J., & Röling, N. G. (1998). The second wing of the Eagle: The human dimension in learning our way to more sustainable futures. In N. G. Roling, & M. A. E. Wagemakers (Eds.), *Facilitating sustainable agriculture: Participatory learning and adaptive management in times of environmental uncertainty*: Cambridge University Press.

PART I

# Rural Community Development and Facilitating Change

CHAPTER 2

# Introduction to Section 1: Rural Community Development and Facilitating Change

*Jim Cavaye*

The vitality of rural communities remains critically important to the social cohesion, economic development and function of regions and indeed nations, including Australia. Communities with quality infrastructure, a diverse and evolving economic base, with good local governance and organisation and with social cohesion and enthusiasm, are desired outcomes of community development processes. Chapters in this section provide examples of rural community development in action including the provision of community infrastructure, changes in agricultural extension and the development of community-based initiatives.

While there are many definitions of rural community development, the concept remains relatively simple. It fundamentally involves the concept of community—a commonality between people, identity, livelihoods, social

J. Cavaye (✉)
University of Tasmania, Hobart, TAS, Australia
e-mail: Jim.Cavaye@utas.edu.au



C. Baldwin and S. van Bommel (eds.), *Rural Development for Sustainable Social-ecological Systems*,
Palgrave Studies in Environmental Sustainability,
https://doi.org/10.1007/978-3-031-34225-7_2

interaction, history and experience, shared values, caring for each other and collective action, through building networks and social capital.

It also involves development—enhancing function, ability, organisation and direction, supported by agencies. Putting the two ideas together—community and development—gives the fundamental concept of community development. A process where communities of people foster what makes a community—their connection to each other, their economic opportunities, organisation, leadership, skills, identity and collaboration. It is a process where they can identify what is important to them, and that values community knowledge.

It often has been described as making progress on the five "capitals" of communities—financial, physical, social, human and environmental. Yet, additional capitals include cultural, political and spiritual assets. It involves enhanced social networks, governance, connection to place, diversity and equity leading to a series of practical actions resulting in improved infrastructure, services and employment.

Elements of rural community development can be seen in many aspects of work involving rural people, their organisations and communities. One of the foundations of functional communities, and the basis for economic development, is access to adequate infrastructure such as roads, power and telecommunications. While it requires the support of agencies, it also relies on a redistribution of power whereby community members lead and implement actions. In Chapter 3, Purwanto describes three case studies of infrastructure provision in Indonesia. The provision of infrastructure itself had variable impacts on local households and communities but what was crucial was communication and engagement between local people and government around improvements to rural infrastructure.

Empowerment and self-direction, supported by appropriate policy and organisational allies, is another core element of community development. This is illustrated in Chapter 5 where Inne Dwiastuti describes the economic empowerment of women in Indonesia and the importance of women's engagement both in terms of creating economic vitality in communities and in fostering gender equity.

Empowerment can also involve challenging the status quo. The civil rights movement and organised rural advocacy reflect elements of the "conflict" approach accepted in community development.

The processes and principles of rural community development are also reflected in the evolution of agricultural extension from a service-oriented technical assistance approach to a social learning and holistic management focus. Indeed, a developing collective and community focus in agricultural extension has moved extension more towards the practices of rural community development. This has led to not only technical improvements in agriculture but also more functional rural communities with greater equity and capacity.

This is shown in Chapter 4, where Severine van Bommel, Jeff Coutts, John James and Ruth Nettle explain the key trends in agricultural extension in Australia. They show how extension practice has changed from technical support for individual producers to seeing agriculture as social and biological systems. In Chapter 7, Chengdong Liao and Yiyu Liu add to this by demonstrating the difficulties and unintended consequences of a single technological extension approach in rural China. They show that a more community-oriented and systemic approach can better engage with landholders and help manage complex issues around land use and food production.

Rural community development fundamentally relies on the empowerment of citizens and self-directed movements. Ideally, these community initiatives are supported by the government and other "allies". A clear example of local empowerment is the Landcare movement in Australia which reflects many principles of rural community development. In Chapter 9, Ken Keith explains the alignment between the origins and operation of Landcare and principles of rural community development. He outlines areas where Landcare, as a rural community development process, could be enhanced. He also suggests important opportunities such as further development of social capital, greater links with First Nations organisations and greater capacity to link with ventures such as carbon farming.

Rural community development both as a practice in communities, and as an approach taken by external agencies, has evolved substantially. The emphasis on asset-based community development from the 1990s has seen an overall strengths-based approach being taken. This has seen communities focusing on assets rather than deficits, and also looking to refine and reframe assets, and combine assets in new ways.

Place-based initiatives have expanded. These have taken comprehensive approaches to community improvement within a locality such as Logan Together, BurnieWorks or the Latrobe Valley Authority. These initiatives often involve leadership development, locally agreed priorities, cross sectorial approaches and greater partnership with government and other agencies. These initiatives often involve collective impact approaches which see community development as system change involving agreed priorities, ongoing communication, coordinated activities and shared evaluation.

Another trend has been a re-emphasis, particularly by government, on community engagement. This has been driven by both greater appreciation of the need to genuinely engage with communities and by the often political effects of clear community disengagement with government. Additional resources, expertise and effort has been invested in engagement with subsequent success. However, lack of coordination, poor follow-through with action and lack of continuity of contact has arguably led to cynicism about engagement processes.

This engagement trend is explored in the two following chapters. In Chapter 6, Sergio Teixeira, discusses community participation in the design of research, development and extension programs in the dairy industry in Australia and Brazil. Close engagement of producers revealed issues of communication and farm management as priorities rather than traditional technical issues of pasture management and genetics. In Chapter 8, in an international development program context, Gomathy Palaniappan emphasises the importance of appropriate local engagement. She reflects on needs being identified broadly by aid agencies, but without close local participation, gender inequities can be entrenched and producers will "go along" with aid projects that do not necessarily meet real needs. Private sector sponsors can also provide incentives that can distort community development processes.

The major recent trend in rural community development is around resilience. Community resilience is the capacity of communities to proactively adapt and manage change, complexity and uncertainty. It fundamentally involves leadership, organisation and community agency and many factors contribute to resilience such as social networks, engaged governance and a diverse economy.

The focus on resilience has largely developed from community and government responses to sudden impacts on communities such as natural disasters such as drought, fire and flood, loss of services and infrastructure, or loss of economic opportunity such as irrigation rationalisation or major business closures. Yet, much of the narrative of community resilience focuses on community stoicism, rather than the adaptive elements of resilience. Many community leaders and policy-makers see resilience as withstanding change or "bouncing back" from a shock. While this is undoubtedly important, true resilience involves anticipating change, managing uncertain systems and building adaptive capacity.

While not necessarily a trend in community development, it is important to note an emerging re-discovery of community as a result of the Covid-19 pandemic. Managing the pandemic has required collective understanding, care and action. Anecdotally, this has led to increased awareness of caring for each other, particularly vulnerable people, and modifying individual actions for collective benefit. The impact of the pandemic has also contributed to a re-focus on health, family and community.

These trends in approaches to rural community development have led to greater sophistication in community-based improvement. Yet, the basic principles of community development remain crucially important. These include community improvement based on values and how people want their community to be, empowerment and self-help, reducing barriers to opportunity and fostering equity in communities, supporting "drivers" and encouraging motivation and passion.

While the many approaches to rural community development have evolved, and many communities have successfully managed change and fostered their capacity, the complex and often "wicked" problems for many communities remain. The forces driving rural decline have seen some single economy and remote communities struggle to adapt. Entrenched disadvantage persists in communities and barriers remain for groups to being fully included in society and economic opportunity.

Volunteer and engagement fatigue limits participation in many communities but also puts greater pressure on engagement processes being more co-designed and appropriate. The provision of infrastructure, particularly high-speed internet access, across dispersed communities remains an important priority together with digital "literacy". The economic and social transition in adapting to climate change offers both challenge and opportunity. The attraction and retention of people in regions, particularly skilled workers who can provide services, continues to be a challenge particularly in remote communities. The engagement of youth, in an age of changing civic participation preferences, also is an important priority.

CHAPTER 3

# Adaptation and Changes in Rural Economic Activities: A Local Community Perspective on the Role of Rural Infrastructure

*Purwanto Purwanto*

## Introduction

Progressive expansion of infrastructure facilities has long been recognized as critical for development, especially in rural areas. It creates better market access, connectivity, trade and job creation, attracts more investment, and more opportunity for new economic activities (Medeiros et al., 2021). Sufficient infrastructure is one of the key factors in connecting rural people to inclusive growth (Hallegatte et al., 2019; World Bank, 2020), and public demand and expectations for improved infrastructure services, both qualitative and quantitative, have increased in line with the growth of social and economic activities in both urban and rural

P. Purwanto (✉)
Research Center for Behavioural and Circular Economics, National Research and Innovation Agency, Java, Indonesia
e-mail: purw013@brin.go.id; purwanto_7@yahoo.com



C. Baldwin and S. van Bommel (eds.), *Rural Development for Sustainable Social-ecological Systems*,
Palgrave Studies in Environmental Sustainability,
https://doi.org/10.1007/978-3-031-34225-7_3

areas. However, many rural areas have experienced slower development progress than urban areas because they receive less attention from governments, resulting in less availability of public services to rural populations (Barrios, 2008). The lack of infrastructure services is therefore a challenge to boost the potential resources for rural economic activities (Agenor, 2008; Pinstrup-Andersen & Shimokawa, 2008).

New provision of rural infrastructure will not create an automatic enhancement in the rural economy. This requires a transitional period supported by government programs, as well as efforts from rural communities, to maximize the benefit of the available infrastructure for rural productive economic activities (Gibson & Olivia, 2010; Rahman, 2014). Assessing the potential opportunity and benefit from available rural infrastructure varies according to rural people's capability to utilize it promptly.

In addition, a proper combination of infrastructure will be more effective through a strong and complementary connectivity than a sole infrastructure facility (Agenor, 2010; Rahman, 2014). For example, the availability of improved roads will support the movement of goods and services, but it will give more benefits if other supportive infrastructure facilities exist, such as telecommunications. The communication devices that connect to internet services to expand the accessibility of information also add electricity to charge the cell phone battery or turn on the television. The associations between various infrastructure facilities are important to support the transitional process of optimizing the rural resources and enhancing the rural economic activities. However, it depends on the response of local people to actively adapt and change their economic activities.

## Statement of the Problem

It is widely known that infrastructure contributes to the delivery of economic growth at the macro level (Aschauer, 1989). However, there is little explanation that shows the impact of infrastructure provision at the micro-level of rural communities and households in rural areas is not. The work of infrastructure at the micro-level requires an understanding between local stakeholders and households to adapt in the transformation process of rural development through changes in the availability of rural infrastructure facilities (Jiao et al., 2017). A better understanding of rural community perceptions of the potential benefits and pitfalls of rural infrastructure can give useful information for government decisions to realize infrastructure needs in rural areas.

It is widely argued that infrastructure will automatically create a sophisticated development outcomes. However, while infrastructure enables development, it does not assure successful development alone. There is also the influence of human factors as the subject or actors of development. It is necessary for the local population to take their role actively in the adjustment and transitional process in order to take advantage of the opportunities offered by their economic activity (Kusumastuti, 2015; Tran & James, 2019). The impacts of infrastructure on rural development varies, either positive, negative, or mixed depending on the responses of different rural areas.

Inspired by the work of my Principal PhD Advisor Prof. Helen Ross on rural community development, this paper aimed to investigate local communities' and households' perspectives on the influence of infrastructure in their economic activity. The following questions were developed to guide this discussion. First, what is the perception of rural communities on the availability of rural infrastructure in their areas? Second, how does the infrastructure facilities influence the activities of rural households and rural business actors in their respective economic activities? The intended outcome is to enrich the discussion from theoretical and empirical insights of the role of infrastructure in rural development at the micro-level, and pose potential ways to maximize the benefits of infrastructure to rural households.

## Methods

To capture comprehensive information on the perspective of local communities on rural infrastructure facilities, this paper draws on field research reports from three selected subdistricts in East Lombok District, Indonesia. East Lombok District is located on the east coast of Lombok Island (West Nusa Tenggara Province) and occupied 33.88% of the island area, see Fig. 3.1. East Lombok district has 1,183,204 inhabitants, counted 23.88% of the total population of West Nusa Tenggara province (Statistics Indonesia, 2018).

The cases were selected to represent: the diversity of available infrastructure; a variety of economic activity; and varying topography (highland, low land, and coastal). The three subdistricts of Sikur, Keruak, and Jerowaru were the fieldwork locations for six focus group discussions

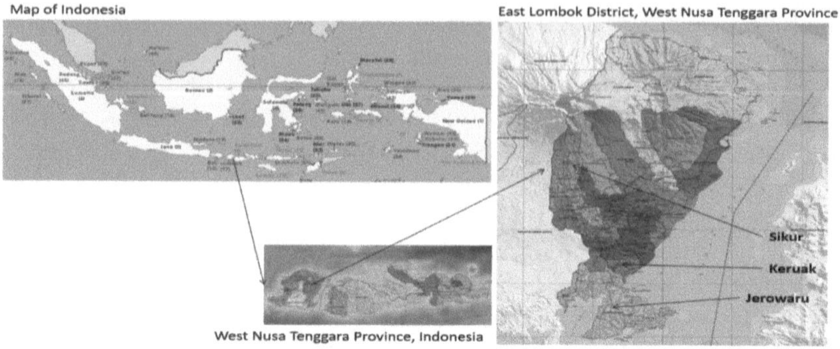


**Fig. 3.1** Map of Indonesia, West Nusa Tenggara Province, East Lombok District

(FGDs) in all subdistricts. FGDs were held with 10–12 participants that divided into two groups of farmer and non-farmer participants in each subdistricts. Sikur represents a highland area near to Rinjani Mountain with a variety of rural economic activities related to farming, business, and tourism; while infrastructure services are available in most areas, there is some problem with road quality, especially in more remote areas. Keruak is located in a lowland area with good infrastructure facilities and people work in farming, business, and those in coastal areas engage in fisheries. Jerowaru is the southernmost area of East Lombok District where some parts experience poor infrastructure service, especially in southern areas with limited to no telecommunication networks, bad roads, and severe effects from drought occasionally. Field observation and interviews of local people were conducted to collect information on the variety of rural economic activities and perceptions of rural communities about the available infrastructure facilities.

## Findings

### *Community Perspectives on the Influence of Infrastructure on Rural Economic Activities*

The first research question aimed to uncover the perception of rural communities about the availability of rural infrastructure in their area. There were different views and perceptions of infrastructure from the communities' perspectives. More than 75% of FGD participants described infrastructure as a government-provided facility that people can utilize it. The example of the participants understanding about infrastructure can be seen from one participant in Sikur's farmer group FGD as mentioned:

> So what we know is that infrastructure is a facility that can support community activities both in terms of economy, agriculture and so on. One example might be opening a road or for a farmer might be irrigation and so on. (FGD I, Farmer Group, Sikur Sub-district, 24 January 2017)

The statement of each FGD participant demonstrated that the availability and condition of infrastructure varies in rural areas. For example participants in Keruak Subdistrict stated they had no problem accessing telecommunication and electricity, while participants in the Jerowaru Subdistrict mentioned that some of their hamlets experience a lack of access to electricity. The problem of a telecommunication network in some hamlets was highlighted by participants from the Sikur Subdistricts as well. The physical conditions of available infrastructure also vary, as explained by participants in all FGDs, from good, reasonably good to bad, mostly caused by poor maintenance and service.

An FGD participant in Sikur Subdistrict expressed his optimism about the progress of infrastructure development in their area. The FGD participant compared the current situations with past conditions, many infrastructure facilities are now available, such as better roads, dams, and telecommunications. However, participants of FGD in Jerowaru Subdistrict regretted the slow progress in infrastructure provision in their area, for example, the southern part of Jerowaru Subdistrict still have no electricity facility. There are three out of four hamlets in Sekaroh Village do not have access to the electricity grid. Some houses use diesel to generate electricity at their own expense. Some other houses received solar panel from the government but the capacity is limited to lighting only and often does not last longer than a year due to the lack of maintenance services.

Regarding telecommunication infrastructure, there are two main telecommunication providers covering most of the area along the roadways. However, those providers are not covering most of the southern region of Jerowaru Subdistrict with a better telecommunication service. People continue to experience weak or a "blank spot" area. This condition is in contrary with the benefit of mobile phone as explained by FGD participants. A farmer FGD participant in Jerowaru Subdistrict gave an example about the way they sell the agricultural products to another island by utilizing telecommunication service. The availability of telecommunications services changes the way of people in promoting their product. It was enough to promote their product by sending pictures through "Whatsapp"[1] service. Another example on the benefit of the internet also explained further in gathering knowledge and information about new techniques/methods, as explained by a farmer in Keruak Sub-district:

> In the past we did not know about what commodities were suitable, how to plant them, and what should we do after harvesting. If you do not have adequate [market] information, most people will be talkative, when the price of a commodity increases, all of them will plant the same commodity. Later, when the harvest takes place, it will cause abundant stock and reduce the selling price, that is market law. (FGD V, Farmer Group, Keruak Sub-district, 8 February 2017)

FGD participants from non-farmer group in Sikur and Keruak Subdistricts described the support of telecommunication and internet facilities for their households in two conditions. Firstly, when a household becomes a producer, they can find some useful information for their business, such as the price of inputs, the availability of new processes/methods, and the dynamic of market demand. Secondly, when a household is a consumer, they can find out about the most affordable price for daily necessities (goods and services). However, FGD participants from non-farmer group in Jerowaru complained about the tendency of overuse of internet services with the possible socio-cultural impact that can lead to negative behaviors. They concerned with the underage internet users who need an intensive supervision and clear guidance in the filtering process from any inappropriate content.

In addition, the availability of internet service supports the online-based governance to improve the rural digital literacy, which is beneficial for village development (Manoby et al., 2021). Village offices need internet access because many administrative matters are currently managed online in terms of public services and making a fast communication and coordination with higher levels of government. Digitizing governance is useful to stimulate and accelerate the preparation and process

[1] Whatsapp is one of the popular social media platforms among Indonesian community that allow people to make a text chat, voice call, and video call as supported by internet service/connection.

of electrification infrastructure, telecommunications, and transportation connectivity between villages.

### *Rural Infrastructure Development in Creating Economic Opportunity*

The second research question focused on how the availability of infrastructure facilities influences the activities of rural households and business actors in their respective economic activities. It can be explained from the view of single and collective infrastructure facilities. Both single infrastructure and collective or a set of infrastructure facilities contribute to rural economic opportunity. It can provide an opportunity for rural people to access more resources through their economic activities and improved communication between rural people (Estudillo & Otsuka, 2015; Gibson & Olivia, 2010). The transformation derived from improved infrastructure can be enhanced by the transmission of macro-level government policy to the meso and micro-level of rural households in the context of economic activity (Gulli, 1998).

"Economic opportunity" is a neutral term to describe the potential or possibility of rural people to gain benefit from the available rural resources. It refers to the equality of opportunity in economic activities and does not reflect the equality of jobs and income, rather income and jobs are part of the opportunity itself (Corneo & Grüner, 2002; Garcia-Valinas & Torgler, 2005). However, the ability to take the opportunity depends on the internal and external factors of the rural economy. It can constrain opportunities to start a possible transformation of their economic activity at the micro-level. Internal factors can include the household's division of tasks and responsibilities in their family. In addition, the individual capability and intention or motivation to take the opportunity varies based on some circumstances, such as educational background, experience, skill, and knowledge. External factors that influence the ability to take the economic opportunity depends on some conditions such as financial capacity, political intervention, environmental issues, and infrastructure limitations (Brett, 2006).

### *The Use of Infrastructure in Rural Economic Activity: Variability in Adaptation and Changes*

*Farming Activities*

Nearly all FGD participants from the farmer groups agreed that they need at least two types of infrastructure to support their activities, namely, farm roads and irrigation. The role of rural infrastructure in farming activities can be illustrated in Table 3.1.

A participant from farmer group in Jerowaru Subdistrict mentioned the far distance of their farm land. In their case, it takes around 30 minutes to one hour to reach their farmland by motorcycle. They combined the use of motorcycle and continued on a walk to reach their farm because there is only a narrow footpath access. This geographical condition also raised a problem in transporting the farming commodity. For example, the cost is around IDR 25,000 (AUD 2.5) per quintal (a unit of weight equal to 100 kg, labor cost included) for moving the paddy from the farm to their house. Since the transport cost is considered as burden to their income, the community is building a farm road by using village funds and other government funds. The farm road is connecting the farmland to the nearest main road. The improvement of the farm road could be done by

**Table 3.1** Farming rice activity and involvement of infrastructure in farming activity

| *Phase/Activities* | *Main infrastructures* | *Supporting infrastructures* | *Periods* |
|---|---|---|---|
| Preparation Phase<br>Land Preparation | Water supply/ Irrigation | Farm road | 10–15 days |
| Vegetative process | Water supply/ Irrigation | Farm road | 30–35 days |
| Generative/reproductive processes | Water supply/ Irrigation | Farm road, Telecommunication | 30–40 days |
| Harvesting time | Farm road, roads | Electricity | 5–10 days |
| Post-harvest activities (collecting, cleaning, drying, milling, packaging, transporting) | | | |
| Consumption (Individual storage) | Electricity | Telecommunication | N.a |
| Marketing for middlemen and industry/end market | Roads, telecommunication | Electricity | N.a |

improving a rural footpath to motorcycle accessible track or upgrading the feeder road around the farmland (Jenkins et al., 2020). If a farm road connects with the highway road (national and provincial highways), it can make farmers save more money and increase farmers' income due to the efficiency of transport cost (Weng et al., 2021). The benefit of farm roads for household income can be calculated from the reduction of transportation costs up to 75% cheaper than before the availability of farm roads. The presence of farm roads also facilitates access for farmers to maintain irrigation along their farmland.

> The experience of the agricultural business from 2003 up to now, we feel supported [by the existence of infrastructure] to improve the economy, income and welfare of the community because of the existence of road infrastructure and irrigation. Five or ten years ago, the road and irrigation infrastructure were in a very, very severe condition. (FGD I, farmer group, Sikur Sub-district, 24 January 2017)

Although paddy farming is considered as the main activity, tobacco commodity is the main source of income for farmers in East Lombok District. Tobacco is usually planted after paddy rice harvest or in the period around March and June. An illustration of tobacco farming activity can be seen in Fig. 3.2.

A farmer from the FGD participant in Jerowaru Subdistrict who is planting tobacco explained the importance of telecommunication infrastructure in negotiating the tobacco contract system with the buyer. Farmers can easily make an exchange information about the best price to negotiate with the buyer before dealing the contract. In contrast with tobacco, a FGD corn farmer participant in Jerowaru Subdistrict mentioned that tobacco contract farming system is not applicable to corn as there is a trend of price hike during the harvest season. Their strategy for selling the corn is by expanding the market and searching the information of current market price that can be found in the internet. The corn farmer can use the telecommunication to contact their farmer's networks or communicate with potential buyers elsewhere.

*Non-Farming Activities: Tourism Villages*

Infrastructure that supports farming may be used by other sectors. Likewise infrastructure developed for tourism purposes can be utilized by other activities such as farming, fishing, or public services such as health

**Fig. 3.2** Some tobacco farming activities. Top left, tobacco farmland; Top right, female farm laborer in land maintenance in tobacco farm

and education. A participant of a FGD non-farmer group in Sikur Subdistrict explained the promising eco-tourism in the northern region of East Lombok District. That region consists of some tourism destination such as mountainous scenery, waterfalls, and a climbing track to Mount Rinjani. The access to the tourism area is already facilitated with better roads. There are several accommodation services available around the villages such as a hotel and homestays. However, internet services is still have bad connection and network problem. For some tourists, mostly domestic tourists, internet service is necessary and it can influence the decision to stay as they planned or leave sooner.

> [telecommunications network] did not cover all of our villages, because there were mountains blocked [the signal], maybe [our area] need two more towers. If you use "Android" [an operating system, but the local people think that "Android" is an internet connection] is very difficult. [We] Can only call and text. For [regions] that are in the south, [even] the 3G network cannot be accessed, only 2G is for old cellular phones [without internet features]. (FGD IV, Non-Farmer group, Jerowaru Sub-district, 27 January 2017)

The potential of tourism also available in the southern part of East Lombok district. FGD participant from the non-farmer group in Jerowaru said that there are agriculture-based tourism (agrotourism) and coastal tourism in the southern region with some attractive landscape of agricultural land. They also have an annual popular cultural festival called "bau nyale",[2] a local ceremony which are combining cultural and tourist elements. However, the non-farmer group participants in Jerowaru explained the problem of the lack of tourism facilities in their area. As a result, even though they have some popular tourism destinations, tourists cannot access the location easily which has burdened the potential of mass tourist to come. The problem of poor facilities in a tourism area was also highlighted by a FGD participant in Sikur Subdistrict.

> Especially in the world of tourism, this road infrastructure in my opinion is the most important [....] Now the [number of] visitor is increasing because of the ease of transportation access. From the hospitality sector, you can now advertise rooms [via] online, etc. The point is that the existence of technology greatly facilitates current economic activities. (FGD II, Non-Farmer group, Sikur Sub-district, 25 January 2017)

> This road infrastructure is the main benchmark in efforts to improve the tourism sector. Tourists often complain, [and] at the same time, [they] give advice to improve the road conditions that currently bad and narrow. (FGD IV, Non-Farmer group, Jerowaru Sub-district, 27 January 2017)

Participants of the FGD non-farmer group in Keruak Subdistrict believed that tourism is beneficial for their economic activity because it can create a mutual linkage to other economic sectors. The livelihood adaptation in village tourism varies due to the different socio-economic backgrounds of the rural household but tourism operators likely have a stronger livelihood adaptability (Li et al., 2020). For example, tourism can provide an alternate market for the farmer to sell their commodity to the tourist or supporting tourism business such as restaurants, and hotels. These activities can be existed because of the adequate road transport and telecommunication service to make a good access between seller and buyer. In addition, as mentioned by a FGD participant who are member

[2] Bau Nyale is an annual cultural festival located along the southern coastal area of Lombok Island. The Nyale Festival, takes place every year in the tenth month of the lunar Sasak calendar and is celebrated in either February or March each year. Anonym, "Bau Nyale Festival: Lombok Annual Cultural Event", available online at https://adventure-lombok.com/news/bau-nyale-festival/.

of Pokdarwis,[3] more supporting facilities are needed by the tourism industry, such as public toilets, garbage collection sites, and parking lots that are not sufficiently available in tourist destination sites.

Further, local participation is essential for village tourism (Priatmoko et al., 2021). Local participation requires good cooperation and connectivity between local actors, not only regarding labor and complementary industries but also to build a sense of urgency between village government and the villager (Meekes et al., 2017). This cooperation is useful to encourage more tourist arrivals through understanding how to promote tourism that aligns with local attitudes, values, and culture.

The support of infrastructure for rural tourism can be illustrated in Fig. 3.3 (Purwanto, 2020). Rural tourism can be improved by optimizing the unique locality, abundant natural assets, and some local attractions such as local tradition, cultural events, and religious ceremony. Such tourism activities involve some rural actors, including farmers, small-scale enterprises, and traders. Some economic activities can be involved in tourism such as accommodation, and transportation as well as supporting services on banking facilities, food and beverage, and local guides. To support the rural tourism, it required a good combination of a set infrastructure facilities.

The expansion of value-adding from tourism will increase the chance of creative economy. FGD participants from farmer and non-farmer groups in Jerowaru Subdistrict believed that improved infrastructure facility (such as road) will significantly support the tourism activities. They pointed out that it is necessary to connecting or integrating the infrastructure facilities. For example, the local and village governments are responsible for building some classification of roads (such as artery, collector, and local roads) that is not covered by the national road project provided by the central government. The governments should also support the community with the implementation of specific program to stimulate the growth of the creative economy, such as improve the quality of handicraft industries, enrich the cultural festivals and increase the tourism services.

[3] Pokdarwis is an acronym of "Kelompok Sadar Wisata" (tourism awareness group), a tourism awareness group/association comprised of local people managing some local tourism activities.

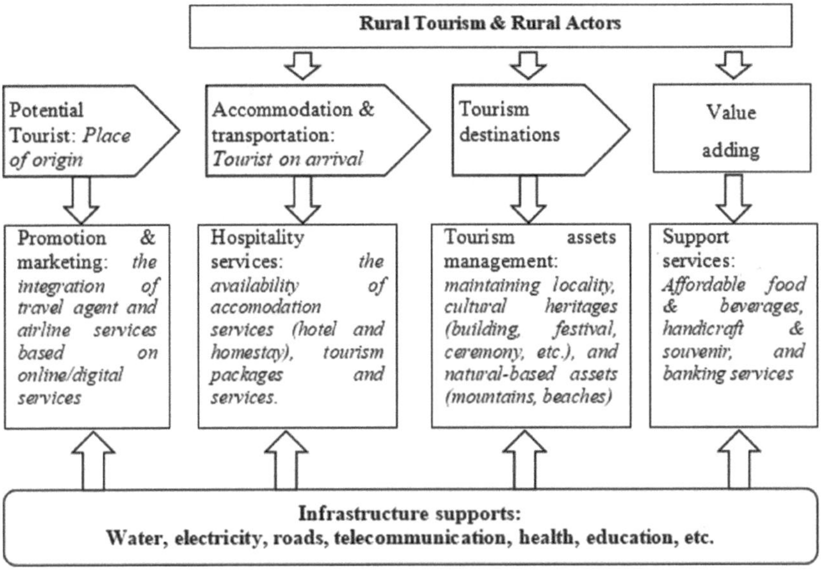


**Fig. 3.3** The support of infrastructure to rural tourism

## Discussion

In line with the work by Prof. Helen Ross, this chapter shows that infrastructure governance needs to link up to local perceptions in order for it to lead to meaningful rural development. This case demonstrates that the availability of infrastructure does not lead to it automatically being utilized by the people. Some types of infrastructure require additional device or equipment. For example, the availability of telecommunication network needs mobile phone and internet services charge. Telecommunication networks will have no effect on people who do not have a telecommunication device, either they do not want to buy one or they have no ability to purchase it. The electricity service will make household to pay the monthly connection fee, as well as buying additional electrical for their house. The process of change in household economic activity is also influenced by the combination of infrastructure available. For example:

1. Improved road and stable telecommunication networks will encourage people to frequent their travel.

2. Road access for tourism is crucial to attract more tourists. It can be matched with a better telecommunication services to enable online tourism promotion, marketing, and bookings accommodation services.
3. The combination of better road and telecommunications infrastructure are supporting effective and efficient trading activities.
4. Farm roads, irrigation, and telecommunication can increase the chance of people to shift their job or having double or multiple jobs.
5. Interaction of a set of infrastructure facilities such as road, telecommunication, and electricity can boost the manufacturing and packaging process to upgrade the business activities.

In response to the research question about the influence of infrastructure on local economic activity, the results suggest that infrastructure availability provides opportunities for some rural households to change their economic activity, but is of little interest or relevance to others. Some people have only ever worked as a farmer of rice, secondary crops, and the tobacco, as a legacy from their parents.

Some people preferred to stay on their current job without having any motivation to shift or change the way they do it, regardless of how good or bad the state of infrastructure in their area is. This is a kind of internal factor that the existence of infrastructure facilities becomes insignificant on people to seize opportunities and make changes (Purwanto, 2020). For example, some farmers insist and remain committed to their current agricultural practice. On the other hand, some households were unable to make a change in their economic activity because they had no relevant skills or adequate knowledge to compete in the job market. Another reason for not making any changes is that they simply had no idea how to utilize the available infrastructure in their area to increase their economic activity.

The opportunity for some households to make adaptations and changes is not an immediate process. The external factors that may influence the length of time needed to make changes in economic activity are the nature of the job itself. For example, agriculture sector depends on the land contour, farm characteristic, commodity price, and the natural resources capacity. The adaptation process will be more effective with a good understanding of the local community perceptions of the benefits for economic activities (Alam et al., 2017; Jha & Gupta, 2021; Kiem &

Austin, 2013). Any improvement in the availability of infrastructure will take different period of time to change households' economic behavior.

Farmer respondents explained how they changed their economic activity in particular ways. For example, some diversified their agriculture by adding new crops (such as tobacco, chilli, and watermelon) and expand their market, while some people just changed their way to travel to their workplace. The local community adaptation and changes are tailored according to local wisdom, social capital, culture, and habits that can be mainstreamed into existing economic activities or community development policy (Jiao et al., 2020).

Non-agricultural activities such as small micro-enterprises will follow a process of adaptation and changes to the availability of infrastructure facilities, where these create opportunities for efficiency in production processes and market expansion. Some of the small, micro-enterprises used online information services to look for suppliers of raw materials and expand their markets. The availability of current infrastructure interacts with other conditions such as employment opportunities, capital, and skills, which can influence people's decision to change jobs or the process of adapting their economic activity (Purwanto, 2020).

People reacted at different speeds to the availability of infrastructure. Some people changed their economic activities quickly, while others responded more slowly, depending on their ability to take advantage. Poor households tended to take fewer opportunities to gain benefits from the available infrastructure due to their circumstances, such as limited capital assets. The type of infrastructure also affected household responsiveness, especially when its services were easy to observe. For example, a new paved road or a water dam was easier for people to observe and utilize it, while the change in telecommunications service, e.g. the improvement in cellular networks has made it harder for people to recognize immediately, especially for those who are not familiar with the internet services.

The availability of infrastructure can change some economic activities. However, every economic activity will follow their process to adapt and change in specific ways (Purwanto, 2020). It can be a change in the production process to increase their capacity by using new machinery/ equipment. Business actor can expand the market for both agricultural and non-agricultural activities by using telecommunication facilities. Rural electricity enables certain small-scale agricultural processing industries to thrive and allows fishermen and fishmongers to purchase cold storage

(Ponce de Leon Barido et al., 2017). The Peruvian solve the problem of telecommunication network coverage by implementing Small Rural Operator (SMO) to reduce the cost of investment and take more benefit from better telecommunication services (Prieto-Egido et al., 2020).

The household decision-making process is also affected by their optimistic or pessimistic views of the prospect of infrastructure facilities. Households with better access to infrastructure facilities tend to get more opportunity to diversify their source of income (Teame & Woldu, 2016). Maya et al. (2019) also concluded that some factors, including education, social networks, and infrastructure, are important to improve rural household's capacity to adapt to any possible changes in their economic activity. Access to information on farming influenced the decision of rural people to make changes and adapt their activities (Khanal et al., 2019).

The FGD results showed that non-farmers can adapt more easily and faster than the farmer group for any changes in infrastructure facilities. The financial benefit gained from utilizing the infrastructure outweighs the cost to access infrastructure facilities. The way infrastructure affects rural household income highlighted the role of infrastructure in rural households' income-generating process. A set of infrastructure works better in supporting the enhancement of rural economy. The establishment of infrastructure facilities created synergies that go beyond the contribution of each infrastructure type when considered individually. Without good roads to access markets and reliable power supply to keep telecommunications going, potential multiple benefits are not realized for rural economic growth.

## Conclusion

The study has provided a more comprehensive picture than previously available of the impact of infrastructure development at micro and household level in rural areas. Inspired by the work of Prof. Helen Ross, this study starts from the perspective of local people in rural areas. As such, the result of this study is a significant departure from previous studies, enriching the analysis by focusing on the process and change at the micro and household level arising from infrastructure availability in rural areas.

The study found that rural people respond differently to the availability of infrastructure. The benefits are promising for people who are able to transform their economic activity, create employment opportunities and increase their income. However, the process of change in economic activity is complex and specific. Such a change requires active decision-making on the part of households in considering all factors to seize opportunities and to derive benefits. The complexity of infrastructure's role in changing economic activities has meant that the ability to take advantage of opportunities is not equal but diverse among households with different backgrounds.

The tourism sector is the sector that has the biggest opportunity to reap the benefits of integrated infrastructure development. Road construction is indeed necessary, but the expansion of telecommunications, internet, electricity, and other networks must well-established. At this point, the completion process can be phased in for all types of infrastructure projects. However, it should be noted that a systematic or targeted plan by the government to improve the rural economy is essential. Each village government can contribute, where possible, by using village funds for small-scale rural infrastructure provision and local economic empowerment.

To sum up, the provision of appropriate infrastructure facilities in rural areas, highlighting the multiple benefits of providing a combination of complementary infrastructure. This study also filled the research gap on success factors for the connection between types of infrastructure facilities in rural areas. Further research is needed because there is still a problem when infrastructure does not meet people's expectations in terms of quality, coverage, and maintenance. This implies the need for an additional or complementary strategy to help people improve their ability to generate income.

**Acknowledgments** This article has been expanded from my thesis titled "The Role of Infrastructure in Improving the Local Economy and Rural Household Income: A Study in Rural Development in Indonesia" (Purwanto, 2020). It was completed under the main supervision of Prof. Helen Ross and I would like to express my deepest gratitude to Prof. Helen Ross for her guidance, advice, and supervision throughout my work.

## References

Agenor, P. R. (2008). Fiscal policy and endogenous growth with public infrastructure. *Oxford Economic Papers, 60*(1), 57–87. https://doi.org/10.1093/oep/gpm018

Agenor, P.-R. (2010). A theory of infrastructure-led development. *Journal of Economic Dynamics & Control, 34*(5), 932–950. https://doi.org/10.1016/j.jedc.2010.01.009

Alam, G. M., Alam, K., & Mushtaq, S. (2017). Climate change perceptions and local adaptation strategies of hazard-prone rural households in Bangladesh. *Climate Risk Management, 17*, 52–63.

Aschauer, D. A. (1989). Is public expenditure productive? *Journal of Monetary Economics, 23*(2), 177–200.

Barrios, E. B. (2008). Infrastructure and rural development: Household perceptions on rural development. *Progress in Planning, 70*(1), 1–44. https://doi.org/10.1016/j.progress.2008.04.001

Brett, J. A. (2006). "We sacrifice and eat less": The structural complexities of microfinance participation. *Human Organization, 65*(1), 8–19. https://doi.org/10.17730/humo.65.1.6wvq3ea7pbl38mub

Corneo, G., & Grüner, H. P. (2002). Individual preferences for political redistribution. *Journal of Public Economics, 83*(1), 83–107. https://doi.org/10.1016/S0047-2727(00)00172-9

Estudillo, J. P., & Otsuka, K. (2015). Infrastructure and the transformation of the rural economy in the Philippines. In J. P. Estudillo & K. Otsuka (Eds.), *Moving out of poverty: An inquiry into the inclusive growth in Asia* (pp. 121–142). Routledge.

Garcia-Valinas, M., & Torgler, B. (2005). *More income equality or not? An empirical analysis of individuals preferences*. http://www.crema-research.ch/papers/2005-23.pdf. Accessed August 5, 2015.

Gibson, J., & Olivia, S. (2010). The effect of infrastructure access and quality on non-farm enterprises in rural Indonesia. *World Development, 38*(5), 717–726. https://doi.org/10.1016/j.worlddev.2009.11.010

Gulli, H. (1998). *Microfinance and poverty: Questioning the conventional wisdom*. Idb.

Hallegatte, S. p., Rozenberg, J., & Rentschler, J. (2019). *Lifelines: The resilient infrastructure opportunity*. World Bank Group.

Jenkins, J., Peters, K., & Richards, P. (2020). At the end of the feeder road: Upgrading rural footpaths to motorcycle taxi-accessible tracks in Liberia. *NJAS—Wageningen Journal of Life Sciences, 92*(1), 1–12. https://doi.org/10.1016/j.njas.2020.100333

Jha, C. K., & Gupta, V. (2021). Farmer's perception and factors determining the adaptation decisions to cope with climate change: An evidence from rural India. *Environmental and Sustainability Indicators, 10*, 100112.

Jiao, X., Pouliot, M., & Walelign, S. Z. (2017). Livelihood strategies and dynamics in rural Cambodia. *World Development, 97*, 266–278. https://doi.org/10.1016/j.worlddev.2017.04.019

Jiao, X., Zheng, Y., & Liu, Z. (2020). Three-stage quantitative approach of understanding household adaptation decisions in rural Cambodia. *International Journal of Climate Change Strategies and Management, 12*(1), 39–58. https://doi.org/10.1108/IJCCSM-01-2019-0004

Khanal, U., Wilson, C., Hoang, V.-N., & Lee, B. L. (2019). Autonomous adaptations to climate change and rice productivity: A case study of the Tanahun district, Nepal. *Climate and Development, 11*(7), 555–563.

Kiem, A. S., & Austin, E. K. (2013). Drought and the future of rural communities: Opportunities and challenges for climate change adaptation in regional Victoria, Australia. *Global Environmental Change, 23*(5), 1307–1316. https://doi.org/10.1016/j.gloenvcha.2013.06.003

Kusumastuti, A. (2015). *Rural infrastructure and gender inequality in agribusiness development (A Case of Batu, East Java Province, Indonesia).* Paper presented at the Food Security and Food Safety for the Twenty-first Century (Proceedings of APSAFE2013).

Li, H., Guo, T., Nijkamp, P., Xie, X., & Liu, J. (2020). Farmers' livelihood adaptability in rural tourism destinations: An evaluation study of rural revitalization in China. *Sustainability, 12*(22), 9544. https://doi.org/10.3390/su12229544

Manoby, W., Afriyanni, A., Fitri, S., Pranasari, M., Setyaningsih, E., Rosidah, R., et al. (2021). Digital village: The importance of strengthening village resilience in the digital age. *Jurnal Bina Praja, 13*(1), 53–63.

Maya, K. A., Sarker, M. A. R., & Gow, J. (2019). Factors influencing rice farmers' adaptation strategies to climate change and extreme weather event impacts in Bangladesh. *Climate Change Economics, 10*(3), 1950012. https://doi.org/10.1142/S201000781950012X

Medeiros, V., Ribeiro, R. S. M., & do Amaral, P. V. M. (2021). Infrastructure and household poverty in Brazil: A regional approach using multilevel models. *World Development, 137*, 105118.

Meekes, J. F., Buda, D. M., & de Roo, G. (2017). Adaptation, interaction and urgency: A complex evolutionary economic geography approach to leisure. *Tourism Geographies, 19*(4), 525–547. https://doi.org/10.1080/14616688.2017.1320582

Pinstrup Andersen, P., & Shimokawa, S. S. (2008). Rural infrastructure and agricultural development. In F. Bourguignon & B. Pleskovic (Eds.), *Rethinking infrastructure for development* (pp. 175–203). Word Bank Publications.

Ponce de Leon Barido, D., Fobi Nsutezo, S., & Taneja, J. (2017). The natural and capital infrastructure of potential post-electrification wealth creation in Kenya. *Energy, Sustainability and Society, 7*(1), 1–15.

Priatmoko, S., Kabil, M., Purwoko, Y., & David, L. D. (2021). Rethinking sustainable community-based tourism: A villager's point of view and case study in Pampang Village, Indonesia. *Sustainability (Basel, Switzerland), 13*(6), 3245. https://doi.org/10.3390/su13063245

Prieto-Egido, I., Valladares, J. A., Muñoz, O., Bernuy, C. C., Simo-Reigadas, J., Quispetupa, D. A., et al. (2020). Small rural operators techno-economic analysis to bring mobile services to isolated communities: The case of Peru Amazon rainforest. *Telecommunications Policy, 44*(10), 102039.

Purwanto, P. (2020). *The role of infrastructure in improving local economy and rural household income: A study in rural development in Indonesia*. PhD Thesis, School of Agriculture and Food Sciences, The University of Queensland. https://doi.org/10.14264/fdde938

Rahman, S. (2014). Impact of rural infrastructure on farm and non-farm enterprise choice and income in Bangladesh. *The Journal of Developing Areas, 48*(1), 275–290.

Statistics Indonesia. (2018). *Statistik Potensi Desa Indonesia 2018* (Village Potential Statistics 2018). Statistics Indonesia.

Teame, G. T., & Woldu, T.-Y. M. (2016). Factors affecting rural households' income diversification: case of Zoba Maekel, Eritrea. *American Journal of Business, Economics and Management, 4*(2), 7–15.

Tran, T., & James, H. (2019). Changing livelihood options as adaptation: A comparative analysis of three flood control schemes in the Vietnamese Mekong Delta. In H. James (Ed.), *Population, development, and the environment* (pp. 137–154). Palgrave Macmillan.

Weng, Y.-Z., Zeng, Y.-T., & Lin, W.-S. (2021). Do rural highways narrow Chinese farmers' income gap among provinces? *Journal of Integrative Agriculture, 20*(4), 905–914. https://doi.org/10.1016/S2095-3119(20)63374-3

World Bank. (2020). *Infrastructure in Asia and the Pacific: Road Transport, Electricity, and Water and Sanitation Services in East Asia, South Asia, and the Pacific Islands*. World Bank.

CHAPTER 4

# Trends in Extension in Australia

*Séverine van Bommel, Jeff Coutts, John James, and Ruth Nettle*

## Introduction

This chapter was written on the occasion of the retirement of Prof. Helen Ross who has played an important role in keeping extension 'alive' as a discipline at the University of Queensland. She drew together the roles and staff of two groups, an agricultural extension research and teaching group at St Lucia campus, founded with the appointment of

S. van Bommel (✉)
School of Agriculture and Food Sciences, Faculty of Science, University of Queensland, Gatton, QLD, Australia
e-mail: s.vanbommel@uq.edu.au

J. Coutts
Coutts J&R, Toowoomba, QLD, Australia
e-mail: jeff@couttsjr.com.au

J. James
Enablers of Change, Hobart, TAS, Australia


C. Baldwin and S. van Bommel (eds.), *Rural Development for Sustainable Social-ecological Systems*, Palgrave Studies in Environmental Sustainability,
https://doi.org/10.1007/978-3-031-34225-7_4

Dr Joan Tully in 1962 and continued by key thinkers such as Assoc. Prof. Shankariah Chamala, Dr Bruce Crouch, Dr Bruce Frank and Ken Keith, and—on its closure—the teaching role of the Rural Extension Centre, a partnership between the university and the Department of Primary Industries, based at Gatton campus. That innovative teaching and research group was led by Prof. Beth Woods then Dr Jeff Coutts, with Dr John James as training manager. In broadening the roles of the merged group—in accordance with her own appointment initially as a research leader—Helen took the critical decision in the early 2000s to retain agricultural extension as a key role in her group, rather than to follow international trends to dissipate use of this term under subsidiary themes such as 'innovation'. Demand from developing country students was a key influence in this decision. She always retained a permanent position in her group which was dedicated to extension, while also embedding extension principles in other teaching and research. She and her group thereby contributed to the retention of agricultural extension as a discipline in Australia.

When agricultural extension in Australia became increasingly privatised in the 1990s, many felt that the field of extension was entering a crisis period, fearing that it would disappear as a profession (Hunt et al., 2012, 2014; Ross & Baldwin, 2021; Vanclay, 1994). The share of public funding for R, D and E dropped from approximately 86% in 1993 to approximately 55% in 2013 (Keogh et al., 2018). Firstly, the decreasing public investment resulted in increased pluralism as more private, industry and other groups stepped into extension roles—an international phenomena (Nettle et al., 2017, 2021). Secondly, it also led to perceived fragmentation, arising from the fact that people delivering extension services commercially—from different organisations—were each involved in their own activities, often duplicating activities or leaving gaps in terms of access to services or types of services (Paschen et al., 2017). However, twenty-five years later, we can conclude that

e-mail: john@enablersofchange.com.au

R. Nettle
Rural Innovation Research Group, University of Melbourne, Melbourne, VIC, Australia
e-mail: ranettle@unimelb.edu.au

agricultural extension in Australia has not died or disappeared in the way that many expected (Ampt et al., 2015; Ross & Baldwin, 2021). In the professional field, we observe that the Australasia-Pacific Extension Network—a network of extension professionals—has maintained a relatively stable membership of approximately 500 members throughout the years (Ross & Baldwin, 2021). In addition to that, we observe that Australian approaches to extension science and practice have been influential internationally, in particular the learning systems philosophies of Hawkesbury college (Bawden—see below)—later introduced at Wageningen University, the Open University of Milton Keynes and the Swedish Agricultural University—and the local and self-organising farmer group model of the Australian Landcare movement (Campbell & Siepen, 1994). This shows that despite initial fears, the science and practice of extension are still very much alive in Australia both in research and in practice (Ross & Baldwin, 2021). Earlier research has described the history of agricultural extension until 2010 (Coutts & Roberts, 2011). This paper provides an elaboration and update of that earlier work, taking as its focus the practices of practitioners in agricultural extension in Australia.

In order to learn what it means to be 'doing' extension in Australia today, the practices of those working in the field of extension (both academically and professionally) and the meaning they generate from this work is important to understand. We therefore conducted a literature review focussing on articles and book chapters that had been published between 2010 and 2021. Firstly, the research team identified 31 key authors to be included. Then using Scopus, any papers published by key Australasian Extension Researchers were downloaded. This resulted in 453 papers. Subsequently the following inclusion/exclusion criteria were used to select the relevant papers: (1) include articles that are relevant specifically to Australia; (2) published between 2010–2021; (3) exclude papers focussing on forestry, impact assessments, economic analysis (cost-benefit) or ones that are highly technical (e.g. new variants of plants or machinery); (4) exclude editorials that introduce journal series and papers that are exclusively general/vague discussions, opinions/perceptions. (5) exclude documents that cannot be accessed online or downloaded (e.g. books or some book chapters). This resulted in 135 papers. Secondly, it was acknowledged that not all extension research is published in academic journals, so to account for this, the publications of the Rural Extension and Innovation Systems Journal (including APEN conference papers)

2010–2021 were downloaded and analysed. This resulted in 205 papers. Thirdly, we used our own experience of working in extension (both extension practice and extension theory) to identify influential and important events, developments and themes to provide context and details to complement the literature.

Our chapter is structured as follows. We start by discussing the meaning of extension in Australia, based on an analysis of what the authors refer to when they talk about extension in their publications. We then present an overview of the evolution of extension in Australia focussing on institutions, funding and extension models. After this, we identify and discuss the three main extension models that underpin extension theory, practice and education in Australia. We end with a reflection on the trends in extension in Australia and we conclude that the pluralistic extension landscape makes it difficult to define a common Australian school of extension. However, considering the extension and advisory system as a whole, the diversity of organisations and people involved in extension is an under-recognised asset to agriculture and the environment in Australia.

## Disentangling the Meaning of Extension

The notion of extension first evolved in Ireland as a result of the outbreak of potato blight in the mid-nineteenth century. To help farmers through the potato famine, appointed lecturers travelled around the country to support smallholder farmers in improving their practices and diversifying their production. These appointed lecturers were referred to as 'practical instructors in husbandry' (Jones & Garforth, 1996). The term 'extension' originated in the UK soon after, where it was used to refer to similar practices of universities 'extending' their teaching beyond university settings. It was introduced by James Stuart of Cambridge University when he referred to his teaching of out-of-college lectures to local associations in the area. This extension work from the UK inspired similar work in the US where land-grant colleges were established to provide informal agricultural education to local farmers (Jones, 1994). The assumption was that new knowledge and technologies developed at universities would automatically benefit farmers and improve their livelihoods (Leeuwis, 2004). Leeuwis (2004, p. 23) put it as follows: 'the basic thrust is that 'the common folk' are to a degree 'living in the dark', and that there is a need for well-educated people to 'shed some light' on their situation by means of educational activities'. So the assumption is that the real world can be known and solutions provided independent of the people that create the world and the problems as we know it (Ison & Russell, 2000). Originally the meaning of 'extension' was very closely related to education and teaching and it had hierarchical and paternalistic connotations (Leeuwis, 2004).

Following Leeuwis (2004), experience showed that despite educational and communicative efforts by extension professionals, farmers were not always able or not always willing to adopt the new technologies or practices that researchers and extensionists were introducing. It became clear that farmers did not always have the same aspirations as the institutions that the extension system was part of. So this led to the insight that extension is not always 'good' or 'beneficial'. Instead it is an active intervention that only works if there is at least a partial overlap between the aspirations of the extension organisations, researchers, policy makers, agricultural industries and the farmers. Instead of a linear process of sending information, extension became to be defined as an interaction in which all stakeholders have important insights to contribute. The process became more like a negotiation or learning process in which changed understandings and practices are constructed in interaction (Leeuwis, 2004). Participants therefore needed to work actively together in order to learn their way out of their own problems (Ison & Russell, 2000; Röling, 2000). The understanding of extension evolved from linear, individual approaches to interactional, collective approaches with interpretive and communicative dimensions (Leeuwis & Aarts, 2011).

This significant change in understanding led some leading authors to completely move away from the concept of extension (e.g. Ison & Russell, 2000; Leeuwis, 2004; Röling & Wagemakers, 1998). These authors argued that the meaning of 'extension' could not be expected to include the interactional, collective approaches. In the Netherlands at Wageningen University the interactive, collective approach was subsequently referred to as 'agricultural innovation' and the word 'extension' became interchangeable with the 'knowledge transfer' approach (Leeuwis, 2004). Similarly, in the UK at the Open University of Milton Keynes a distinction was made between 'first order R&D' and 'second order R&D' with the former referring to the knowledge transfer approach and the latter referring to the interactive approach (Ison & Russell, 2000, 2011; Leeuwis, 2004). In contrast to this, Australia and most developing countries continued to use the 'extension' term (Ampt et al., 2015).

Funding bodies, governments and farmers still use the term and in the 2010–2020 decade, attempts of extension professionals to change the language to terms such as 'capacity building' or 'innovation/knowledge brokering' were unsuccessful. Despite the initial expectation that this group was likely to erode (Leeuwis, 2004), extension practitioners ended up stretching the meaning of extension and now use it as an umbrella term to refer to knowledge transfer as well as participation, dialogue, social learning, capacity building or co-innovation. As a result, in Australia the notion of extension was redefined, yet retained (Ampt et al., 2015; Jennings et al., 2011).

When we take a closer look at what it is that people do when they refer to their practices as 'extension' in Australia then we see two different 'types' of extension (see also Marsh & Pannell, 2000; Ross & Baldwin, 2021). An analysis of articles published in the Rural Extension and Innovation Systems journal—the journal associated with the Australasia-Pacific Extension Network—during the last decade, reveals an extension to mean 'agricultural extension' in addition to 'environmental extension'. First of all, we can distinguish articles on 'agricultural extension' that focus on productivity and control of plant diseases. Issues discussed are for example agricultural technologies (Achora et al., 2018; James et al., 2015; Kamruzzaman et al., 2019; Phillips et al., 2021; Stitzlein et al., 2020), gender issues in agriculture (Bhattacharjee et al., 2018; Creelman, 2018; Hilmiati et al., 2017; Howard et al., 2015; Rogers & Grieves, 2017; Sarker et al., 2011; Walters et al., 2018), agricultural value chains (Creelman, 2018; Hill & Hathaway, 2017; Jayasena et al., 2019), mental health in agriculture (Botha & White, 2013) or agricultural extension in conflict or post-conflict areas (Koyenikan & Vincent-Osaghae, 2019; Vock & Carusos, 2019). Second, we can distinguish articles on 'environmental extension' that focus on land degradation and integrated management of water catchments. Issues discussed include for example social resilience (Collins, 2013; Keith, 2017; Wegscheidl et al., 2013); bush fire recovery (Roberts et al., 2011; Young et al., 2017); Natural Resource Management practices (Barbi et al., 2015; Braddick, 2016; Maskey et al., 2017; Murray & Hasselman, 2013; Pike, 2013) including Reef protection (Maher et al., 2017; McCosker & Northey, 2015; Rouse & Davenport, 2017; Royle & Di Bella, 2017; Wegscheidl et al., 2013, 2015) and climate change (Collins, 2013; Graymore et al., 2016).

So taking this together, the meaning of 'extension' in Australia includes both knowledge transfer as well as the interactional approaches that evolved in response to it. It also includes agricultural extension as well as environmental extension. When comparing the use of the term 'extension' and the associated practices in Australia relative to the Netherlands and the UK, two stark differences emerge. First, environmental extension in the Australian context has the closest parallel with 'area-oriented policy' in the Netherlands and social learning in natural resource management or 'second order R&D' (Ison & Russell, 2011) used by the Systems group at the Open University of Milton Keynes (UK). Second, agricultural extension in the Australian context is sometimes described in terms reflective of Agricultural Innovation systems thinking at Wageningen University in the Netherlands. The differentiated practices of agricultural and environmental extension in Australia are a unique attribute internationally in terms of the meaning and practices of extension. To examine how extension has evolved in Australia, the next section examines the institutions of extension in Australia and the way different approaches have emerged.

## The Evolution of Extension in Australia

Following Hunt et al. (2012) the evolution of extension in Australia shares some similarities with the way in which extension evolved in the US and in Europe, but there are also some remarkable differences. Extension in Australia began as a government instrument during WWII during which it was used by the Commonwealth to increase agricultural production to feed the soldiers and increase export (Hale & Ashton, 2002). Producers had to adapt their agricultural practices and technologies to environmental circumstances that were very different from the European ones that they were used to (Ross & Baldwin, 2021). Extension became increasingly important in the 1950s and 1960s when Australian agricultural policy focussed on increasing production (Hunt et al., 2012) and Australian agriculture jumped on the 'agricultural treadmill' (Cochrane, 1958). Extension practice was very much based on Transfer of Technology thinking as developed by Everett Rogers in the US in the 1950s and aimed at changing the knowledge and skills of farmers. Australia did not pursue a Land Grant University style system as in the US and different to Europe where the extension was mostly driven by National government agricultural departments, in Australia extension was organised at the State level. Compared to the US, the role of Australian universities was relatively limited and mostly focussed on training agricultural science graduates who increasingly took up positions as extension practitioners (Hunt et al., 2012). At the University of Queensland, Dr Joan Tully developed the first postgraduate course on agricultural extension in Australia in the 1960s (Crouch, 2002). At the University of Melbourne, an agricultural extension research unit was established by Peter Salmon and Stuart Hawkins and undergraduate and postgraduate subjects in extension were initiated (Salmon & Hawkins, 1973).

In the 1970s, both in Australia and overseas, it became clear that technologies and practices from 'station based research' (such as on government or university farms) were not always easily transferrable to local farming contexts (Bawden, 1978; Leeuwis, 2004; Petheram & Clark, 1998). New genetic varieties, for example, that would perform well under the controlled conditions of the research station, would not always show the same increase of productivity in the variable conditions of a farmers' field. This led to the insight that the local farming context needed to be taken into account in the development of new technologies and practices. Farming System Research (FSR) responded to the challenges and short comings of the earlier Transfer of Technology (ToT) model and local producers became actively involved in defining research priorities as well as in conducting on-farm research (Petheram & Clark, 1998). Initially extension practice was characterised by pointing farmers in the direction of technical solutions or informing them about different sorts of management practices (Coutts & Roberts, 2011). Similar to ToT, FSR was aimed at generating knowledge and information to be used by farmers, however FSR was significant in being demand driven (farmers in the lead) instead of supply driven (researchers in the lead) (Hunt et al., 2012).

In the 1980s, based on experiences from Farming Systems Research, group learning and farmer-led discussion groups were developed and practised in Australia and New Zealand. This required collective action, facilitation and social learning which are now quite common but which were pioneered by extension practitioners. Alongside these trends in extension practice, extension science evolved. Similar to elsewhere in the world, Hawkesbury Agricultural College (HAC) recognised that agriculture is a complex system with social-economic elements as well as biophysical elements. Breaking out of the knowledge transfer tradition, systems thinking, experiential learning and adult learning principles were introduced into an innovative undergraduate 'Systems Agriculture' degree. Post-graduate training was developed for mid-career professionals (Hunt et al., 2012). Meanwhile increased awareness of environmental degradation led to initiatives that were intended to promote environmental stewardship. In 1986, the Landcare programme was launched in Victoria to assist farmers in learning their way out of natural resource management problems such as land degradation, soil erosion or feral pests. Farmers were encouraged to form Landcare groups at the catchment or sub-catchment level to tackle problems that none of them would be able to solve on their own (Campbell et al., 2017). Alongside the momentum for group-based and farmer focussed models of extension, public investment in agricultural extension was increasingly under scrutiny. In 1982, the New Zealand government privatised their extension service and in Australia, most state governments began reducing investment in extension, often by not replacing staff who retired or left.

In the early 1990s, the University of Queensland, following the lead of HAC, partnered with the Queensland Department of Primary Industries (DPI) to form the UQ Rural Extension Centre in Gatton which offered postgraduateextension training to mid-career professionals. As extension was mostly delivered by government departments, extension practitioners working for the Department of Primary Industry were the main target audience (Hunt et al., 2012). The Executive Director of the agriculture group of the Department of Primary Industries (DPI) saw this as a potential trigger to follow the USA Land Grant System but it never worked out that way in practice. Although initially it was very successful, privatisation of extension meant that the number of extension professionals working for the Department greatly diminished over the years. This resulted in a lower number of student enrolments for the UQ Rural Extension Centre until eventually led to the demise of the Centre.

In 1993 the Australian Pacific Extension Network (APEN) was founded to raise the profile of extension as a profession (Bourne, 1999). With a bi-annual conference at its core, it became a vibrant network with a relatively stable membership of around 500 people (Coutts & Roberts, 2003). In the meantime, the success of the Landcare program in Victoria was recognised by the Commonwealth government and Landcare was embraced as a national initiative (Campbell et al., 2017; Curtis & de Lacy, 1998). The 1990s were designated as the 'decade of Landcare' (Lockie, 1999) with nearly two thirds of the Australian farmers being involved in Landcare groups (Curtis & de Lacy, 1998; Lockie, 1999).

Inspired by Soft Systems thinking (Checkland, 1981) and the 'Farmers First' movement (Chambers et al., 1989), extension practice was characterised by group-oriented social learning and facilitation of interactive processes (Coutts & Roberts, 2011). State-based agricultural extension however was on the decline, with continued disinvestment.

In the late 1980s, The Primary Industries and Energy Research and Development Act 1989 (PIERD Act) provided for the establishment of rural Research and Development Corporations (RDCs) and the legislative basis for regulating their funding and administration. After these reforms in the 1980s and 1990s industries, organised by the agricultural sector, began to invest in their own Research and Development. Research and Development Corporations (RDCs) were formed for each agricultural commodity group and were funded through a co-funding arrangement in which farm levies are matched by Federal Government funding (Ampt et al., 2015). The late 1990s can be characterised by privatisation of agricultural extension as well as a rise of public environmental extension. Reflecting the trend in privatisation of agricultural extension in other parts of the world (Leeuwis et al., 2006), public funding for agricultural extension was reduced. It was argued that individual farmers were getting the benefit of productivity advances from extension support and therefore extension should be funded privately. In addition to this, the implicit assumption was that private and corporate actors would be more effective and efficient in providing demand-driven extension services and therefore private actors should replace public extension practitioners (Marsh & Pannell, 2000; Paschen et al., 2017) The RDC model, is still considered as the largest structural change to agricultural RD&E in Australia and the positioning of extension relative to the research effort differed in each RDC, whereby some RDCs co-invested with government for extension services, others established their own industry extension service and others did not invest in extension at all. Meanwhile many former public extension officers re-established themselves as private consultants. So in contrast to New Zealand or Europe where agricultural extension was completely privatised, Australian extension gradually evolved into a pluralistic system (Hunt et al., 2012) with multiple actors—public, industry (farmer levy paid) and private—delivering extension services (Hunt et al., 2014; Nettle et al., 2017; Paschen et al., 2017). There is no central coordination of extension activities in the form of policies like there is in the EU (Nettle et al., 2021). In response to this lack of coordination, in 2006 the State Extension Leaders Network (SELN) was formed under the auspices of APEN. This enabled extension managers representing each of the state and territory governments to meet and discuss the approaches used in each of their agencies (State Extension Leaders Network, 2006).

As a result of this increased privatisation, extension professionals—especially the private consultants—experienced an increasing 'projectification' of activities whereby targets for adoption or practice change in defined areas made it difficult to advocate for or maintain a 'farming systems' approach nor develop long-term relationships with producers or stay up to date through training (Allan, 2012). This lack of visibility and support for the extension as a profession and discipline by government and other organisations was reflected in lower enrolments in undergraduate and postgraduate extension programmes. Some universities retained subjects and courses on extension or community and stakeholder engagement as part of broader agricultural programmes (Ampt et al., 2015).

In contrast, environmental extension remained publicly funded (Ross & Baldwin, 2021). The success of the Landcare movement led to the creation of Natural Resource Management bodies which continued to support the environmental stewardship of landholders (Campbell et al., 2017). As environmental stewardship is a public good, this legitimised continued public funding of environmental extension activities. Environmental extension required extension professionals with skills in capacity building and community engagement but now environmental extension practitioners new to the job had to deliver this without proper training (Pannell et al., 2006). Many environmental science graduates were unfamiliar with farming systems and integrating environmental stewardship with the issues and challenges faced by farmers in maintaining a profitable and sustainable farm business.

Since 2010, the trend of privatisation of extension continued but the degree of privatisation varies depending on the sector. In some sectors, such as cotton, sugar, horticulture or pork, extension is almost fully provided by private consultant through fee-for-services models. In other sectors, such as Dairy, Meat and Livestock or Grape and Wine, extension is provided by the industry through a mix of private consultants and public extension officers. In the NRM and Biosecurity field, extension is still almost entirely delivered by publicly funded extension officers (Nettle et al., 2018, 2021). Sectors with shorter and more intensive production cycles (often less remote), such as annual cropping and intensive animal production, seem to have embraced privatisation more than sectors such as grazing enterprises which have longer and more extensive production cycles (often more remote) (Wellington et al., 2021).

However if we 'follow the money' instead of following the actors, privatisation is less pronounced than when we just look at the type of actor that delivers the service. Farmers, for example, may access subsidised private sector advice which was funded by government or industry in a project (Nettle et al., 2021). Based on self-report of the income streams of over 700 Australian farm advisers, support to farmers is funded 50:50 from the public and private sector (Nettle et al., 2018), reflecting a public-private partnership in the agricultural extension system. This is quite different from the situation in the US or the Netherlands where the investment of the private sector in extension now exceeds public investment (Geerling-Eiff et al., 2014). The challenge with the increasing privatisation of extension is in how to address longer-term complex, industry-wide problems such as land degradation or climate adaptation that extends beyond the provision of information and advice for the short-term. Natural Resource Management groups—mostly government agencies either statutory or non-statutory—have taken over some of the responsibilities from State Departments to do some of this, for example through voluntary, publicly funded Best Management Practice programs (BMP)[1] (Moravek et al., 2017), while State Departments have retained responsibility for regulation of biosecurity and the environment (Wellington et al., 2021).

During the last decade, new approaches to extension science and practice have emerged. On the one hand, there is a revitalisation of farmer-led and peer-to-peer learning in agricultural extension practice (Knook & Turner, 2020). In recent years, for example, the Reef program invested in research on behaviour change based on principles of behavioural psychology for the sugar industry (Pickering et al., 2020). And as part of the Queensland Government's Reef Water Quality Program, a farmer-centred extension model of practice was developed (Williams et al., 2021). On the other hand, agricultural innovation systems and co-innovation are also coming up in extension discourse and practice (Fielke et al., 2018; Paschen et al., 2021). The privatisation of the extension in Australia has been evaluated in several assessments during the last decade (Hunt et al., 2014; Keogh et al., 2018; Nettle et al., 2018). These show that

[1] Not all BMP programs are run by NRM Groups. BMP such as cotton and wine are run by industry groups and RDCs.

firstly, producers are not engaging with private extension professionals as much as had been expected and limits their ability to make complex decisions e.g. with regard to digital agriculture or with regard to new market requirements such as animal welfare or environmental credentials (Nettle et al., 2018). Secondly, the split between agricultural extension and environmental extension has led to framing of land degradation and environmental problems as externalities of the farming system instead of taking a whole farming approach and thus seeing these as a pathology of the system (Nettle et al., 2018) and that makes it easier for farming operations to live off environmental capital (Wellington et al., 2021). Thirdly, there is a high turnover among private extension professionals associated in part with high overhead costs with short project cycles as well as high costs of professional development in staying up to date, especially for those working across various industries (Nettle et al., 2018).

The Australian Government is now concerned that the market failures associated with the privatisation of extension services are starting to impact productivity and innovation in the agricultural sector (Nettle et al., 2018). Recent investments by the Australian government in resourcing innovation platforms and extension roles reflect a turning point in thinking about the importance and role of extension. In early 2021, in establishing a bi-partisan 'Future Drought Fund', a $64 million program for 8 Drought resilience Adoption and Innovation hubs across Australia was announced (Australian Government, 2022b). These Hubs are intended to support farmers and communities in their drought preparedness. Each hub employs a knowledge broker, Regional Soil Coordinators and Adoption Officers who will work with farmers and communities as well as agricultural experts to improve the on-farm Drought resilience (Australian Government, 2022a). These seem to be a way for the Australian government to directly resource innovation and extension independent from the industry and state government pathways. Within an agricultural innovation system discourse, the Australian Government is re-investing in publicly funded agricultural extension. It is too early to say what the impact will be on the extension and the advisory system and the private sector response.

## Current Approaches to Extension in Australia

A wide range of approaches to extension have emerged over the past 50 years in Australia (see Fig. 4.1). Following Rickards et al. (2018), we distinguish three main, often overlapping, approaches to extension that are used in Australia today, namely producer-focussed approaches, interaction focussed approaches and system focussed approaches.

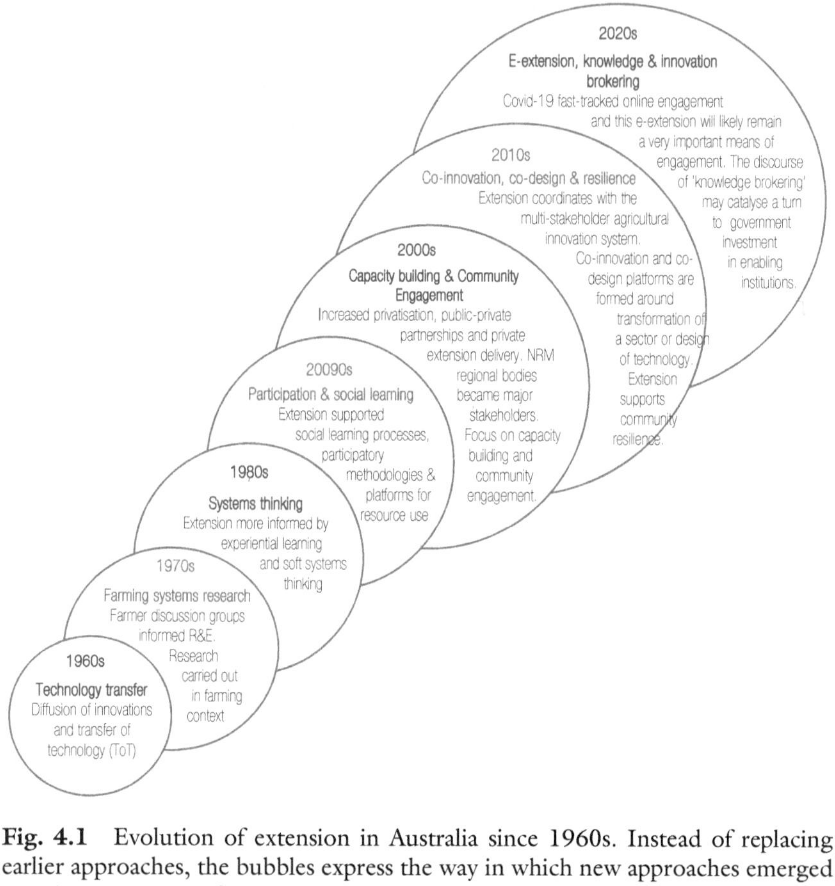


**Fig. 4.1** Evolution of extension in Australia since 1960s. Instead of replacing earlier approaches, the bubbles express the way in which new approaches emerged overtime on top of, and in addition to earlier approaches (Adapted from Coutts & Roberts, 2011)

### *Adoption and Diffusion*

The adoption and diffusion of innovations model, originally developed by Everett Rogers, is arguably the most well-known theory about how extension works. In many countries, the adoption and diffusion model became the leading policy model for how innovation in agriculture should be promoted. It emphasised Transfer of Technology and free markets as conditions for agricultural development (Roling, 2009). In Australia it is influential among researchers, RDC and policy makers (but less so among practitioners). The Australian agricultural R, D&E system is predominantly based on this model in which knowledge and technology that is developed by researchers (in one form or another), is distributed to producers and then adopted by producers. Examples include improved genetic crop varieties that are developed in the lab or slow release fertiliser as well as new digital technologies that are developed often with limited producer input. There is also a proliferation of decision support systems (expert driven) with limited uptake (McCown et al., 2009).

The most recent application of the diffusion and adoption theory in Australia is the ADOPT tool which was developed by CSIRO. ADOPT is an online, digital tool that predicts the farmer uptake of new agricultural practices based on (1) the characteristics of the innovation; (2) the characteristics of the target population; (3) the relative advantage of the innovation to the user; and (4) the awareness of the relative advantage of the innovation. Users are presented with a number of questions. Based on 22 variables, the ADOPT tools predicts the adoption curve and thus also predicts the time until the peak in adoption is reached (Kuehne et al., 2017). Although the original tool was developed for use in Australia, an adapted version of the tool had been developed for developing country contexts (Brown et al., 2016). In early 2022, the ADOPT tool was mostly used by policy makers (50%). Other users included higher education (25%) and business or private actors (25%). Although 75% of the users came from Australia, 25% of the users came from abroad (Rural Practice Change, 2022).

One could argue that the Best Management Practice programs, which were launched to improve land degradation in various agricultural industry sectors, also draw on adoption and diffusion thinking. BMP programs are voluntary programs in which producers can benchmark themselves against industry standards and improve their practices should they want to do so. Access to good information is seen as a key factor determining uptake. For example, in 2012 the Queensland government launched BMPs in Reef catchments to reduce erosion and run-off to protect the Great Barrier Reef. Among others, Grazing BMP was launched by the Beef industry whereas Smartcane BMP was launched by the Sugar industry. The uptake of BMPs has been limited with, for example only 35.8% of the grazing land managed under a BMP in 2018 and only 33% of the cane growers operating under BMPs in 2020 (Rural and Regional Affairs and Transport References Committee, 2020). As a result, industries are rethinking how to promote further uptake of BMPs and some industries are turning to behaviour change approaches based on social and organisation psychology to improve uptake. In the sugarcane industry, the government invested in the development of a behaviour change programme for cane growers to improve the uptake of BMPs. And in 2019, the Queensland government commissioned behavioural psychology training for extension professionals working in the sugarcane industry, grazing industry and the horticultural industry based on the work in the sugar industry (Pickering et al., 2020).

### *Interaction and Learning*

Although initially Landcare pioneered the development of social learning methods that later became mainstream worldwide, the institutionalisation of the model and the associated bureaucracy has been suggested to have led to an erosion of the community-based model that Landcare was based on (Campbell et al., 2017; Keith, 2019). There is concern that few Landcare groups are thriving, while others are surviving or have disappeared altogether (Keith, 2017). Despite this, the collaborative and interactive approaches, based on adult learning principles, have remained popular among producers and extension practitioners in practice. Field days and group activities that include peer-to-peer learning are often used by extension practitioners (Wellington et al., 2021). In addition to this, farmers—especially growers in Western Australia—started to self-organise into extension-based community groups. Some grower groups emerged in response to a perceived lack of extension at the local level. Other groups have arisen in response to specific needs or issues such as no-till associations, implementation of precision agriculture or grain quality issues (Anil et al., 2015). Currently more than 5000 farm enterprises across Australia are now part of grower groups with 40% of these located in Western Australia. To develop the capacity of these groups and to coordinate the links between the grower groups and other stakeholders, overarching alliances were formed. Although the Landcare groups can be seen as the forerunners of the grower groups there are also remarkable differences. Whereas Landcare groups were formed to deal with environmental issues, the grower groups are focussed on-farm productivity and profitability: 'the highly organized member-driven grower groups are the recipients of major research and development funding and, in some cases, invite the traditional research organizations and researchers to become participants, in a distinct contrast to earlier models where agencies engaged farmers' (Anil et al., 2015, p. 1105).

Furthermore, after Landcare the deployment of collaborative, participatory approaches and social learning in policy development and implementation also evolved further in relation to co-management of natural resources in partnership with indigenous groups (Ross et al., 2009). While not without challenges (Bardsley et al., 2021; Enqvist et al., 2018), Australia has also been a frontrunner in co-management of Natural Resources (Ross et al., 2009). Co-management allows Aboriginal people to form partnerships with other public and private actors for the collaborative management of terrestrial or marine resources and areas. It is a political process that is closely linked to land rights. It includes social learning and capacity building in which the development of trust and negotiation of power are very important elements (Nursey-Bray & Rist, 2009). There are various arrangements in Australia, ranging from Indigenous Protected Areas (IPAs) which are governed by Aboriginal people themselves, to joint management arrangements in which power is shared in a more equitable way between Aboriginal communities and other stakeholders (Ayre et al., 2021; Zurba et al., 2012).

### *Agricultural Innovation Systems and Systems Innovation*

Theoretically extension science has always been influenced by systems studies. We see three different system focussed approaches in extension studies in Australia. First of all, social learning was influenced by soft systems thinking (Checkland, 1981 in Klerkx et al., 2012) in the sense that it pays attention to how system boundaries are drawn, by whom and with what sorts of consequences. The idea is that when stakeholders recognise their interdependence in relation to resource use, they will redraw the system boundaries more appropriately and that will open up space for change (Klerkx et al., 2012). This evolved into further thinking about 'system innovation' or 'systemic innovation' for sustainable development (Ison, 2016; Ison & Straw, 2020). Resource dilemmas and competing claims are seen as a systemic failure that requires system transformation. In the late 1990s social-ecological systems thinking began to influence the thinking and practice of environmental extension. Social-ecological systems thinking is based on the idea that social systems and ecological systems are structurally coupled and thus co-evolve together (Berkes, 2009; Folke et al., 2005). Actions and technology can mediate the relationship between the social system and the ecological system (Renaud et al., 2010). From this perspective, land degradation is an emergent property of a social system that has lost its correspondence with the environmental system. When the system reaches a tipping point, it can move from one regime into a different, usually less desirable, regime. Soils, for example, may lose their capacity to filter due to changes in their biophysical and/or biochemical structure (Renaud et al., 2010). Therefore a lot of academic work is focussed on understanding and increasing the social-ecological resilience (Folke, 2006) for example by means of social learning or collective learning and co-management (Berkes, 2009; Folke, 2006; Gunderson et al., 1995; Plummer & Armitage, 2007). In Australia, the concept of social-ecological resilience was developed further from 2010 onwards in work on co-management (Maclean et al., 2013, 2017), community resilience (Berkes & Ross, 2013, 2016; Leite et al., 2019); human values (Jones et al., 2016; Ross et al., 2018) and agriculture (Ayre & Nettle, 2017; Nettle et al., 2015).

Third, in the early 2000s, Agricultural Innovation Systems thinking was developed alongside social learning and social-ecological systems thinking. Following Klerkx et al. (2012), Agricultural Innovation Systems thinking conceptualises agricultural innovation as a co-evolutionary process of hardware (technology), orgware (institutions) and software (people's perceptions and practices). Whereas systemic innovation and social-ecological system mostly work with problems on regional level, Agricultural Innovation Systems tends to work with problems on national level or agricultural sector level. It is therefore well suited to address problems that surpass the regional level such as zoonotic diseases, climate change or animal welfare issues. Agricultural Innovation Systems thinking is still 'under development' and different approaches co-exist. Some approaches are more structural, some are more process-oriented and some are more functionalist (Klerkx et al., 2012). The work in Australia has been influenced mostly by the Multi-Level Perspective framework (MLP) and the process-oriented view of Agricultural Innovation Systems (see Klerkx & Nettle, 2013; Murphy et al., 2013; Paschen et al., 2021) in which innovation is seen as the emergent property of a network of actors and their interactions which are influenced by circumstances (Leeuwis & Aarts, 2011). Niches are seen as the space in which actors experiment with new innovations. When the conditions are right—the incumbent regime has become instable due to wider external 'landscape' pressures—these niche experiments can grow and eventually they can break through and change the incumbent regime (Geels, 2006; Klerkx et al., 2012). In Australia, Agricultural Innovation Systems thinking was picked up and further developed by the Rural Innovation Research Group of the University of Melbourne. Since 2010, the agricultural innovation systems approach has been used in research in the Dairy industry (Klerkx & Nettle, 2013; Murphy et al., 2013), research on regional water productivity (Ayre et al., 2016) as well as research on the Australian extension system itself (Paschen et al., 2017, 2021). It is now also taught as part of the Agricultural Science Master at Melbourne University.

## Discussion

In this chapter, the evolution of extension in Australia has been explored with a focus on developments in extension science and practice relative to extension science scholarship in the Netherlands and the UK. Our findings show that Australian extension science and practice is unique in three different ways. First, extension in Australia includes both agricultural extension and environmental extension. Where the shift from extension to agricultural innovation elsewhere in the world has led to an apparent 'blind spot' for environmental extension, the practice of environmental extension has remained part and parcel of the way in which meaning has been given to extension in Australia. This can perhaps be explained by the way that the government retained extension practices in the context of Natural resource management (NRM) or environmental programs. Second, Australian extension science and practice is characterised by a diversity in thematic focus (both agriculture and environment), empirical practice (public and private), funding sources, the modes of employment of practitioners and methodologies. Whereas pluralism in agricultural advisory and extension services is more commonly defined only on the basis of the number and type of organisations involved in providing extension, the Australian extension system is pluralistic across these multiple dimensions. While this may be supporting flexibility and tailoring of extension to demand on one hand, coordination is difficult (Paschen et al., 2017, 2021) and there are risks of duplication or missing important stakeholders and groups (Norton & Alwang, 2020). Third of all, extension science and practice are characterised by social learning and interactive, systems thinking, reflecting second-order R&D and Agricultural Innovation Systems thinking. This is likely due to the strong Landcare legacy and the ongoing influence that Landcare as a philosophy and model still has in the Australian extension landscape.

Although extension has managed to survive under conditions of reduced public funding and with little central coordination—admittedly with various degrees of success—extensionists are struggling to keep their knowledge and skills up to date. Extensionists are under the constant pressure to improvise and learn their way out of new challenges such as digital agriculture, mental health issues of producers and impacts of natural disasters (such as drought, bushfires and floods) which are further exacerbated by climate change. Also the agricultural landscape is changing. Alternative farming systems, such as regenerative agriculture, are on the rise and markets are increasingly demanding agricultural products that come with environmental credentials (such carbon balance, biodiversity stewardship and/or de-forestation free credentials). We also see an increase in 'lifestyle' landholders, people in rural areas that are relatively new to farming and that deliberately chose sustainable—part time—farming as a lifestyle (Ross & Baldwin, 2021). There is huge potential for extension professionals to contribute to these new global challenges in agriculture and natural resource management (Ampt et al., 2015). A whole system approach is needed in which coordination by the government allows extension professionals to become true catalysts for the change people expect.

## Conclusion and the Contributions of Prof. Helen Ross

Coincidentally, or by design, rethinking extension was one of the important contributions of Prof. Helen Ross's work. At the University of Queensland, she kept extension alive by always having a position dedicated to 'extension and rural development' in her group. She argued that in light of new global challenges, we needed to re-invigorate extension to include not just food production but also 'food security, food sovereignty and food justice, empowerment and adaptive management capacity' (Ampt et al., 2015, p. 162). She argued that we needed to equip extensionists with 'a deeper understanding of change as a very personal, emotional process' (Ampt et al., 2015, p. 162) in order for them to be able to support people that are going through change processes. She always worked tirelessly and passionately to equip students in agriculture with some of these people skills during their studies and for many years she coordinated the Master of Rural Development which provided students and mid-career professionals with skills and capacities in facilitation, conflict management, community development and leadership. Prof. Helen Ross's legacy—both in research and teaching—lives on and will continue to contribute to 'retaining, refining and reinvigorating' extension in agriculture and natural resource management in order to allow 'for the field to continue to evolve in addressing contemporary and future global and local challenges' (Ampt et al., 2015, p. 163).

## References

Achora, J. C., Sseguya, H., Kyazze, F., Mkomwa, S., & Okello, D. (2018). ICTs for conservation agriculture: Influence of actor positioning in knowledge networks in Laikipia and Machakos counties, Kenya. *Rural Extension and Innovation Systems Journal, 14*(1), 24–33. https://doi.org/10.3316/informit.563342011055881

Allan, C. (2012). Rethinking the 'project': Bridging the polarized discourses in IWRM. *Journal of Environmental Policy & Planning, 14*(3), 231–241. https://doi.org/10.1080/1523908X.2012.702012

Ampt, P., Cross, R., Ross, H., & Howie, B. (2015). The case for retaining, redefining and reinvigorating extension in agricultural innovation systems. *Rural Extension and Innovation Systems Journal, 11*(1), 157–164. https://doi.org/10.3316/informit.661157849570442

Anil, B., Tonts, M., & Siddique, K. (2015). Grower groups and the transformation of agricultural research and extension in Australia. *Agroecology and Sustainable Food Systems, 39*(10), 1104–1123. https://doi.org/10.1080/21683565.2015.1081857

Australian Government. (2022a). *Drought resilience adoption and innovation hubs.* https://www.awe.gov.au/agriculture-land/farm-food-drought/drought/future-drought-fund/research-adoption-program/adoption-innovation-hubs. Accessed 5 May 2022.

Australian Government. (2022b). *Future drought fund.* https://www.awe.gov.au/agriculture-land/farm-food-drought/drought/future-drought-fund. Accessed 5 May 2022.

Ayre, M. L., & Nettle, R. A. (2017). Enacting resilience for adaptive water governance: A case study of irrigation modernization in an Australian catchment. *Ecology and Society, 22*(3), 1. https://doi.org/10.5751/ES-09256-220301

Ayre, M., Nettle, R., Erazo Bobenrieth, M., & Klerkx, L. (2016). Doing water research differently for innovation in regional water productivity in Australia. *Australian Journal of Water Resources, 20*(1), 39–52. https://doi.org/10.1080/13241583.2016.1162763

Ayre, M. L., Yunupingu, D., Wearne, J., O'Dwyer, C., Vernes, T., & Marika, M. (2021). Accounting for Yolŋu ranger work in the Dhimurru Indigenous Protected Area Australia. *Ecology and Society, 26*(1), 24. https://doi.org/10.5751/ES-12149-260124

Barbi, E., Denham, R., & Star, M. (2015). Do improved grazing management practices lead to increased levels of ground cover? *Rural Extension and Innovation Systems Journal, 11*(1), 114–121.

Bardsley, D. K., Bardsley, A. M., Moskwa, E., Weber, D., & Robinson, G. M. (2021). Challenges to the co-management of biodiversity in a reflexive modernity. *Geographical Research, 59*(3), 362–377. https://doi.org/10.1111/1745-5871.12468

Bawden, R. J. (1978). A perspective for parasite management. *Agriculture and Environment, 4*(1), 43–55.

Berkes, F. (2009). Evolution of co-management: Role of knowledge generation, bridging organizations and social learning. *Journal of Environmental Management, 90*(5), 1692–1702. https://doi.org/10.1016/j.jenvman.2008.12.001

Berkes, F., & Ross, H. (2013). Community resilience: Toward an integrated approach. *Society & Natural Resources, 26*(1), 5–20. https://doi.org/10.1080/08941920.2012.736605

Berkes, F., & Ross, H. (2016). Panarchy and community resilience: Sustainability science and policy implications. *Environmental Science & Policy, 61*, 185–193. https://doi.org/10.1016/j.envsci.2016.04.004

Bhattacharjee, D., Afrad, M. S. I., Haque, M. E., & Barau, A. A. (2018). Gender mainstreaming and participation in shifting cultivation among 'Chakma' tribe: A comparative study in selected areas of Bangladesh and India. *Rural Extension and Innovation Systems Journal, 14*(1), 11–18. https://doi.org/10.3316/informit.563211580257074

Botha, N., & White, T. (2013). Distress and burnout among NZ dairy farmers: Research findings and policy recommendations. *Extension Farming Systems Journal, 9*(1), 160–170. https://doi.org/10.3316/informit.409040801665872

Bourne, J. (1999). Australasia-Pacific Extension Network Inc. (APEN): Promoting extension as a discipline. *The Journal of Agricultural Education and Extension, 6*(2), 131–134. https://doi.org/10.1080/13892249985300241

Braddick, L. (2016). WA Wheatbelt landholders' perception of support for adoption of natural resource management practices. *Rural Extension and Innovation Systems Journal, 12*(1), 12–22. https://doi.org/10.3316/informit.210495120529482

Brown, P. R., Nidumolu, U. B., Kuehne, G., Llewellyn, R., Mungai, O., Brown, B., et al. (2016). *Development of the public release version of Smallholder ADOPT for developing countries.* Australian Centre for International Agricultural Research.

Campbell, A., Alexandra, J., & Curtis, D. (2017). Reflections on four decades of land restoration in Australia. *The Rangeland Journal, 39*(6), 405–416.

Campbell, A., & Siepen, G. (1994). *Landcare: Communities shaping the land and the future.* Allen & Unwin.

Chambers, R., Pacey, A., & Thrupp, L. A. (1989). *Farmer first: Farmer innovation and agricultural research.* Intermediate Technology Publications.

Checkland, P. (1981). *Systems thinking, systems practice.* Wiley.

Cochrane, W. W. (1958). *Farm prices: Myth and reality.* University of Minnesota Press.

Collins, H. (2013). Resilience through change: 'Beyond reasonable drought.' *Extension Farming Systems Journal, 9*(1), 20–27. https://doi.org/10.3316/informit.408724041154482

Coutts, J., & Roberts, K. (2003). *Extension models and best practice in extension.* Australasian Pacific Extension Network Forum.

Coutts, J., & Roberts, K. (2011). Theories and approaches of extension: review of extension in capacity building. In *Shaping change: Natural resource management, agriculture and the role of extension* (pp. 22–31). Australia-Pacific Extension Network (APEN).

Creelman, Z. (2018). Sharing the load: Improving technical skills and competencies of women on farms. *Rural Extension and Innovation Systems Journal, 14*(1), 147–151. https://doi.org/10.3316/informit.563751936423562

Crouch, B. (2002). Tully, Joan (1907–1973). In *Australian Dictionary of Biography.* National Centre of Biography, Australian National University. https://adb.anu.edu.au/biography/tully-joan-11891/text21297. Accessed 5 May 2022.

Curtis, A., & de Lacy, T. (1998). Landcare, stewardship and sustainable agriculture in Australia. *Environmental Values, 7*(1), 59–78. https://doi.org/10.3197/096327198129341474

Enqvist, J. P., West, S., Masterson, V. A., Haider, L. J., Svedin, U., & Tengö, M. (2018). Stewardship as a boundary object for sustainability research: Linking care, knowledge and agency. *Landscape and Urban Planning, 179*, 17–37. https://doi.org/10.1016/j.landurbplan.2018.07.005

Fielke, S. J., Botha, N., Reid, J., Gray, D., Blackett, P., Park, N., et al. (2018). Lessons for co-innovation in agricultural innovation systems: A multiple case study analysis and a conceptual model. *The Journal of Agricultural Education and Extension, 24*(1), 9–27. https://doi.org/10.1080/1389224X.2017.1394885

Folke, C. (2006). Resilience: The emergence of a perspective for social-ecological systems analyses. *Global Environmental Change, 16*(3), 253–267. https://doi.org/10.1016/j.gloenvcha.2006.04.002

Folke, C., Hahn, T., Olsson, P., & Norberg, J. (2005). Adaptive governance of social-ecological systems. *Annual Review of Environment and Resources, 30*(1), 441–473. https://doi.org/10.1146/annurev.energy.30.050504.144511

Geels, F. W. (2006). Multi-level perspective on system innovation: Relevance for industrial transformation. In X. Olsthoorn & A. Wieczorek (Eds.), *Understanding industrial transformation* (pp. 163–186). Springer. https://doi.org/10.1007/1-4020-4418-6_9

Geerling-Eiff, F. A., Linderhof, V. G. M., & Poppe, K. J. (2014) *Study on investment in agricultural research: Review for The Netherlands*. Wageningen University & Research. https://library.wur.nl/WebQuery/wurpubs/491076

Graymore, M., Schwarz, I., & Brownell, B. (2016). Identifying farming styles to encourage farmer engagement in climate adaptation and mitigation programs. *Rural Extension and Innovation Systems Journal, 12*(1), 32–42. https://doi.org/10.3316/informit.210569652414515

Gunderson, L. H., Holling, C. S., & Light, S. S. (Eds.). (1995). *Barriers and bridges to the renewal of ecosystems and institutions*. Columbia University Press.

Hale, P., & Ashton, P. (2002). *Raising the nation: A history of Commonwealth departments of agriculture, fisheries and forestry, 1901–2001*. Department of Agriculture Fisheries & Forestry Australia.

Hill, M., & Hathaway, S. (2017). Using the 'Market, message, means of communication framework' to guide design of grape rootstock extension. *Rural Extension and Innovation Systems Journal, 13*(1), 19–30. https://doi.org/10.3316/informit.192713579759025

Hilmiati, N., Budiwiranto, B., & van de Fliert, E. (2017). Gender, ethnicity and engagement: Uptake strategies for smallholder cattle farming innovation in West Nusa Tenggara, Indonesia. *Rural Extension and Innovation Systems Journal, 13*(2), 61–69. https://doi.org/10.3316/informit.187160954324081

Howard, K., Stelling, A., & Mahoney, C. (2015). Building women's participation in the wool industry. *Rural Extension and Innovation Systems Journal, 11*(1), 103–113. https://doi.org/10.3316/informit.660896987972827

Hunt, W., Birch, C., Coutts, J., & Vanclay, F. (2012). The many turnings of agricultural extension in Australia. *The Journal of Agricultural Education and Extension, 18*(1), 9–26. https://doi.org/10.1080/1389224X.2012.638780

Hunt, W., Birch, C., Vanclay, F., & Coutts, J. (2014). Recommendations arising from an analysis of changes to the Australian agricultural research, development and extension system. *Food Policy, 44*, 129–141. https://doi.org/10.1016/j.foodpol.2013.11.007

Ison, R. (2016). What is systemic about innovation systems? The implications for policies, governance and institutionalization. In J. Francis, L. Mytelka, A. van Huis, & N. Röling (Eds.), *Innovation systems: Towards effective strategies in support of smallholder farmers* (pp. 37–52). Technical Centre for Agricultural and Rural Cooperation (CTA) and Wageningen University and Research (WUR)/Convergence of Sci-ences-Strengthening Innovation Systems (CoS-SIS).

Ison, R., & Russell, D. (2000). *Agricultural extension and rural development: Breaking out of knowledge transfer traditions*. Cambridge University Press.

Ison, R., & Russell, D. (2011). The worlds we create: Designing learning systems for the underworld of extension practice. In J. Jennings, R. G. Packham, & D. Woodside (Eds.), *Shaping change: Natural resource management, agriculture and the role of extension* (pp. 44–55). Australasia-Pacific Extension Network (APEN).

Ison, R., & Straw, E. (2020). *The hidden power of systems thinking: Governance in a climate emergency*. Routledge.

James, J., Coutts, J., & Gururajan, R. (2015). It's all about the benefits: Why extension professionals adopt Web 2.0 technologies. *Rural Extension and Innovation Systems Journal, 11*(1), 72–82. https://doi.org/10.3316/informit.660841089059052

Jayasena, N., Wood, B. A., Reid, J. I., & Ramilan, T. (2019). Smallholder participation in milk markets in Sri Lanka: A livelihood analysis. *Rural Extension and Innovation Systems Journal, 15*(1), 40–48. https://doi.org/10.3316/informit.940486295637312

Jennings, J., Packham, R. G., & Woodside, D. P. (2011). *Shaping change: natural resource management, agriculture and the role of extension*. Australasia Pacific Extension Network.

Jones, G. E., & Garforth, C. (1996). The history, development, and future of agricultural extension. In R. Swanson, P. Bentz, & A. J. Sofranko (Eds.), *Improving agricultural extension: A reference manual* (pp. 2–12). Food and Agriculture Organization.

Jones, H. S. (1994). Federal agricultural policies: Do black farm operators benefit? *The Review of Black Political Economy, 22*(4), 25–50. https://doi.org/10.1007/BF02689978

Jones, N. A., Shaw, S., Ross, H., Witt, K., & Pinner, B. (2016). The study of human values in understanding and managing social-ecological systems. *Ecology and Society, 21*(1), 15–15. https://doi.org/10.5751/ES-07977-210115

Kamruzzaman, M., Chowdhury, A., Odame, H. H., & Sarapura, S. (2019). How do extension agents of DAE use social media for strengthening agricultural innovation in Bangladesh? *Rural Extension and Innovation Systems Journal, 15*(1), 10–19. https://doi.org/10.3316/informit.940430396723537

Keith, K. (2017). Thriving or just surviving: Social capital and resilience of long-term Landcare groups in Queensland. *Rural Extension and Innovation Systems Journal, 13*(2), 11–21.

Keith, K. (2019). *Re-visioning landcare to deliver ecosystem services West of the divide*. The Royal Society of Queensland. http://www.royalsocietyqld.org/wp-con-tent/uploads/documents/Stewardship/Rangelands%20Policy%20Dialogue/Rangelands%20Briefs/Keith_K_Landcare_delivery_of_ecosystem_services.pdf. Accessed 1 April 2022.

Keogh, M., Heath, R., Henry, M., & Darragh, L. (2018). *Enhancing private-sector investment in agricultural research development and extension (R, D&E) in Australia*. Australian Farm Institute.

Klerkx, L., & Nettle, R. (2013). Achievements and challenges of innovation co-production support initiatives in the Australian and Dutch dairy sectors: A comparative study. *Food Policy, 40*, 74–89. https://doi.org/10.1016/j.foodpol.2013.02.004

Klerkx, L., van Mierlo, B., & Leeuwis, C. (2012). Evolution of systems approaches to agricultural innovation: Concepts, analysis and interventions. In I. Darnhofer, D. Gibbon, & B. Dedieu (Eds.), *Farming systems research into the 21st century: The new dynamic* (pp. 457–483). Springer Netherlands. https://doi.org/10.1007/978-94-007-4503-2_20

Knook, J., & Turner, J. A. (2020). Reshaping a farming culture through participatory extension: An institutional logics perspective. *Journal of Rural Studies, 78*, 411–425. https://doi.org/10.1016/j.jrurstud.2020.06.037

Koyenikan, M. J., & Vincent-Osaghae, G. N. (2019). Gender perspectives in coping and resolution strategies in violent communal conflicts experienced in communities of Edo and Ondo States, Nigeria. *Rural Extension and Innovation Systems Journal, 14*(2), 12–22. https://doi.org/10.3316/informit.248737238142298

Kuehne, G., Llewellyn, R., Pannell, D. J., Wilkinson, R., Dolling, P., Ouzman, J., et al. (2017). Predicting farmer uptake of new agricultural practices: A tool for research, extension and policy. *Agricultural Systems, 156*, 115–125. https://doi.org/10.1016/j.agsy.2017.06.007

Leeuwis, C. (2004). *Communication for rural innovation: Rethinking agricultural extension*. Blackwell Science.

Leeuwis, C., & Aarts, N. (2011). Rethinking communication in innovation processes: Creating space for change in complex systems. *The Journal of Agricultural Education and Extension, 17*(1), 21–36. https://doi.org/10.1080/1389224X.2011.536344

Leeuwis, C., Smits, R., Grin, J., Klerkx, L. W. A., Mierlo, v. B. C., & Kuipers, A. (2006). Equivocations on the post privatization dynamics in agricultural innovation systems. In *The design of an innovation-enhancing environment* (pp. 3–5). Transforum Agro & Groen.

Leite, M., Ross, H., & Berkes, F. (2019). Interactions between individual, household, and fishing community resilience in southeast Brazil. *Ecology and Society, 24*(3), 2. https://doi.org/10.5751/ES-10910-240302

Lockie, S. (1999). The state, rural environments, and globalisation: 'Action at a distance' via the Australian Landcare Program. *Environment and Planning A, 31*(4), 597–611. https://doi.org/10.1068/a310597

Maclean, K., Ross, H., Cuthill, M., & Rist, P. (2013). Healthy country, healthy people: An Australian Aboriginal organisation's adaptive governance to enhance its social–ecological system. *Geoforum, 45*, 94–105. https://doi.org/10.1016/j.geoforum.2012.10.005

Maclean, K., Ross, H., Cuthill, M., & Witt, B. (2017). Converging disciplinary understandings of social aspects of resilience. *Journal of Environmental Planning and Management, 60*(3), 519–537. https://doi.org/10.1080/09640568.2016.1162706

Maher, E., Roberts, A., Park, G., Knapp, A., & Sweatman, C. (2017). Evaluating a new model for the integrated delivery of Reef water quality outcomes. *Rural Extension and Innovation Systems Journal, 13*(2), 141–146. https://doi.org/10.3316/informit.187533613749245

Marsh, S. P., & Pannell, D. (2000). Agricultural extension policy in Australia: The good, the bad and the misguided. *The Australian Journal of Agricultural and Resource Economics, 44*(4), 605–627. https://doi.org/10.1111/1467-8489.00126

Maskey, R., Murdoch, H., Pike, R., & O'Halloran, N. (2017). Assessing the impact of beyond soilcare project: Using a goal attainment scaling technique. *Rural Extension and Innovation Systems Journal, 13*(2), 52–60. https://doi.org/10.3316/informit.187123688381565

McCosker, K., & Northey, A. (2015). Paddock to reef: Measuring the effectiveness of large scale investments in farm management change. *Rural Extension and Innovation Systems Journal, 11*(1), 177–183. https://doi.org/10.3316/informit.661251014426733

McCown, R. L., Carberry, P. S., Hochman, Z., Dalgliesh, N. P., & Foale, M. A. (2009). Re-inventing model-based decision support with Australian dryland farmers. 1. Changing intervention concepts during 17 years of action research.

*Crop and Pasture Science, 60*(11), 1017–1030. https://doi.org/10.1071/CP08455

Moravek, T., Nelson, B., Anderson, A., & Reid, D. (2017). Quantifying the effectiveness of extension delivery methods on practice change—The experience of the Grazing BMP extension support project. *Rural Extension and Innovation Systems Journal, 13*(2), 70–75. https://doi.org/10.3316/informit.187198220266598

Murphy, C., Nettle, R., & Paine, M. (2013). The evolving extension environment: Implications for dairy scientists. *Animal Production Science, 53*(9), 917–923. https://doi.org/10.1071/AN12347

Murray, A., & Hasselman, L. (2013). A solution to rigid government NRM planning requirements through adaptive management. *Extension Farming Systems Journal, 9*(1), 290–296. https://doi.org/10.3316/informit.409804753487458

Nettle, R., Ayre, M., Beilin, R., Waller, S., Turner, L., Hall, A., et al. (2015). Empowering farmers for increased resilience in uncertain times. *Animal Production Science, 55*(7), 843–855. https://doi.org/10.1071/AN14882

Nettle, R., Klerkx, L., Faure, G., & Koutsouris, A. (2017). Governance dynamics and the quest for coordination in pluralistic agricultural advisory systems. *The Journal of Agricultural Education and Extension, 23*(3), 189–195. https://doi.org/10.1080/1389224X.2017.1320638

Nettle, R., La, N., & Smith, E. (2018). Research Report C: The advisory and extension system. In *Stimulating private sector extension in Australian agriculture to increase returns from R&D. (June, 2018). A project of the Department of Agriculture and Water Resources (DAWR) Rural R&D for profit program.* University of Melbourne.

Nettle, R., Morton, J. M., McDonald, N., Suryana, M., Birch, D., Nyengo, K., et al. (2021). Factors associated with farmers' use of fee-for-service advisors in a privatized agricultural extension system. *Land Use Policy, 104*, 105360. https://doi.org/10.1016/j.landusepol.2021.105360

Norton, G. W., & Alwang, J. (2020). Changes in agricultural extension and implications for farmer adoption of new practices. *Applied Economic Perspectives and Policy, 42*(1), 8–20. https://doi.org/10.1002/aepp.13008

Nursey-Bray, M., & Rist, P. (2009). Co-management and protected area management: Achieving effective management of a contested site, lessons from the Great Barrier Reef World Heritage Area (GBRWHA). *Marine Policy, 33*(1), 118–127. https://doi.org/10.1016/j.marpol.2008.05.002

Pannell, D. J., Marshall, G. R., Barr, N., Curtis, A., Vanclay, F., & Wilkinson, R. (2006). Understanding and promoting adoption of conservation practices by rural landholders. *Australian Journal of Experimental Agriculture, 46*(11), 1407–1424. https://doi.org/10.1071/EA05037

Paschen, J.-A., Ayre, M., King, B., Reichelt, N., & Nettle, R. (2021). Shaking it up: The realities of 'doing' co-innovation in a privatised agricultural advisory and extension system. *Journal of Rural Studies, 87*, 338–351. https://doi.org/10.1016/j.jrurstud.2021.09.003

Paschen, J.-A., Reichelt, N., King, B., Ayre, M., & Nettle, R. (2017). Enrolling advisers in governing privatised agricultural extension in Australia: Challenges and opportunities for the research, development and extension system. *The Journal of Agricultural Education and Extension, 23*(3), 265–282. https://doi.org/10.1080/1389224X.2017.1320642

Petheram, R. J., & Clark, R. A. (1998). Farming systems research: Relevance to Australia. *Australian Journal of Experimental Agriculture, 38*(1), 101–115. https://doi.org/10.1071/EA96055

Phillips, T., McEntee, M., & Klerkx, L. (2021). An investigation into the use of social media for knowledge exchange by farmers and advisors. *Rural Extension and Innovation Systems Journal, 17*(2), 1–13.

Pickering, J., Jenner, A., Haantera, K., Moore, S., Iseppi, C., Markey-Towler, B., et al. (2020). Why behavioural science matters in extension. *Rural Extension and Innovation Systems Journal, 16*(1), 24–30. https://doi.org/10.3316/informit.365299420143463

Pike, R. (2013). A story 50 years in the making: The Benwell surface water management system. *Extension Farming Systems Journal, 9*(1), 250–257. https://doi.org/10.3316/informit.409599790803618

Plummer, R., & Armitage, D. (2007). A resilience-based framework for evaluating adaptive co-management: Linking ecology, economics and society in a complex world. *Ecological Economics, 61*(1), 62–74. https://doi.org/10.1016/j.ecolecon.2006.09.025

Renaud, F. G., Birkmann, J., Damm, M., & Gallopín, G. C. (2010). Understanding multiple thresholds of coupled social-ecological systems exposed to natural hazards as external shocks. *Natural Hazards (Dordrecht), 55*(3), 749–763. https://doi.org/10.1007/s11069-010-9505-x

Rickards, L., Alexandra, J., Jolley, C., Farhey, K., & Frewer, T. (2018) *Review of agricultural extension*. Australian Centre for International Agricultural Research (ACIAR).

Roberts, K., Healy, D., Apted, R., & Cotter, M. (2011). Restoring the landscape after Black Saturday. *Extension Farming Systems Journal, 7*(2), 118–122.

Rogers, A., & Grieves, T. (2017). Chicks in the sticks—Supporting rural women to lead in agriculture and natural resource management. *Rural Extension and Innovation Systems Journal, 13*(2), 181–186. https://doi.org/10.3316/informit.187813108318118

Roling, N. (2009). Pathways for impact: Scientists different perspectives on agricultural innovation. *International Journal of Agricultural Sustainability, 7*(2), 83–94. https://doi.org/10.3763/ijas.2009.0043

Röling, N. G. (2000). Gateway to the global garden: Beta/gamma science for dealing with ecological rationality. In J. Pretty (Ed.), *The earthscan reader in sustainable agriculture* (pp. 179–205). Routledge.

Röling, N., & Wagemakers, M. A. E. (Eds.). (1998). *Facilitating sustainable agriculture: Participatory learning and adaptive management in times of environmental uncertainty*. Cambridge University Press.

Ross, H., & Baldwin, C. (2021). The quiet rise of environmental extension. *Australasian Journal of Environmental Management, 28*(2), 75–80. https://doi.org/10.1080/14486563.2021.1923434

Ross, H., Grant, C., Robinson, C. J., Izurieta, A., Smyth, D., & Rist, P. (2009). Co-management and Indigenous protected areas in Australia: Achievements and ways forward. *Australasian Journal of Environmental Management, 16*(4), 242–252. https://doi.org/10.1080/14486563.2009.9725240

Ross, H., Witt, K., & Jones, N. A. (2018). Stephen Kellert's development and contribution of relational values in social-ecological systems. *Current Opinion in Environmental Sustainability, 35*, 46–53. https://doi.org/10.1016/j.cosust.2018.10.007

Rouse, A., & Davenport, C. (2017). Project catalyst: An innovation project for cane growers in the Great Barrier Reef catchment. *Rural Extension and Innovation Systems Journal, 13*(2), 167–171. https://doi.org/10.3316/informit.187738576433086

Royle, A., & Di Bella, L. (2017). Targeted sugarcane grower extension improves reef water quality. *Rural Extension and Innovation Systems Journal, 13*(2), 187–192. https://doi.org/10.3316/informit.187831741289377

Rural and Regional Affairs and Transport References Committee. (2020). Rationale for the regulation of farming practices and the adoption of best practice. In The Senate (Ed.), *Identification of leading practices in ensuring evidence-based regulation of farm practices that impact water quality outcomes in the Great Barrier Reef*. Commonwealth of Australia. https://www.aph.gov.au/Parliamentary_Business/Committees/Senate/Rural_and_Regional_Affairs_and_Transport/GreatBarrierReef/Report/section?id=committees%2freportsen%2f024386%2f73346

Rural Practice Change. (2022). *Adopt*. http://www.ruralpracticechange.net/tools. Accessed 5 May 2022.

Salmon, P. W., & Hawkins, H. S. (1973). *The agricultural extension research unit: Its origins, philosophy and research program*. University of Melbourne.

Sarker, M. A., Chowdhury, A. H., Miah, M. A. M., Aurangozeb, M. K., & Peloschek, F. A. (2011). Participatory rural video centre in fostering women's voices—A model from Bangladesh. *Extension Farming Systems Journal, 7*(2), 27–32.

State Extension Leaders Network. (2006). *Enabling change in rural and regional Australia: The role of extension in achieving sustainable and productive futures.* https://www.academia.edu/8372595. Accessed 15 May 2022.

Stitzlein, C., Fielke, S., Fleming, A., Jakku, E., & Mooij, M. (2020). Participatory design of digital agriculture technologies: Bridging gaps between science and practice. *Rural Extension and Innovation Systems Journal, 16*(1), 14–23. https://doi.org/10.3316/informit.365373952028496

Vanclay, F. (1994). A crisis in agricultural extension. *Rural Society, 4*(1), 10–13.

Vock, N., & Carusos, E. (2019). Retooling agricultural extension for conflict-affected areas of the southern Philippines. *Rural Extension and Innovation Systems Journal, 14*(2), 1–11. https://doi.org/10.3316/informit.248662706257265

Walters, J., Milne, R., & Thompson, H. (2018). Online farm trials: A national web-based information source for Australian grains research, development and extension. *Rural Extension and Innovation Systems Journal, 14*(1), 117–123. https://doi.org/10.3316/informit.570292109335190

Wegscheidl, C., Coutts, J., & West, A. (2013). The role of agricultural extension in improving the health and resilience of the Great Barrier Reef. *Extension Farming Systems Journal, 9*(1), 196–203. https://doi.org/10.3316/informit.409338929206003

Wegscheidl, C., Trendell, P., & Coutts, J. (2015). Evaluating the role of extension in helping to improve water quality in the Great Barrier Reef. *Rural Extension and Innovation Systems Journal, 11*(1), 1–10. https://doi.org/10.3316/informit.660710658260245

Wellington, M., Mclean, I., & Holmes, P. (2021). *Towards effective extension services in Australia.* www.aginstitute.com.au. Accessed 1 April 2022.

Williams, A., James, J., & Prichard, P. (2021). Developing an extension model of practice to guide and empower extension practitioners. *Rural Extension and Innovation Systems Journal, 17*(1), 10–20.

Young, M.-A., Tothill, M., Drew, J., & Dohle, L. (2017). Experiences in supporting primary producers' recovery from the Pinery bushfire, South Australia. *Rural Extension and Innovation Systems Journal, 13*(2), 135–140. https://doi.org/10.3316/informit.187514980777987

Zurba, M., Ross, H., Izurieta, A., Rist, P., Bock, E., & Berkes, F. (2012). building co-management as a process: Problem solving through partnerships in aboriginal country, Australia. *Environmental Management, 49*(6), 1130–1142. https://doi.org/10.1007/s00267-012-9845-2

CHAPTER 5

# Promoting Women's Economic Empowerment in Achieving Rural Economic Development: Case Study in Lombok Island, Indonesia

*Inne Dwiastuti*

## Introduction

The relationship between women's empowerment and economic development is a central issue in research development and policy today. Women's Economic Empowerment (WEE) is now recognized as one of the main driving forces of well-being and global economic growth. Women's empowerment is an important policy discussion in development economics and modernization theory. It can lead to greater equality in relation to the human resources capacity for economic development

I. Dwiastuti (✉)
School of Agriculture and Food Sciences, The University of Queensland, Brisbane, QLD, Australia
e-mail: i.dwiastuti@uq.net.au



C. Baldwin and S. van Bommel (eds.), *Rural Development for Sustainable Social-ecological Systems*, Palgrave Studies in Environmental Sustainability,
https://doi.org/10.1007/978-3-031-34225-7_5

(Pal et al., 2021). Dominguez et al. (2022) stated that the empowerment concept helps to understand the processes of influence and control over decisions, quality of life, the ways organizations function and expectations of the various scopes of human experience, whether individual, community or organizational. Empowerment can be seen as a process of behaviour, in which individuals or communities gain control of the decisions that give impact to their lives, to achieve common welfare and positive changes (Zimmerman, 2000). Community empowerment has been seen as a tool for making people more powerful, active and able to participate in the growth of their local economy (Gralewiex, 2022). The community empowerment concept translates into both power to create something individually for themselves or collectively for others, and collective power over the local economy (Brophy et al., 2017).

This chapter analyzes women's economic empowerment to achieve rural economic development in Central Lombok and West Lombok, Indonesia. It identifies the process by which women empower themselves and their communities by running small businesses in rural areas, and explores the problems and challenges in achieving their economic development goals. It also elaborates on government interventions to empower them. The chapter concludes that women's economic empowerment has a positive impact on rural economic development, but needs more substantial, and ongoing, support from local, provincial and national governments. Investing in women's economic empowerment not only sets a direct path towards gender equality, but also for poverty eradication and inclusive economic growth.

## Context

A characteristic of Indonesia, including the study area, Lombok Island, is a large proportion of the population dispersed in rural areas. Lombok, a relatively small island, under West Nusa Tenggara Province (only 26 percent from total 20,012 square kilometres of West Nusa Tenggara Province), population 3,758,631, according to the Statistic Agency of West Nusa Tenggara Province (2020a). Based on a master file of villages in West Nusa Tenggara Province, Lombok island has 518 villages in 4 districts (Statistic Agency of West Nusa Tenggara Province, 2020b).

Indonesia is strongly committed to rural empowerment, and emphasizes empowerment specifically within its strategies for rural development.

Among its policies and programs, it distributes about 1 billion Indonesian rupiah (about $95,000 AUD) to every village in Indonesia for the development of the village. These are discretionary funds, which can be invested in different ways, allowing each village "bottom-up" opportunities to decide development directions. The purposes of these funds are:

1. To empower the rural people (they make their own decisions about use of these funds)
2. To expand the local economy, based on principles of local welfare.
3. Improving access to natural resources.
4. The availability and maintenance of rural infrastructure.
5. Optimizing the use of rural funds based on their needs (Republic of Indonesia, 2015).

One of the channels for community empowerment in Indonesia is by increasing the level of activity by small to medium enterprises (SMEs). The discretionary funds allocated to villages offer them the option of investing in their economic development via SMEs, though not all recognize this opportunity. SMEs have the capabilities and capacities to generate income and employment which can provide people fair opportunities for owning, accessing and controlling resources, provided conditions are conducive for business operations to grow (Firdausy, 2005; Miranda & Miranda, 2015).

## Economic and Other Forms of Empowerment

Wu (2013) proposes a set of four domains to represent economic empowerment in the community (see Fig. 5.1). First, "agency" is defined as capabilities, knowledge, individual will, skills and confidence to pursue one's own interests, and access assets, services and needed support. This may include the power of decision-making and control to adopt new strategies and technologies to enhance productivity and income. Second, the "institutional environment, norms, recognition and status" are defined as systems of values, norms, institutions and policies that shape the economic and social environment and condition people's choices. This includes access issues in the social and physical environment, in relation to rights and use of assets and services, opportunities and expectations. Third, "social relations, accountability and networks" influence power relations and networks that help people, including women, to achieve their potential and negotiate for their rights and interests. This may involve processes of bargaining, negotiation, decision-making, collaboration and collective action. Fourth, "economic advancement" includes income, assets, resilience and return on labour. This framework offers an important foundation for considering economic and other forms of empowerment, in the context of women's entrepreneurship.

Silva et al. (2014) claimed that increasing women's access to earning opportunities, productive resources and decision-making power can result in a more productive use of resources, more investment in children's welfare and more representative public institutions. Hence, female economic empowerment contributes to economic growth and poverty reduction, while jobs for women stimulate development. Similarly, empirical findings by Doepke and Tertilt (2019) suggest that empowering women may promote overall economic development and ultimately lead to faster economic growth.

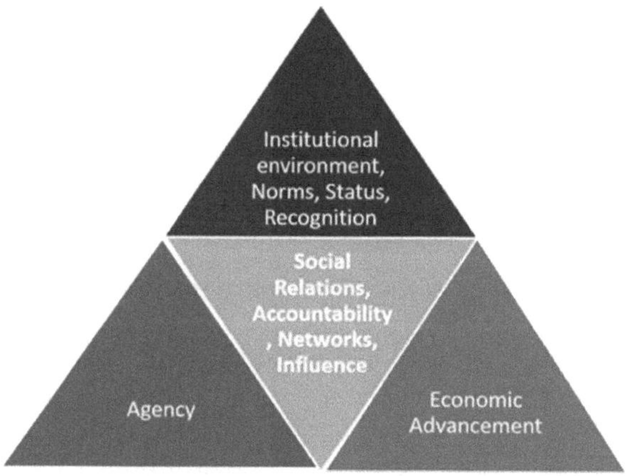


**Fig. 5.1** Economic empowerment (*Source* Wu [2013])

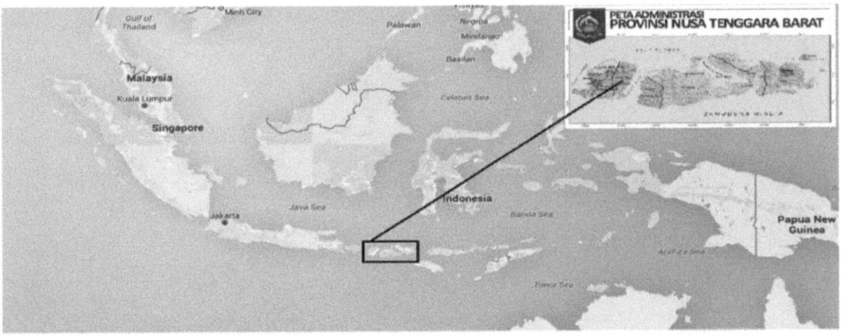


**Fig. 5.2** Map of Indonesia, Lombok (*Source* Google Maps 2016, Statistics Indonesia, 2016)

## Purpose and Location of Study

This study examines women's economic empowerment using two case studies: Sukarara village in Central Lombok district and Batu Mekar village in West Lombok district, West Nusa Tenggara (WNT) province, Indonesia. The villages were chosen as case studies because of their locational opportunities for growth, because both are targeted by the Indonesian government for growth, and in these villages most of the economic development is driven by women. Both have economies focused on local crafts which require high levels of skill. The locations of Lombok within Indonesia, and Lombok's districts and the two villages, are shown in Figs. 5.2 and 5.3.

## Methods

A pragmatic paradigm is used in this case study approach. The case study design used mixed quantitative and qualitative methods. A survey of SME owners, workers and people involved closely in village community groups and associations was conducted, with 87 respondents in Sukarara village and 83 in Batu Mekar village in Batu Mekar. In-depth interviews were conducted with 20 SME owners in Sukarara Village and 20 SME owners in Batu Mekar village, and 18 key decision makers at different levels of government who are able to comment on government policies and their implementation (see Table 5.1). Meanwhile observation was conducted throughout fieldwork in each village, including observation of premises

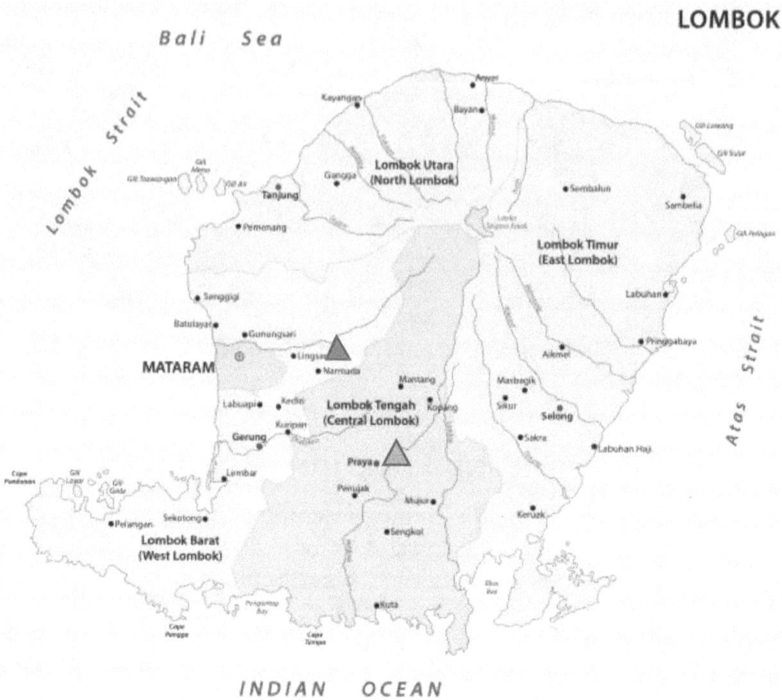


**Fig. 5.3** Map of Lombok Island (*Note* Sukarara Vil-lage, Jonggat district at Central Lombok (green triangle). Batu Me-kar village, Lingsar district at West Lombok (blue triangle). *Source* Istock [2017])

and activities while visiting SMEs for recruitment and interview, observation around each village and participation in village events when these occurred, such as a major day of weaving demonstration in Sukarara village. These data sources were supported by secondary data from Statistics Indonesia. The summary of the research method is presented in Table 5.1.

**Table 5.1** Research methods

| *No.* | *Method* |
|---|---|
| 1 | Qualitative<br>Interviews with key decision makers at 4 key institutions, namely, Ministry of National Development Planning; Ministry of Village, Development of Disadvantaged Region and Transmigration; Ministry of Cooperatives and Small Medium Enterprises; and Ministry of Industry. These have offices at national, provincial and local levels. The total number of organizations for in-depth interview was 18. By government level, at the national level in Jakarta, capital city of Indonesia (4), at province level based in Mataram, capital city of Lombok (4), Central Lombok District (4), West Lombok District (4), and a rural leader in Sukarara village (1) and in Batu Mekar village (1)<br>Interviews with SME owners<br>In-depth interviews were conducted with 40 SME owners (20 in Sukarara and 20 in Batu Mekar)<br>Participant observation<br>Secondary data from Statistics Indonesia |
| 2 | Quantitative<br>A survey of SMEs was conducted, with 87 respondents in Sukarara village and 83 in Batu Mekar village. Nearly 68% of the respondents in Sukarara are SME owners, 25.28% are SME workers, and the remainder (6.6%) are involved in village organizations but not members of SMEs. In Batu Mekar the proportions are 65% SME owners, 20.48% workers and the remaining 14.46% in other roles |

## Sukarara Village

Sukarara village is 21 km, about 30-minutes drive, from Mataram, capital of West Nusa Tenggara Province. However there is no public transport to Mataram, or to the subdistrict and district capitals, both of which are seven km away but nevertheless difficult to reach without a car.

### *Village Profile*

The population of 10,060, 5,181 females (51.5%) and 4,879 males (48.5%) comes mostly from the Sasak Tribe, the original cultural group of Lombok (Sukarara Village Profile, 2018). The total number of households was 3,561 in 2018 (Sukarara Village Profile, 2018). The village covers an area of 755 hectares, of which 77.5% (585 hectares) is devoted to irrigated rice fields. It is a lowland area, on average 102 metres above sea level. The average daily temperature is 30 degrees Celsius, and annual rainfall is 142 mm, mainly concentrated in six months of the year.

Education levels in Sukarara village are shown in Table 5.2. Nearly forty-four per cent of the population has only primary (elementary) school education, while a further 10.5% have a junior high school. A proportion with degrees (6%, in which women predominate) reflects rising education levels among the younger generation in the village, some of whom attended university in Mataram but continue to live in the village.

Most people in Sukarara are weavers (2520 people) and the second main occupation is farmer (see Table 5.3).

**Table 5.2** Sukarara village population with education levels

| *Education* | *Males* | *Females* | *Total (%)* |
|---|---|---|---|
| Elementary school | 837 | 799 | 1636 (43.7) |
| Junior high school | 250 | 145 | 395 (10.6) |
| Senior high school | 478 | 739 | 1217 (32.5) |
| Diploma—1 | 48 | 71 | 119 (3.2) |
| Diploma—2 | 48 | 32 | 80 (2.1) |
| Diploma—3 | 28 | 40 | 68 (1.8) |
| Bachelor degree | 84 | 142 | 226 (6) |
| Masters degree | 0 | 0 | 0 (0) |
| Total | 1773 | 1968 | 3741 (100) |

*Source* Sukarara village profile, 2018

**Table 5.3** Occupations in Sukarara village, 2018

| *No* | *Occupation* | *Total (person)* |
|---|---|---|
| 1 | Farmer | 2032 |
| 2 | Farm workers | 1558 |
| 3 | Private employee | 60 |
| 4 | Public official | 125 |
| 5 | Weaver | 2520 |
| 6 | Trader/merchant | 110 |
| 7 | Stock farmer | 750 |
| 8 | Police and army | 22 |
| 9 | Mechanic | 9 |
| 10 | SMEs | 75 |

*Source* Sukarara village profile, 2018

### *Songket Weaving as a Women's Enterprise*

Sukarara is the largest production centre in Lombok of the famous local weaving craft called songket. Songket is a woven fabric which is made by adding gold, silver or other-coloured synthetic yarn to the weft. It is very important in the village and island economy, and is also exported overseas. The woven fabric is usually produced at home (i.e. it is a home industry); almost every house has its own loom. All of the weavers are women, while the men mostly work as farmers. The head of the Cooperatives and Small Medium Enterprises Agency at the provincial level explained at interview that Sukarara village is targeted by the government of West Nusa Tenggara as a supporting village which can provide raw materials, products and labour for the Special Economic Zone of Mandalika, Kuta Mandalika resort, Central Lombok district. This is an opportunity for SMEs.

For Sasak women, being able to weave traditional woven cloth is a must. They are not allowed to marry if they cannot weave. If a Sasak woman cannot weave but want to get married, she must pay a fine in the form of rice field, money, or rice. Young women of Sukarara are taught this old-age tradition before they reach puberty, so most Sasak women can weave by the time they are 15 years old. One of them said she learned to weave when she was nine. Usually their mothers teach the young women the techniques, motifs and patterns to make this traditional hand-woven cloth.

In the village, we observed how the Sasak women dye their threads using natural ingredients, which are paler and more subtle than the very bright artificially coloured threads of machine-made competitors, and how they weave the cloth on hand looms. Tourists can observe, and also try to weave under the Sasak women's guidance. Weaving traditional songket cloth is difficult and takes time. As a consequence, the price is quite expensive.

The economic structure associated with songket weaving and marketing involves micro-enterprises—individual women or family members working in and sometimes selling from their own homes—and a few SMEs which operate on a larger scale with showrooms and more formal marketing. Most of these SMEs evolved from household weavers, and there is a strong symbiosis between them. When an SME receives an order that is too large for their own workforce and stocks, the women in charge call on their own networks of individual weavers, usually comprising weavers whose quality they trust. These relationships evolve into semi-formal women's groups, which meet socially as well as for production purposes. When the SMEs have many orders from the customer, they can ask the women weaving group to help them to fulfil the order under their supervision, especially in terms of quality control. Meanwhile the leaders of the SMEs tend to be conduits to training offered by the Cooperatives and Small Medium Enterprises Agency at both provincial and district levels, to improve weaving techniques, new designs and maintain good quality. SME leaders may lobby to have training provided, using their contacts and reputations, or if the training is initiated by the agency, they can help to recommend attendees. Thus the SME leaders help themselves by helping the smaller enterprises, in a mutual positive interaction, in which the women's weaving groups are pivotal. Thus economic empowerment, through economic opportunity, relates closely to other forms of empowerment, related to building social capital and contacts, building skills and building commercial and other forms of confidence. There is mutual support between SMEs and other community institutions. For instance the SMEs donate to the religious community group for the main annual religious and cultural festival, and in return the religious group members wear the SMEs products at the event, to promote them.

The annual cultural festival in Sukarara village, *Nyesek Begawe*, is initiated by the Sukarara village government. In this festival, large numbers of women belonging to the community weaving groups (around 1500 weavers) gather and demonstrate weaving and processing near the Sukarara village office. Representatives from the Indonesian Ministry of Industry and from local Industrial Agencies, local and foreign tourists, attend the festival.

### *Weaving Production Process*

There are two types of woven fabric, *Rang Rang* and *Songket*. Production of Rang Rang has a shorter process than Songket, which requires two additional steps which made it more expensive than Rang Rang. First, the raw material of white cotton yarn is bought from Mataram. Second, the yarn is coloured using natural dyes. Third, it is spun into yarn once it is dry. Fourth, the process of making base for the woven fabric, "*ngane*", involves determining the width and the length of the woven fabric. This process needs about 2–3 hours for a 2-metre width of woven fabric, whereas for Songket woven fabric this process needs 4–5 hours. Fifth, the yarn is put into traditional weaving tools, "*suri*" which are used to arrange and control the yarn so that the size of the base stays the same until the end of the process. Sixth, rolling the yarn, "*begulung*", makes the woven fabric soft and solid. Seventh, the weaving process commences by putting the yarn into the initial base. The woven fabric stages 1–6 are the same for both types of cloth, but songket requires two additional steps to create a pattern using a wood stick and then rope before proceeding to the final weaving process. The last process of Songket weaving usually uses gold, silver or copper yarn. The steps of Rang Rang and Songket woven fabric production are shown in Fig. 5.4.

### *Value Chain and Marketing*

The value chain associated with weaving involves considerable empowerment for the women. The larger SME owners recruit others to help them whenever they have an order which is too large for their own team to handle. To do this they first look to their weaver community group to find individual weavers who can assist, and whose work they trust.

In marketing the SMEs owners use one or both of two types of marketing, namely, offline marketing and online marketing. Offline marketing includes sales to "collectors" in the village (people who specialize in gathering a product and organizing sales); sales to a larger enterprise such as an art shop in the village, i.e. a larger SME which has a proper store to sell weaving; sales to a cooperative in the village; sales to a larger art shop in Central Lombok district or in Mataram city; and exhibitions. Exhibitions for example are held by the Industry agency at province level, Special Economic Zones at Kuta Mandalika, Lombok international airport and also at Ina Craft, an annual craft festival held in Jakarta Convention Center. Online marketing is through media such as Instagram, Facebook and Whatsapp. The value chain of weaving is shown in Fig. 5.5.

Weavers who work individually will usually rely on their own sources of raw material and sell the woven fabric to a collector. There is also a group of weavers who work at their own place but receive the raw materials from an SME owner and receive wages. This group of weavers is categorized as the labour of the SME, even though they work from home. This is partially because workshop facilities to gather all the women weavers at the same place are limited. The SME owners will collect the woven products from the individual weavers for further sale.

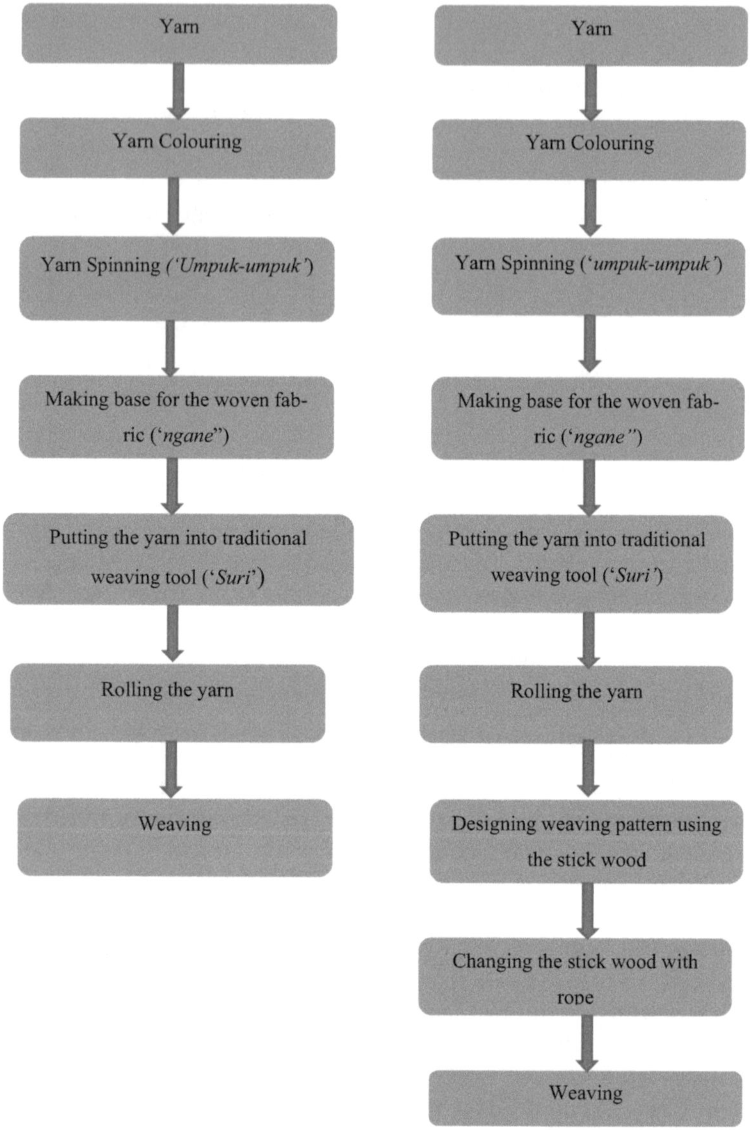


**Fig. 5.4** Sukarara—Rang Rang Process and Sukarara—Songket Process

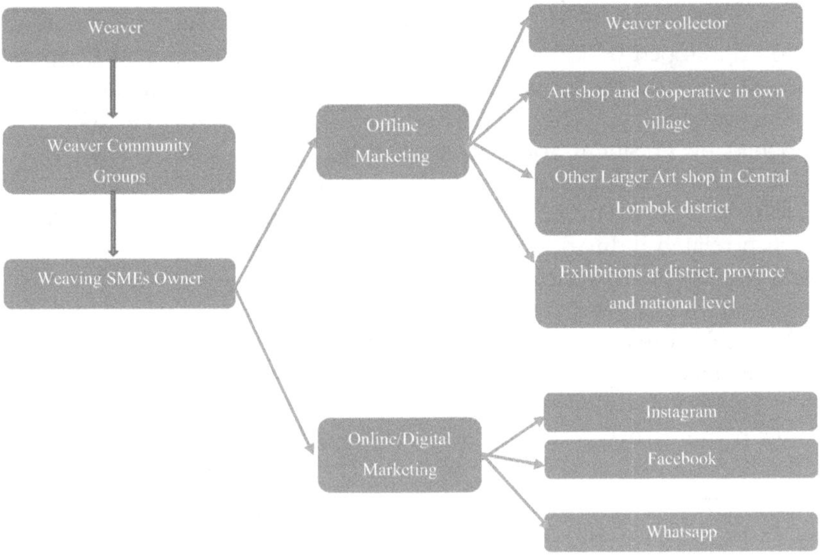


**Fig. 5.5** Value chain of weaving

### *Opportunities*

Sukarara has several advantages. First, it has an abundance of local labour. Second, the non-farm workers of Sukarara village are mostly women who learnt their weaving skills from their mothers and other female relatives, from a young age. Third, Sukarara village is strategically located with relatively good road infrastructure. It lies along the main road between Lombok International Airport and the capital city of Mataram. Fourth, local champions are owners of successful medium-sized enterprises who export products internationally and have relatively large stores in the village. Fifth, many of the owners of the larger SMEs in Sukarara village are already familiar with digital marketing for promoting and selling their final products through Whatsapp, Facebook and Instagram accounts. Sixth, the synergy between SME owners, weavers and the weaving community groups in the village is relatively good; they can support each other to meet the demands of the customers.

Other opportunities arise from regional development. Sukarara village is located in Central Lombok district approximately 26 km away or 34 minutes drive away from the Kuta Mandalika ecotourism tourist destination developed by ITDC (Indonesian Tourism Development Corporation). This is the developer that built Nusa Dua—a world class tourist destination in Bali. The USD 3 Billion project Kuta Mandalika was launched by President Jokowi on May 2017. It is being developed on 1175 hectares of land and is the key project driving interest in this region. It is in a Special Economic Zone (SEZ) which will consist of tourism-related industries and supporting industries, over 10,000 luxury hotel rooms (Pullman, Paramount resort, Novotel, Grand Mercure, Club Med and more), and an international road racing circuit (Moto GP 2021), golf courses, a yacht marina and more. Local government has a major focus on supporting the locals through the comprehensive Indigenous People's Development Plan (IPDP) of the Kuta Mandalika Urban Tourism Infrastructure Project. Therefore, the Small Medium Enterprise and Cooperative Agency at the province level have targeted the area for the development of SME products and skills to support the Kuta Mandalika Project in providing high-quality souvenirs for tourists. Thus Sukarara village has a competitive advantage and opportunities due to its relatively close location to this major ecotourism destination.

### *Constraints*

However, there are also some challenges identified at Sukarara village. First, there is a lack of additional skills and training support from the local government. Since the businesses in Sukarara village were established a long time ago, they now attracted less attention from local government in continuously improving their skills such as creating new designs and processing of natural colours to meet market tastes. Second, there is a lack of sophisticated technology. Currently, the weavers only use traditional weaving tools, yet improvements would help productivity. They need support from the government in terms of overcoming technology limitations. In an interview, an SME owner explained that the local government had once provided newer weaving tools and aids to the SME owners. However, because of a lack of communication between the local government and SME owners in terms of the kinds of tools they really needed to meet the standards to produce good quality woven fabrics, the tools offered unfortunately did not meet their standards and they could not use them. Third, there was a lack of workshops and showrooms. At the time of research, only three SME owners had a suitable workshop for weaving in, and showroom for marketing and sales. Most of the SME owners use their houses as the location of their businesses. Individuals can weave at home but they cannot weave in groups, as they could in a workshop. Some of them have small display cabinets for buyers who come to their houses, but buyers, especially tourists, are more likely to come to showrooms especially if the location is prominent. Most cannot afford workshops and showrooms because of the cost to buy centrally located land and to build.

## Batu Mekar Village

Batu Mekar has natural resources in the form of large agricultural and plantation land that is suitable for various fruits including Rambutan, Mango and Mangosteen. Batu Mekar also has a rich source of water, and there is a famous rafting activity which attract as tourist destination.

Batu Mekar produces *ketak* (similar to rattan), which can be shaped into many forms of craft such as bags, jewellery, pencil cases and jars. Most of the SME owners and workers for *ketak* production are women.

Batu Mekar village is 20 km from Praya, the capital city of Central Lombok district, with approximate travelling time of 1.5 hours. However, there is only minimal public transportation. The distance to Mataram, the province capital city is 17 km, with an approximate travelling time of 1 hour, but there is no public transportation so people have to rely on their own transportation such as motorcycles, if they have them.

### *Village Profile*

Batu Mekar is one of ten villages that are located in Lingsar subdistrict, part of the West Lombok district. It has a total population of 9,882 consisting of 4,953 females and 4,929 males, with 3,141 households. As in Sukarara, the majority of the population are Sasak (8,312) and of Balinese background (1,570). Most of the people have Islam as their religion (8,398) or Hindu (1,227). The village has an area of a total of 841.97 hectares, which consists of rice fields (233 hectares); forest areas (366.45 hectares), plantation areas (185.15 hectares), dry land areas (48.18 hectares), public facilities (9.19 hectares) and the remainder consists of public housing, office buildings and other facilities.

Batu Mekar village has an altitude of 150–750 metre above sea level. It has 200–300 mm/year of rainfall mainly within the 6 month period of the rainy season, and average daily temperature of 30 °C.

The population of Batu Mekar village based on educational background is shown in Table 5.4. This suggests that despite low education levels, many of the population are able to support themselves with their traditional skills.

In Batu Mekar, major employment is in farming (farmers and farm workers) followed by SME owners, indicating SMEs are a major contributor to the economy (see Table 5.5).

There is mutually positive interaction between SMEs and community groups in relation to producing *ketak* handicrafts. When training is given to the SMEs by the SMEs Agency, those trained can further give training to the community groups, for example about improved equipment, new product design and maintaining high quality from processing raw material to packaging. This is a common practice by the larger SME owners. As in Sukarara, when the larger SMEs have many orders from customers from local and foreign countries (mostly from Japan, and European countries) they can ask the community groups involved in *ketak* production to help them to fulfil the orders under their supervision in terms of quality control.

**Table 5.4** Batu Mekar village population by education levels

| *Education* | *Males* | *Females* | *Total (%)* |
|---|---|---|---|
| Elementary school | 1,071 | 1,176 | 2247 (54) |
| Junior high school | 192 | 206 | 398 (10) |
| Senior high school | 98 | 100 | 198 (5) |
| Diploma—1 | 386 | 308 | 694 (17) |
| Diploma—2 | 306 | 235 | 541 (13) |
| Diploma—3 | 0 | 0 | 0 |
| Bachelor degree | 2 | 0 | 2 |
| Master degree | 0 | 0 | 0 |
| Doctoral degree | 25 | 20 | 45 |
| Total | 2080 | 2045 | 4125 (100) |

*Note* The high proportion with doctoral degrees reflects young people being given opportunities to take degrees at the University of Mataram, then continuing to study for higher degrees overseas. Young people in Lombok are targeted to get scholarships to study overseas. Under the method of data collection, statistics are kept on those born and raised in the village, not necessarily currently living there

*Source* Batu Mekar village profile, 2018

**Table 5.5** Occupations in Batu Mekar village

| *Occupation* | *Total (person)* |
|---|---|
| Farmer | 887 |
| Farm worker | 773 |
| SMEs | 825 |
| Livestock farmer | 135 |
| Trader/merchant | 112 |
| Clerical worker (private sector) | 54 |
| Police and army | 45 |
| Public official | 41 |
| Manual worker (manufacturing or private sector) | 31 |
| Mechanics | 5 |

*Source* Batu Mekar village profile, 2018

## The Process of Making *Ketak* Handicrafts

*Ketak* handicraft is an art and tradition inherited through generations of Sasak in Lombok. *Ketak* is a type of wild plant; it is not cultivated. They process the *ketak* plant, especially the grass part, to make tools that are useful in everyday life and have artistic value.

The process of making *ketak* handicraft is complicated. For a start, collecting the *ketak* plant is not easy. The craftswomen must go to the forest area to select and sort high-quality *ketak* grasses. This requires special skill and experience as only certain types of *ketak* grass can be used to make good handicrafts. The harvested *ketak* grass must be used immediately, otherwise it will become difficult to process and the quality of handicraft will suffer. To begin the weaving process, each *ketak* plant, especially the branch part, must be cracked manually. At this point, careful steps are necessary since the parts of the *ketak* plant now become quite sharp.

There are three main stages in the process of making *ketak* handicraft, starting with making the basic material, then making a model and weaving the object, and finally the process of preservation (see Fig. 5.6).

> *First stage*: the *ketak* stem is broken into three to six parts which are smoothed using a *ketak* drawstring. The smooth part must also be further broken into smaller sizes. This part is referred to as the *ketak terekan* or *ketak sek* by the craftswomen. There are also some parts

**Fig. 5.6** Batu Mekar—Production process for ketak handicrafts

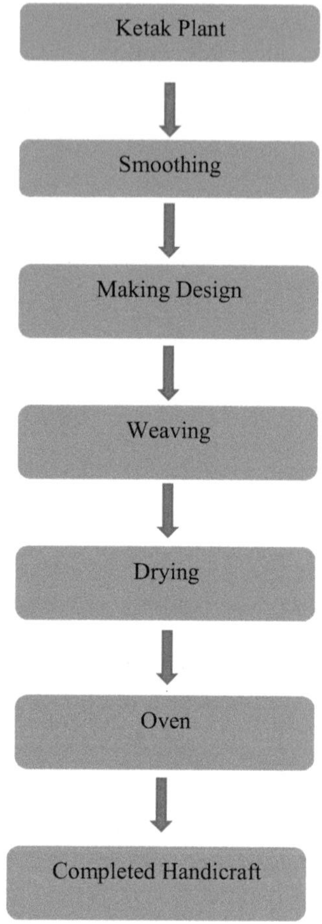


from the *ketak* stem that are left unbroken, which are called *jelujuh*. At the later stage, this will be used as the basic model.

*Second stage*: a high-level of creativity and ability is required to draw or form the design for a client to approve, then to make a sample for the SME workers to follow as they produce quantities of the craft objects. The construction process looks rather complicated from a lay perspective. Each *ketak* stick needs to be measured, formed and then arranged very carefully before the object is woven.

*Third stage*: the ***ketak*** handicraft must then undergo a preservation process, starting with the drying of the base ingredients until they are completely dry, then in a special wood-burning oven for three days and three nights to get the best results. This makes ***ketak*** handicrafts durable and termite resistant. Figure 5.6 summarizes the process.

In terms of the value chain of ***ketak*** handicrafts, SME owners recruit some individual ***ketak*** craftswomen and/or community groups *of* ***ketak*** craftswomen. Then, the ***ketak*** SME owners have two types of marketing, similar to those for weaving, namely, offline marketing and online marketing. Offline marketing includes sales to a ***ketak*** collector in the village; sales of product at a house shop; sales to an art shop in Mataram; and exhibitions. Exhibitions are held by the Industrial Agency at the province level. Exhibitions in Jakarta are initiated by the Indonesian Ministry of Industry and the same ministry partners in international exhibitions. Online marketing is conducted through media such as Instagram, Facebook and Whatsapp. The value chain of ***ketak*** handicrafts is shown in Fig. 5.7.

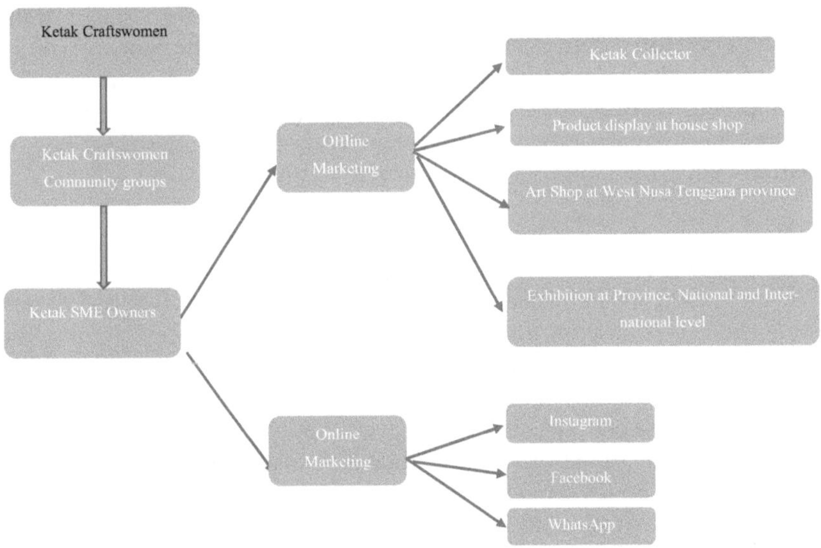


**Fig. 5.7** Value chain of ketak handicraft

### *Opportunities*

Strengths in the production and sales of *ketak* are identified as follows. *First*, Batu Mekar village has an abundance of local labour. *Second*, the people of Batu Mekar village who are involved in k*etak/*rattan SMEs are mostly female. *Third*, Batu Mekar village has a strategic location which has relatively good road infrastructure. *Fourth*, there is an SME owner in Batu Mekar village who has won awards for excellence from the national government (Indonesian Ministry of Industry) and from an international organization (World Craft Council). This national and international awareness of *ketak* handicraft products should encourage the other SME owners in Batu Mekar and have a spillover effect on the local economy especially to the SMEs in the village. *Fifth*, many of the SME owners in Batu Mekar village are already familiar with digital marketing in promoting and selling their products through Instagram, Whatsapp and Facebook. *Sixth*, the synergy between SME owners, *ketak* craftswomen and *ketak* community groups in the village is relatively good; they can support each other to meet the demands of the customers. Many of these are similar to the advantages regarding weaving production in Sukarara.

Meanwhile Batu Mekar village -is approximately 50 km or one hour 24 minutes car drive away from Kuta Mandalika, the ecotourism tourist destination. Therefore, Batu Mekar can take advantage of the Small Medium Enterprise and Cooperative Agency at the Province level targeting the development of the SMEs around Kuta Mandalika. Like Sukarara, Batu Mekar village has a competitive advantage and opportunities regarding its relatively close location to the ecotourism destination.

### *Constraints*

However, some challenges are identified at Bau Mekar village as follows. *First*, there is a lack of additional skills and training support from the local government for the same reasons as in Sukarara. *Second*, there is a lack of more sophisticated technology. Currently, they only use traditional tools. They need support from the government in terms of overcoming their technology limitations. *Third*, they lack workshops and showrooms from which to run their businesses and market their products. At present, only one SME owner has an adequate workshop and showroom. As with Sukarara, most SME owners use their house as the location of their business, which is far from ideal.

## Government Intervention for Economic Empowerment Through SMEs

Government has a serious interest in promoting SMEs, as an important avenue of local level economic development. At the national level the main ministries involved are the Ministry of Cooperatives and Small Medium Enterprises, and the Ministry of Industry. At the local level (provincial and district governments) there are counterpart institutions, namely Cooperatives and SMEs Agency and Industrial Agency. Village governments also play roles. The Head of Cooperatives and SMEs Agency at the provincial level stated at interview that as each SME business becomes developed, capital and labour will increase. It will also create a multiplier effect where the workers will want to open their own businesses, and be capable of doing so. Where women are active in SMEs, this empowers women and entire communities.

The Head of Cooperative and SMEs Agency at the provincial level explained that they have taken specific actions to encourage the establishment of SMEs and rural community empowerment, as follows. The Central and Provincial governments provide assistance in the form of training and capital. Capital assistance is provided through loans of Kredit Usaha Rakyat (KUR) by West Nusa Tenggara Province (Nusa Tenggara Barat, NTB) towards SMEs, with distribution by local banks. By 31 Oct 2017, the amount of KUR loans issued had reached IDR 903 billion for small businesses (about $85 million AUD) and IDR 580 billion (about $55 million AUD) for micro businesses. KUR loans for SMEs were channelled through 12 banks in NTB Province, with low interest rates. Other forms of capital assistance for SMEs are provided through, the Agency for Revolving Fund Management for Cooperatives and Small Medium Enterprises (*Lembaga Pengelola Dana Bergulir Koperasi dan UMKM*/LPDB) administered by the Ministry of Cooperatives and Small and Medium Enterprises and delivered through NTB province banks, with the total amount of IDR 180 billion (about $17 million AUD). As these funds are supposed to beallocated to SMEs, it is unfortunate if they are not taken up because the interest rate is relatively small, only 6% per year. The government's Cooperatives and SMEs Unit also provides assistance in the form of a salary bonus of IDR 50,000 ($5 AUD) for the officers in the sub-districts for every permit issued to SMEs. The Cooperatives and SMEs Unit also encourages the development of cooperatives in rural villages so that the SME entrepreneurs do not need to borrow from banks but can borrow capital through the cooperative.

Capital assistance for SMEs is also available in the form of Corporate Social Responsibility (CSR) assistance from state-owned enterprises such as from Angkasa Pura (state-owned airport management company), Telkom (state-owned telecommunication company) and PLN (state-owned electric company). Angkasa Pura gives the SMEs loans up to a maximum of IDR 60 million (about $5,500 AUD) with a fairly low-interest rate of 3%. CSR is also provided in the form of loans, training and guidance to SMEs, or support with equipment. There is a low-interest rate on loans through the CSR fund. The Cooperatives and SMEs Agency also facilitates exhibitions and bazaars for SMEs, and encourages public officials to use local products ranging from food, drinks and woven batik with Lombok's pattern. They also recommend the launch of small retail shops and chains such as *Kios 21* mart of Sharia Cooperative that sells SMEs' products. In addition, the Cooperative and SMEs Agency also provides training for management, accounting and bookkeeping for SMEs with allocated funds of IDR 2.5 billion (about $235,000 AUD) per year.

The Cooperative and SMEs Agency also collaborates with other ministries, agencies and institutions. *First*, the provincial government collaborates with the Ministry of Cooperatives and SMEs to provide salary allowances for field consultant workers in each district who are assigned to assist the SMEs. There are around 50 of these workers based at the Cooperative and SMEs Unit, with their salaries paid by the Cooperatives and SMEs Ministry. This arrangement fosters the SMEs. Since 2018 the Cooperative Office from the Cooperatives and SMEs unit has also collaborated with KEK Mandalika by providing kiosks for SME entrepreneurs to sell their products, and provides managerial training for SMEs. *Second*, the Cooperatives and SMEs Agency at NTB province also partners with the Ministry of Cooperatives and SMEs to establish an integrated business service centre at the provincial level to support the development of the SMEs. The same partnership also facilitates the enhancement of the SMEs in the Mandalika special economic zones. In addition, they have also built a people's market in Lombok for SME product marketing, also in order to encourage the growth of the SMEs.

The challenges and constraints in developing SMEs according to the Head of the Cooperative and SMEs Agency include:

- business persistence of the SME owners;
- business competition from more modern retailers;
- the emergence of modern retailers that have their own products (not local products) and place local products in a less strategic location in the store; and
- more expensive pricing of the SME's products, hence consumers often choose the cheaper products over the hand-made ones.

## Conclusion

Women's economic empowerment is central to the awareness and promotion of gender equality and acknowledging women's rights. It enables access to and control over productive resources, access to decent work, women's ability to participate equally in existing markets, and meaningful participation in economic decision-making at all levels from the household to international institutions. Investing in women's economic empowerment not only sets a direct path towards gender equality, but also for poverty eradication and inclusive economic growth. In terms of economies, women make enormous contributions, whether in businesses, on farms, as entrepreneurs or employees, or by doing unpaid care work at home.

The case studies of two villages, Sukarara village in Central Lombok and Batu Mekar village in West Lombok, showed that economic development involving weaving and *ketak* handicrafts generated income, mostly by women, in those villages. The process of economic empowerment is particularly interesting, as it involves other forms of empowerment. The owners of the larger SMEs, who tend to be more experienced (and often have workshops and showrooms) organize with individual weavers and *ketak* craftwomen, and through groups of these craftspeople, to fulfil orders. This requires strong social relationships among the craftspeople, and strengthens them. Meanwhile the owners of the larger SMEs assist the smaller SMEs with quality control for their orders, and organize training opportunities for the groups. Together the women learn technological improvement, too. Thus economic empowerment is a strongly social process, which enhances economic opportunity, social relations and technological improvement. All of this helps build the women's confidence, and standing in their families and community. Despite low levels of education, female SME owners in those villages already use digital marketing to promote and sell their products. This gives them more power to sell their products independently, and having this skill has given them more confidence to build their businesses.

In terms of government intervention, local government claimed that they support SMEs by giving training, additional equipment and also provide market places for SMEs in Kuta Mandalika ecotourism tourist destination in Central Lombok. However, the SME owners in both villages stated that there is a lack of support for additional skills and training from the local government. In addition, they have technological limitations and a lack of capital to increase their business capacity, in spite of the various opportunities offered by government and funding agencies.

These two village case studies show that women are economically empowered when they have both the power to make and act on economic decisions and the ability to advance and succeed economically. Female entrepreneurs in both villages also represent economic empowerment by having four important foundations; agency, institutional environment, networks and economic advancement (Wu, 2013). Female entrepreneurs in both villages have skills as a cultural heritage, and they exert agency especially in organizing among themselves, using their social networks. They also have a supportive institutional environment (government policies that shape social and economic environment), although they noted limitations in the extent of practical help from national and local government. They form and use social networks around business interests, through the weaver and craft groups in which individuals interact with larger SMEs for mutual benefit. Economic advancement through additional household income has been giving them and their families more power and confidence.

The women's economic empowerment in Sukarara village and Batu Mekar village has thus promoted rural economic development in each village. The process of empowerment is holistic, involving social and technological empowerment arising from economic self-organizing processes.

**Acknowledgements** Prof. Helen Ross is the primary PhD adviser for this study, and has assisted with editing this chapter.

## References

Brophy, P., Weissbourd, R., & Beideman, A. (2017). *Transformative economies: Emerging practices for aligning growth and inclusion*. Federal Reserve Bank of Philadelphia.

Doepke, M., & Tertilt, M. (2019). Does female empowerment promote economic development? *Journal of Economic Growth, 24*(4), 309–343.

Dominguez, I. G., Alfonso, M. J. P., Cruz, A. D., & Ortiz, H. T. (2022). Rural community-based tourism management in Central Mexico: A view from the empowerment theory. *Cuadernos de Turismo, 50*, 71–96.

Firdausy, C. (2005). Roles, problems and policies of the Indonesian small and medium enterprises in globalization. In C. Tisdell (Ed.), *Globalization and world economic policies: Effects and policy responses of nations and their grouping* (pp. 249–272). Serials Publications.

Gralewiex, P. (2022). Community empowerment as a tool to reduce unemployment: Contrasting cases of Iceland and Ireland. In T. Chaiechi & J. Wood (Eds.), *Community empowerment, sustainable cities, and transformative economies*. Springer. https://doi.org/10.1007/978-981-16-5260-8_33

Istock. (2017). "Map of Lombok island", Road Map Of The Indonesian Island Of Lombok Stock Illustration—Download Image Now —Lombok, Map, Asia—iStock. https://www.istockphoto.com

Miranda, A. T., & Miranda, J. L. F. (2015). Empowering rural communities through Small and Medium Enterprises (SMEs) in Samar Island, Philippines. The Asian Conference on Society, Education & Technology 2015, Official Conference Proceedings.

Pal, M., Gupta, H., & Josie, Y. C. (2021). Social and economic empowerment of women through financial inclusion: Empirical evidence from India. *Equality, Diversity and Inclusion: An International Journal, 41*(2), 294–305. https://doi.org/10.1108/EDI-04-2021-0113

Republic of Indonesia. (2015). Rules of Village Minister no 21/2015, about priority decision of rural funds for year 2016. Permendesa PDTT No. 21 Tahun 2015 tentang Penetapan Prioritas Penggunaan Dana Desa Tahun 2016 [JDIH BPK RI].

Silva, S. J. D., Paci, S., & Posadas, J. (2014). *Lessons learned and not yet learned from a multicountry initiative on women's economic empowerment*. The World Bank.

Statistic Agency of West Nusa Tenggara Province. (2020a). *Population census result 2020 West Nusa Tenggara* (pp. 8–9).

Statistic Agency of West Nusa Tenggara Province. (2020b). *Master file of village in West Nusa Tenggara province, Semester 1* (p. 19).

Wu, D. (2013). *Measuring change in women entrepreneur's economic empowerment: A literature review* (Working Paper), The Donor Committee for Entrepreneurship Development (DCED).

Zimmerman, M. A. (2000). Empowerment theory: Psychological, organizational and community levels of analysis. In J. Rappaport & E. Seidman (Eds.), *Handbook of community psychology* (pp. 43–63). Kluwer Academic/Plenum.

CHAPTER 6

# Getting the Priorities Right: Stakeholders' Involvement for a Holistic View of Research and Extension Needs in the Australian and Brazilian Dairy Industry

*Sérgio Rustichelli Teixeira*

## Introduction

Agricultural scientific communities work to control many variables affecting a particular technology under development; but without a vision shared with stakeholders of how it can be applied in the production systems, which have an even greater number of variables. As a result, new technologies are unlikely to be adopted successfully. Researchers and academics mostly concentrate on technologies that can result in publications, as this is the way their work is evaluated by most Research and

S. R. Teixeira (✉)
School of Agriculture and Food Sciences, University of Queensland, St Lucia, QLD, Australia
e-mail: rusti0305@gmail.com



C. Baldwin and S. van Bommel (eds.), *Rural Development for Sustainable Social-ecological Systems*,
Palgrave Studies in Environmental Sustainability,
https://doi.org/10.1007/978-3-031-34225-7_6

Development (R&D) organisations. However, dairy research corporations need to increase their understanding of regional production systems as a whole and in the design of Research, Development and Extension (R,D&E) programmes to meet industry needs comprehensively.

The objective of this research was, therefore, to develop a methodology to involve a broad range of dairy stakeholders in the identification of dairy industry priorities in order to help R,D&E personnel to understand production realities and develop appropriate research and actions.

## Review: Evolution of the Extension Framework

If agricultural R,D&E organisations do not provide Solutions for agricultural industries, sooner or later they will not have their support. Many extension experts and researchers in different countries, such as Australia and Brazil, have developed technologies that helped their dairy industries to achieve world importance while also developing methods to involve the stakeholders in the design of R,D&E activities. As a result of the identification of priorities over the last four decades of development of extension models, the participation of stakeholders in R,D&E activities is improving. The following presents an evolution of relevant methods, theories and approaches.

Agricultural extension originated in Oxford and Cambridge Universities in the nineteenth century to extend the benefits of university education to ordinary people. Coutts (1994) classifies four paradigms of extension:

(1) technology transfer, providing information, opportunity and persuasion;
(2) problem solving, assisting individuals to find Solutions to technological or management problems. The adoption of new technology or management practices is an indirect, though 'inevitable', consequence of this process;
(3) educational, pro-active informal education assisting individuals to better understand their situation, and so enable them to make choices and take action to improve their condition; and
(4) human development, to facilitate and stimulate individuals and communities to take the initiative in problem definition and in seeking Solutions to individual and societal concerns/opportunities.

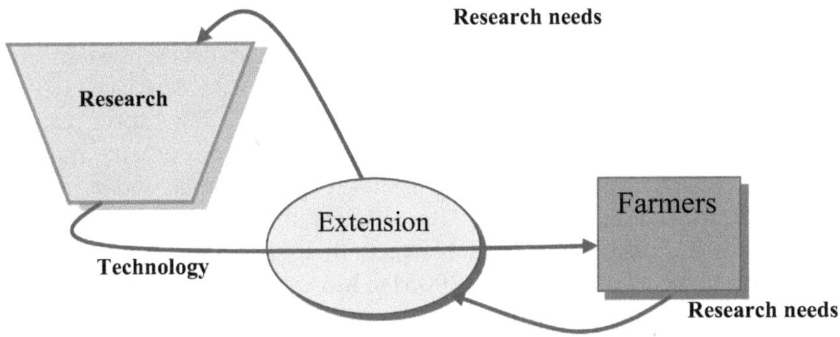


**Fig. 6.1** The role and relations of R,D&E and farmers

Agricultural extension is changing, going beyond technology transfer (Coutts, 2000) towards greater interaction with the production system. According to Freire (1992), extension work is related to the interaction between the extension agent, farmers and community. That means that the extension agent should have an attitude of being both teacher and student at the same time. The outcome of agricultural science cannot be considered finished before its insertion into the production system (Macedo, 1997), and achieving a practical result. Thus, extension is seen as an effective instrument when it stimulates participation and combines efforts with other segments of the agricultural industry chain, such as research. Participatory approaches involving stakeholders are seen as a way to improve the relations between research and agricultural industries in order to improve the performance of the extension role (AFFA, 2002; Dairy Australia, 2003; USDA, 2003; Vilela & Bressan, 2002). In this study, extension is treated as a 'bridge' (Fig. 6.1) to help the transfer of agricultural research findings into production systems and as a feedback instrument to drive new research directions and activities.

## Paradigms Influencing Extension

Extension has been influenced strongly by both positivist and constructivist paradigms, with a shift from the former to the latter, though both now co-exist. In the seventeenth century, the Cartesian paradigm, usually termed rationalism or positivism, dominated scientific investigation (Pretty, 1994). A traditional approach of positivism is to reduce or break complex systems into smaller parts in order to investigate the nature of a property or phenomenon independent of the system and then return the results to the whole system (Petheram & Clark, 1998). By the late nineteenth century, the modern scientific revolution realised that phenomena are transformed in the act of measurement and the observer is never really independent of the phenomena under investigation (Robinson et al., 2001). Researchers make sense of what they observe through the interaction between their own 'interpretive framework' and those of others (Levy, 2003). Thus constructivist epistemology rests on the assumption that people construct meaning actively within the contexts of social interaction in which they are situated. Both positivist and constructivist researchers are shaping and being shaped by phenomena. They should therefore be engaged in merging their expertise (Cupchik, 2001).

## A Brief History of Evolution of Extension Approaches and Models

1960s: The Transfer of Technology approach (TOT) was a top-down process whereby scientists developed products and methods promulgated by extension agents, and farmers were expected to adopt. Extension agents considered farmers who failed to adopt new techniques to be recalcitrant and irrational (Vanclay & Lawrence, 2001). The premise was that scientists know best, new technology is better than old, technology is needed, innovators will transfer information to laggards and many people are not information seekers (Chamala, 1999). This is a 'linear' model in which the farmers themselves were seldom involved. The positivist paradigm and reductionist approach featured little feedback, a single-innovation focus and the purpose of increasing production (Chamala, 1999). As a result, while extension people had much optimism about progress, they had few proven results (Russell et al., 1989).

1970s: The Farming Systems approach evolved in the post 'Green Revolution' period (Chamala et al., 1999). The World Bank developed strategies such as the Training and Visit system (Benor & Baxter, 1984; Coutts, 1997) which, although not strictly a farming systems approach, was another effort for better interaction with farmers for more effective transference of technology. Farming Systems Research and Extension (FSR/E) emerged in the late 1970s in reaction to the prevailing TOT model in an effort to drive research according to farmers' needs.

However, FSR/E was based on some assumptions derived from a TOT approach. Researchers continued to keep control over research agendas (King, 2000). The Participatory Rural Appraisal (PRA) approach emerged to facilitate and contribute to the empowerment of local people to take control over their own appraisal of problems and opportunities, and to select appropriate action (Reid, 1996).

1980s: Approaches to understand the production community better saw a transition from one-way R,D&E informing farmers to two-way communication between R,D&E and farmers. New methodologies such as the Problem-Census technique (Crouch, 1983), Farmer Researcher (Silva, 1984), MIDDA model (Chamala, 1987) and Agricultural Knowledge and Information Systems—AKIS (Röling, 1988) were developed. The Farmers First model raises the concept that most R,D&E problems arose from attempting to transfer technologies that did not relate well to farmers' own capacities and priorities. This philosophy generated much interest and caused a revolution in agricultural extension science. It promoted the view that farming communities are complex, diverse and risk-prone environments, which share common goals and access to resources, and that local knowledge is unitary, systematised and available for assimilation and incorporation with scientific knowledge.

1990s: Participatory models hold that development communication is not a vertical process of information transmission from the knowledgeable to the less knowledgeable, but rather a horizontal process of information exchange: an interaction. Proponents of this approach stress the model of dialogue, as a catalyst for individual and community empowerment, adopted from the work of Paulo Freire (1992). The participatory approach in essence lies in working with citizens to determine their needs and to design and implement programmes to address these needs, rather than imposing an intervention on a community. That means that people are regarded as agents rather than objects. Many methods were developed, such as Strategic Extension Campaign, Rapid and Interactive Model, The DART model, Clark: Six Steps for Better Practice and The PAM model (Chamala et al., 1999).

2000s: Engagement approaches go beyond simple participation because there is pressure for greater stakeholder involvement and engagement in decision-making processes. In United States Department of Agriculture (USDA), participation means business in 'The Technology Transfer Information Center' (TTIC). This process establishes R&D in a professional and technological relationship with agricultural industries, centred on each industry's needs. At the Australasia Pacific Extension Network (APEN) forum in 2003, it seemed that one word pervaded the meeting and the goals of extension agents: engagement. According to Paine and Nettle (2003), extension agents had failed to adequately represent the stakeholders. The following statement by McClelland and Eyres (2001) represents this trend: '*Tell me and I forget, show me and I may remember, involve me and I will understand*'.

## Key Extension Theories

From the spread around the world of extension models and methods stressing the transfer of technology paradigm, until the present, when research and extension seek the engagement of people in the production systems through R,D&E design, extension science has increasingly used knowledge developed in other branches of science to deliver better results. Communication theory contributes more than mass media, group methods and individual methods. In practice, communication is not just about packaging information in easy-to-understand language and putting it into popular communication channels; it is not a linear process whereby a message is sent and received. It is also about understanding the social, professional and institutional contexts in which the communication occurs, the social environment. It is also necessary to consider leaders' opinions and the communication network in a community. Other factors, such as behaviour, must be considered since humans are both rational and emotional beings and social science studies about human behaviour are helpful to achieve effective stakeholder participation. Trust, motivation, community knowledge/development and ownership of ideas, power and empowerment are particularly important in building relationships with people.

*Systems theory* became an important part of this study since it describes how a designated set of parts has no meaning unless the parts are connected together to become one functional dynamic unit (Edwards, 2000). The term 'system' itself was not emphasised until the middle of the twentieth century when the 'system' emerged as a key concept in scientific research and was studied as an entity rather than a conglomeration of parts (Bertalanffy, 1973). Checkland (1981) defines a system as the interactions, and lessons from managing the interactions, between the relevant combinations of parts (sub-systems) that make up the complex whole (the system). Dillon (1992) describes a system as an organised unitary whole.

It is a set of interrelated elements, each of which relates directly or indirectly to every other element, and no subject of which is unrelated to any other subject. Thus, anything that affects one part of the system has repercussions throughout. In addition, systems interact with other systems in a complex way and are generally part of a larger system.

*Farming and environmental systems*—internally, each farm is a unique and purposeful sub-system related to: (a) technology and resources, (b) formal organisation structure, (c) informal relations, goals and values and (d) management. In other words, each farm has the interaction of four production factors: earth, work, capital and technology. External to the farm sub-system is its environment, which exerts both general and specific influence on any particular farm sub-system. Major influences in the environmental supra-system are cultural, technological, market, educational, political, legal, climatic, demographic, sociological and economic factors. These broad environmental influences, along with intra-system determinants, help decide and constrain the goals of the farm sub-system. Farmers use many different sources to obtain the knowledge and information they need to manage their farms, and new knowledge is developed not only by research institutes, but also by many different stakeholders. It is useful to improve the understanding of the farming systems to analyse how all these sources supplement and support each other in a systemic way. Few farmers can have direct contact with researchers, especially in less industrialised countries where the number of farmers is large compared with the number of agricultural researchers (Ban & Hawkins, 1996). When one considers the different production technologies, and the different sources for obtaining knowledge, and how they are working and interacting together, it is easy to understand that a researcher and the research will only have a real impact on agricultural production if there are others, working synergistically, acting as effective communicators between researchers and farmers.

Christiansen and Hunt (2000) report that R,D&E agents can in fact learn much from farmers about multi-disciplinary decision-making. Farmers are constantly making decisions that take into account physical, environmental, technical, economic and social aspects (Fig. 6.2). Development of good Solutions requires contributions from researchers who have a deep knowledge of certain aspects of the problem, from extension agents who should have a broad knowledge and understanding of the farming system, and from the stakeholders in the communities who transform resources into products. It is the interaction of these people and

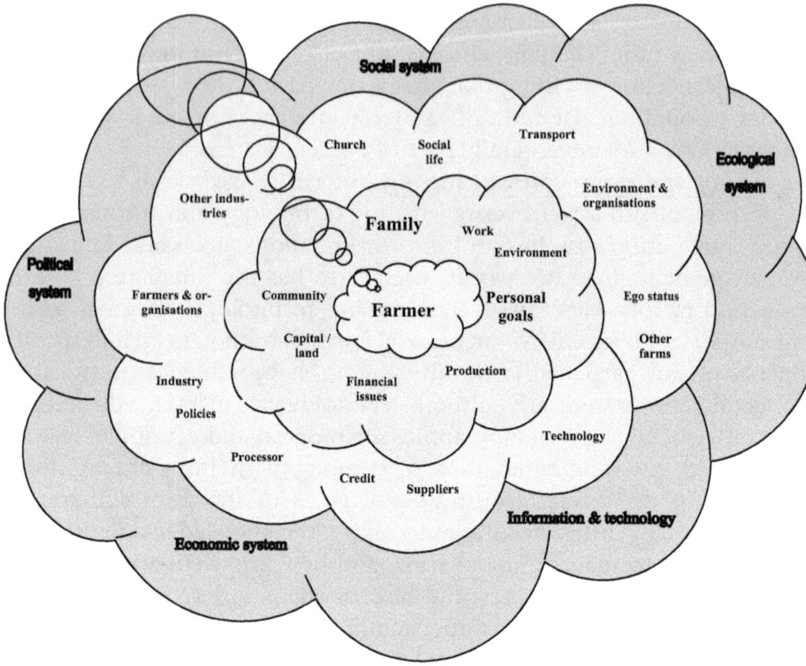


**Fig. 6.2** Illustration of farm/farmer's place in several systems

exchange of ideas among them that directs the actions that they undertake within the production community; that needs to be understood by R,D&E, before planning their research projects and extension activities.

## Why a New Model or Methodology for Better Involvement of Stakeholders in R,D&E?

The first activities related to extension are almost as old as science since they started with the Renaissance in the fifteenth and sixteenth centuries (Swanson & Claar, 1989). The late nineteenth-century evolution of science brought qualitative research but only in the last three decades of the twentieth century has it exerted real impacts in agricultural extension with stakeholders becoming part of extension teams instead of merely clients. An additional explanation for research and extension interest in higher engagement with stakeholders is that the public funds for R,D&E are being reduced and private funding, mainly from agricultural industries, is very welcome and, in Australia, encouraged through government incentives.

In the past, scientists themselves were rarely expected to play the role of an interventionist (Davies, 2000). The adoption of participatory approaches requires 'opening up' of scientific practice to stakeholder involvement. This could place R&D in a negotiating rather than simply a controlling role, even when R&D people consider themselves as the experts. In addition, engaging in dialogue with stakeholders and clients rather than simply communicating results challenges the basis by which scientific judgements are made (Stefanson & Fulton, 1997). Walker et al. (2001) conclude that the challenge is to develop approaches by which data, knowledge and scientific judgements of the data, can all be made available for integration into a negotiation process that attempts to deal with inherent uncertainty.

*The model for this study*—Fig. 6.3—represents two strategies to develop research, both supported by literature, but with different approaches to research problem identification and experimentation. The increasing participation and financial support of agricultural industries in research and extension enable working in closer contact with, and responding to the agricultural industries' demands. Extension tends to occupy an information management role. So, there are two ways to identify and meet demands for research (Fig. 6.3). It requires R,D&E organisations to widen their perceptions of the priorities of production communities in order to have a broader understanding of the R,D&E role and its usefulness in the development of agricultural industries. That gives impetus to a study focused on higher engagement of researchers in the understanding of the diversity of production community issues, and on how R,D&E needs can be identified, involving stakeholders in developing a holistic view of their realities. In other words, two-way engagement is essential in the identification of the production communities' priorities.

## Research Design and Methodology

The methodology for this study emerged from what was missing in other approaches as well as the advice of experienced research and extension people. The academic challenge was to make good connections between the extension-related theory, a logical evolution of the methodology,

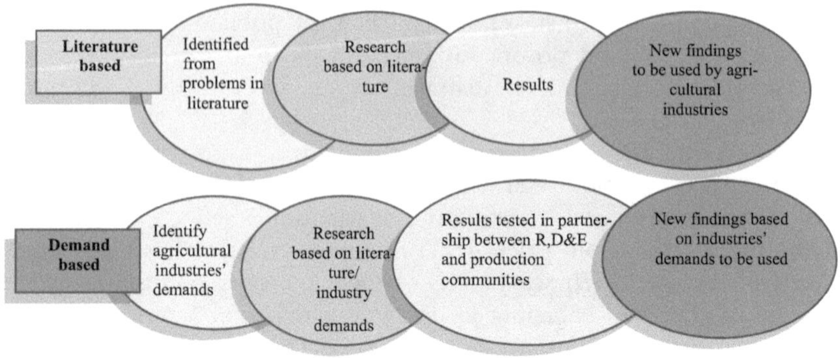


**Fig. 6.3** Two ways to identify and meet demands for research

research activities and the field research data. Trust and staying and working on a farm were fundamental to get a closer understanding of communities and commitment by farmers to the study.

The research design was based on five components working together: purpose, conceptual context, research questions, methods and validity (Maxwell, 1996).

While both qualitative and quantitative methods are important tools when employed properly, standardised methods make it impossible to capture the richness and individuality of subjects' perspectives (Patton, 2002), as required in this study. Qualitative research provides important and often unexpected insights that would almost certainly have been missed by quantitative methods (Kozel, 1999).

Case study selection—a case study is the method of choice when the phenomenon under study is not readily distinguishable from its context, featuring complex multivariate conditions and not just iSulated variables (Yin, 2003).

Dairy industries in both the Australian northeast coast of the State of New South Wales and Brazilian south in the northwest of the State of Rio Grande do Sul have a year-round pasture-based dairy production system. The similarities and differences between the regions were expected to return rich contributions to the study. Working and networking in Embrapa (*Empresa Brasileira de Pesquisa Agropecuaria*) Dairy gave me a clear picture of the industry in the South of Brazil, in which I selected two regions for investigation: Santa Rosa and Alto Jacuí. In Australia, the overall picture was obtained through reports, discussion with researchers from Australia, and websites. Networking with a diversity of Australian dairy stakeholders helped to give more insight into the research problem during the definition of the research. The farmers and the New South Wales Department of Agriculture office in Kyogle cooperated with the study. Ultimately, I chose two case study regions in Rio Grande del Sul, and one in the northeast of New South Wales.

### *The Consultation Strategy*

The consultation strategy is shown in Fig. 6.4 and described below. The literature review and observations from networking suggest the need for a person with some years of experience and understanding of farming system and community issues to interview the stakeholders and implement this strategy.

***1st Stage—Build trust***—the first initiative was to present the study to some stakeholders in each region including a selection of farmers, R,D&E specialists, advisors and others that could be highly interested in the study.

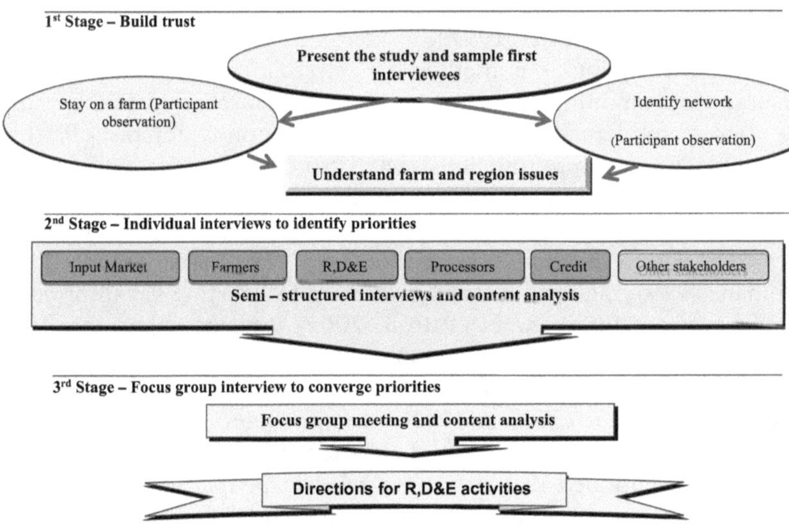


**Fig. 6.4** Consultation stages: a framework to identify the R,D&E needs of stakeholders

At this stage, one farm in each region of Rio Grande do Sul was identified where I could stay and work, using participant observation to identify the farmers' network, and build an understanding of farms and the region. Some farmers looked at me as a researcher and wanted to show me the farm instead of putting me to work with him, but at the end I was highly involved with the farm work and even slept at their farms.

***2nd Stage***—*Semi-structured interviews to identify priorities through purposive and snowball sampling*—face-to-face semi-structured interviews offer the possibility of modifying one's line of enquiry, following up interesting responses and investigating underlying motives in a way that survey or other self-administered questionnaires cannot (Robson 2002). The interviewer asks questions in such a way as to obtain valid responses and to record the responses accurately and completely. Freire (1992) stated that the idea is challenge stakeholders to think about problems recognising themselves as able to transform their realities. Besides creating a relaxed atmosphere for the respondent, the interviewer should state briefly the purpose of the interview but should avoid giving too much information about the study which could bias the respondent. That requires that the interviewer be familiar with the questions, their sequence, and have done a pilot test beforehand, so that the questions can be asked in a conversational tone (Burns, 1997).

Purposive sampling is a method to select information-rich people, organisations, communities, cultures, events or critical incidents, because they offer useful manifestations of the phenomenon of interest (Patton, 2002). Participants were initially selected from the six groups of stakeholders described below, using a set of desired characteristics. As it was not possible to know all stakeholders that could be important to be interviewed within the interconnected network of people or organisations (Neuman, 1999), snowball sampling was used for locating key informants through previous interviewees (Patton, 2002), usually when more than one interviewee recommended that person.

Criteria to select types of participants: Interviewees were selected because of their participation and influence in milk production in different parts of the dairy supply chain. Others were interviewed whenever the interviewees mentioned that they were relevant people to be consulted, validated by the repetition of that suggestion by other interviewees. Table 6.1 reports the number of participants by group.

**Table 6.1** Distribution of the groups of interviewees per studied region

| *Groups* | *Australians* | | *Brazilians St Rosa* | | *Brazilians Alto Jacuí* | | *Total* | |
|---|---|---|---|---|---|---|---|---|
| | *N* | *%* | *N* | *%* | *N* | *%* | *N* | *%* |
| Input market | 3 | 8 | 1 | 2 | 1 | 2 | 17 | 4 |
| Farmers | 20 | 56 | 20 | 48 | 27 | 59 | 67 | 54 |
| R,D&E people | 5 | 14 | 7 | 17 | 7 | 15 | 65 | 16 |
| Processors | 3 | 8 | 3 | 7 | 3 | 6 | 30 | 7 |
| Credit providers | 2 | 6 | 4 | 9 | 3 | 6 | 30 | 7 |
| Others | 3 | 8 | 7 | 17 | 5 | 11 | 51 | 12 |
| Totals | 36 | 100 | 42 | 100 | 46 | 99 | 423 | 100 |

Farm input market is those consulted when selling products to farmers. They hear the opinions and problems of different farmers.

Farmers are the main stakeholders of the dairy production systems, and the knowledge and technology that they use in their farms are the focus for R,D&E.

R,D&E people included consultants and veterinarians who provide professional and technical services to different farmers.

Processors are the natural link between production, processing and a bit of market.

Credit system includes those who make judgements about the financial feasibility of technologies and the financial health of farmers because of their interest in rates of return.

Off-farm stakeholders include every stakeholder identified at interviews with other participants as those who could provide useful contributions for the study and the region.

During the field research, I considered the following advice:

1. Most farmers are very smart and sensitive people. They will interpret what you want to hear from them, and at the end you will not have their real thoughts, unless you have created a relaxed atmosphere so that they trust you.
2. Dairy farmers always tend to 'cry' about the industry, they will 'kidnap' the meeting and you will lose it.

To meet the objectives of this study a mix of closed and open-ended questions was required, the latter allowing respondents to answer in their own words. Pilot testing of interview questions was conducted during the interview design stage to fix order, wording of questions, clarity, approach to the interviewees and procedures for processing the data.

*The main interview questions* were organised to match the objectives of the study. Some of the questions were presented with arm, hand and even body movements or comparison with other industries to help interviewees to understand the question. The interviewer drew on Freire's (1992) advice, to challenge the participant to participate and think, in a two-way communication process between researcher and participant. Avoid stressing the interviewee too much, questions that challenge the interviewees were presented in between those that appear to be in a one-way communication. Some trends were presented to challenge interviewees to see the dairy industry beyond the farm gate and of his/her region and country.

*Interview questions*—to ease the environment prior to the interview, some talks and magazines about the dairy industry were shown. An advisor told me that farmers like to manipulate possible answers, so I made a set of cards related to some questions. The questions covered differences among the regions studied, regional main profile and characteristics of the interviewees such as: gender, age, years in the dairy industry, major income of farms, characteristics of the farms such as size, milk production, number of cows and people involved. Comprehensive questions were asked of the farmer and about the farm. They included:

What made them select the dairy business? What are the goals in the dairy business and personal goals? What are the good and bad things that happened to your dairy farm business and for the dairy industry as a whole? What has potential to work well? What are the priorities of each dairy region from a broad range of stakeholders' perspectives? How can one involve the range of dairy stakeholders that influence the industry, and exchange information and ideas within a dairy community, to identify a comprehensive and relevant set of priorities for the design of R,D&E activities? With whom do you discuss dairy technological issues first? What is your main source of information? Who are the persons you consult to discuss your dairy issues? Which are the sub-systems of a dairy farm that you consider of high concern? If you were a manager of a R,D&E organisation what would be your priority issues to address? Who else should I consult for the study?

*3rd Stage—Focus group to converge priorities*—a Focus Group Meeting (FGM) is a method originating from market research (Kiel, 1999; Mitchell, 1979) involving inviting a small group (8–12 people) for about three hours of duration.

In this study, the facilitator explained the purpose of the meeting, the procedure and the need for recording the conversation of the group, then presented the content analysis data from the individual interviews. The aim was to provide an opportunity to share the findings so far, help in the identification of more issues, and discuss and converge the main priorities.

The facilitator should be skilled to conduct the discussion smoothly, pleasantly around a few questions or points, avoiding conflicts, not letting the discussion escape the objective of the meeting and involve everyone (Barbour, 1999; Morgan et al., 1998a, 1998b). The facilitator must keep in mind that this is not a passive process, that it is a two-way street and works better when the research team and participants are equally interested.

In this study, the first question was about the strengths and weaknesses about the dairy industry in each region. The second question was to nominate priorities for the region and potential activities. There was then an interval in which the priorities for the region were listed on butchers' paper. After that each participant was asked to pretend that they had an imaginary $100,000 to invest in the listed priorities and they distributed the money wherever he/she decided. Finally, the investments were summed to rank the shared priorities for investment.

Recruitment of participants with relevant characteristics and the agenda are critical for a successful FGM (Morgan et al., 1998b; Stewart & Shamdasani, 1990). In group meetings, people can answer differently from individual interviews, especially if they feel themselves in an inferior position or if the group discussion prompts new thoughts. Thus, analysing the FGM in conjunction with individual interviews is useful because in individual interviews, shy people may say things that they would not say in a group meeting Michell (1999).

*Feedback content analysis*—after the FGM a first content analysis (explained below) of data collected in the FGM was made. Each participant of the FGM received feedback on the FGM findings. The feedback complemented the process rather than added new important information. Nevertheless, the validation of the process was very useful.

The project received ethical clearance from the University of Queensland Committee of Research Involving Human Participants.

## Results

*Brief description of interviewees and farms sampled*—the 124 interviewees of the study were grouped according to their role in the dairy supply chain. The groups were Input market (suppliers), farmers, R,D&E people, processors, credit providers and others. The distribution of the groups of interviewees per case study is presented in Table 6.1 Distribution of the groups of interviewees per studied region.

## Individual Interviews Results

One main result was to have the interviewees' priorities if they were managers of an R,D&E organisation. Results would be compared with opinions from FGMs. Participants were invited to "wear the shoes" of a manager of a R&D organisation. The answers rarely came immediately after the question. Many times, after a vague answer, such as 'research in pasture', I asked: 'Can you be more specific?' The combined results of the three regions show Pasture and Communication as main priorities for the three regions (see Fig. 6.5).

## Focus Group Interviews

The data from the individual interviews introduced and facilitated the discussion. The combined results of the three regions show Communication and Organization as main priorities for the three regions (see Fig. 6.6).

Content analysis of individual interviews—content analysis refers to words, meanings, pictures, symbols, ideas, themes or any message that can be communicated (Neuman, 1999). Twenty-two codes were identified. The same coding frame was used for the individual and group interviews (FGMs). The data was coded twice with an interval of over two months to test the consistency of interpretation and coding.

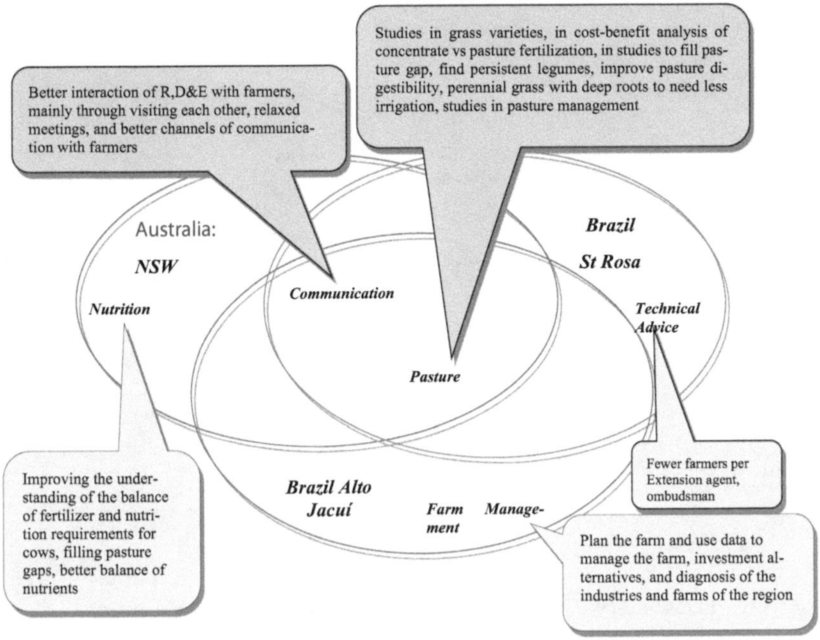


**Fig. 6.5** Interviewees' priorities as if they were R,D&E organisation managers

## Conclusions

Some learning and recommendations from the study include:

1. Agricultural R,D&E should involve a greater number of stakeholders in the design of activities in order to achieve their mission and have greater support of the agricultural industries for projects.
2. Australian and Brazilian dairy R,D&E organisations can be improved by involving a diversity of stakeholders that can contribute a more holistic view of the regional dairy industry providing a sense of the importance of production technologies for a production region.
3. The deeper contact between the farming system and production community through day-by-day relationships helped to change the image that technology advisors went to talk only about production technologies.

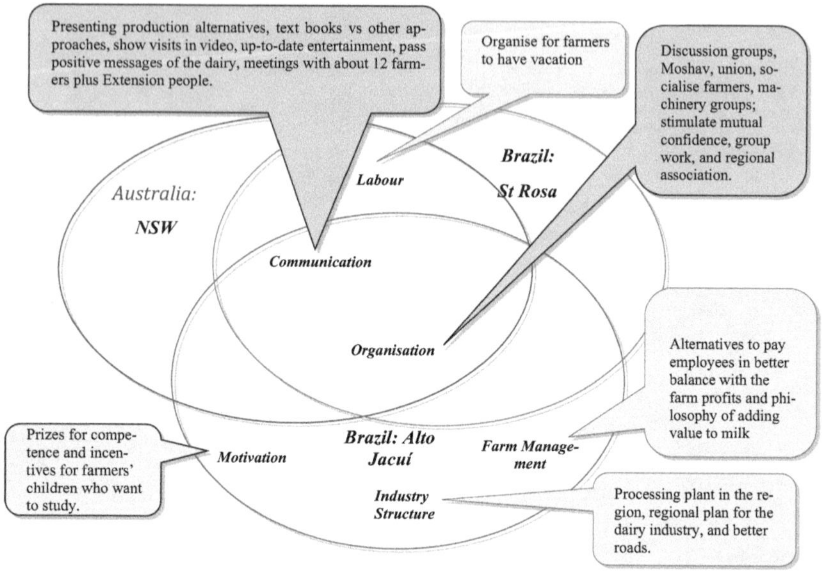


**Fig. 6.6** FGM interviewees' suggestions of any nature for their regions

4. The use of in-depth interviews with stakeholders revealed more farm concerns and pointed to a wider view of priorities. That gave R,D&E the opportunity to understand the broad context before selecting specific activities and matching with present studies. It showed the need of more studies such as to improve labour productivity.
5. The final report to participants of the group interviews was a good method for validation of findings and feedback. Participants complained several times that most researchers collect data but did not return results.

The study requires a long time (about 3 months) and the need of a researcher with a broad view of dairy business. A Solution is higher involvement of local agricultural extension agents.

**Acknowledgements** I wish to acknowledge the fundamental help by the advisory team, mainly my principal advisor, Professor Helen Ross, who influenced my systems approach. Her experience during her PhD research, whereby she lived with Aboriginal people to understand the reasons for their decisions on housing, in the context of the broader environment influenced my research involving living and working on farms. I also acknowledge the support of Associate advisors Professor Tom Cowan and Dr Shankariah Chamala. I also acknowledge the financial support from Embrapa and the Coordination for Professional Development of Graduates Foundation (CAPES). The Central Cooperative of Rio Grande do Sul State also provided logistical and financial support. In Australia, the University of Queensland provided the major theoretical and financial support for the study in Australia. By networking, I obtained an agreement for co-operation in the study from the Dairy Research Development Corporation (now Dairy Australia), the Subtropical Dairy Program (SDP), the Australian Tropical Dairy Institute (now The Tropical Dairy Research Centre) and the Department of Agriculture of New South Wales.

## References

AFFA. (2002). *Rural policy and innovation—The rural research & development corporation model*. www.affa.gov.au. Accessed 23 October 2002.

Ban, A. W. V. D., & Hawkins, H. S. (1996). *Agricultural extension*. Blackwell Science.

Barbour, R. S. (1999). *Developing focus group research: Politics, theory and practice*. SAGE.

Benor, D., & Baxter, M. (1984). *Training and visit extension*. The World Bank.

Bertalanffy, L. (1973). *General systems theory*. George Braziller.

Burns, R. B. (1997). *Introduction to research methods*. Addison Wesley Longman.

Chamala, S. (1987). Adoption processes and extension strategies for conservation farming. In PSCJE Pratley (Ed.), *Tillage: New direction in Australia* (pp. 400–419). Inkata Press.

Chamala, S. (1999). Factors affecting diffusion and adoption processes. In S. R. Raine (Ed.), *Research, development and extension in irrigation and water use efficiency: A review for the rural water use efficiency initiative* (Vol. 1, pp. 58–71). National Centre for Engineering in Agriculture Publication USQ, Toowoomba.

Chamala, S., Coutts, J., & Pearson, C. (1999). *Innovation management: Participatory action management methodologies for R,D,E & Industry stakeholders*. Canberra, Australia.

Checkland, P. (1981). *Systems thinking, systems practice*. Wiley.

Coutts, J. (1994). *Process, paper, policy and practice: A case study of the introduction of a formal extension policy in Queensland Australia* (Unpublished PhD Thesis). University of Wageningen.

Coutts, J. (1997). *Changes in extension—An Australian perspective, the coming of age of extension*. Paper presented to Conference Managing Change—building knowledge and skill, Albury, Australia, November.
Coutts, J. (2000). *Five domains of extension*. Paper presented to Extension in Australia Conference, Melbourne, 26–27 October 2000.
Crouch, B. (1983). The problem census: Farmer-centred problem identification. In *Training for agriculture and rural development* (pp. 27–38). FAO.
Cupchik, G. (2001). *Constructivist realism: An ontology that encompasses positivist and constructivist approaches to the social sciences*. Paper presented to Forum Qualitative Social Research, February 2001.
Dairy Australia. (2003). Australian dairy industry in focus 2003. In *Focus*. Dairy Australia.
Davies, M. (2000). *The Blackwell encyclopaedia of social work*. Blackwell Publishers.
Dillon, J. L. (1992). *The farm as a purposeful system*. The University of New England and Miscellaneous Publication.
Edwards, 2000Edwards, R. L. (2000). *Encyclopaedia of social work*. Washington.
Freire P. (1992). *Extensão ou Comunicação (Extension or Communication)*. Paz e Terra S/A.
Kiel, G. (1999). *An introduction to research techniques in action learning*. Brisbane, Australia.
King, C.A. (2000). *Systemic processes for facilitating social learning* (Unpublished Ph.D. Thesis). Swedish University of Agricultural Sciences.
Kozel, V. (1999). *New approaches, new methods: The need for cross disciplinary research on poverty*. Paper presented to WDR (World Development Report) on poverty and development 2000/01, Washington DC, 6–8 July 1999.
Levy, P. (2003). A methodological framework for practice-based research in network learning. *Instructional Science, 31*, 87–109.
Macedo, M. M. C. (1997). *The process of agricultural technology generation in Brazil: A social audit* (Unpublished PhD thesis). University of Sussex.
Maxwell, J. A. (1996). *Qualitative research design: An interactive approach*. Sage.
McClelland, I., & Eyres, T. (2001). *Farmers determining their own destiny*. Paper presented to 10th Australian Agronomy Conference, Hobart, Australia, 1 February 2001.
Michell, L. (1999). Combining focus group and interviews: Telling how it is; Telling how it feels. In R. S. Barbour & J. Kitzinger (Eds.), *Developing focus group research* (pp. 36–46). SAGE.
Mitchell, G. D. (1979). *A new dictionary of social sciences*. Aldine Publishing Company.
Morgan, D. L., Krueger, R. A., & King, J. A. (1998a). *Focus group kit*. SAGE.
Morgan, D. L., Krueger, R. A., & King, J. A. (1998b). *Involving community members in focus groups, Focus group kit 5*. SAGE Publications.

Neuman, W. L. (1999). *Social research methods—Qualitative and quantitative approaches*. University of Wisconsin.
Paine, M., & Nettle, R. (2003). *Best practice extension: Mediating learning relationships*. Paper presented to APEN forum Extending Extension: Beyond Traditional Boundaries, Methods and Ways of Thinking, 26–28 November 2003, Hobart, Australia.
Patton, M. Q. (2002). *Qualitative research and evaluation methods*. Sage.
Petheram, R. J., & Clark, R. A. (1998). Farming systems research: Relevance to Australia. *Australian Journal of Experimental Agriculture, 38*, 101–115.
Pretty, J. (1994). Alternative systems of inquiry for a sustainable agriculture. *IDS Bulletin, 25*, 37–48.
Reid, J. I. (1996). *Farming systems research: A background paper to the Farmer First research project at Massey University*. Massey University.
Robinson, J. B., Freebairn, D. M., Huda, A. K. S., & Cawley, S. T. (2001). *A view of science, agricultural science and farming systems research in these postmodern times*. Paper presented to 10th Australian Agronomy Conference, Hobart, Australia, 1 February 2001.
Röling, N. (1988). *Extension science*. University of Cambridge.
Russell, D. B., Ison, R. L., Gamble, D. R., & Williams, R. K. (1989). *A critical review of rural extension theory and practice*. University of Sydney.
Silva, A. R. (1984). A hora de decidir (The time for decision). *A granja*, 42–44.
Stefanson, B., & Fulton, M. (1997). *New generation co-operatives: Responding to changes in agriculture*. Centre for the Study of Co-operatives, viewed 7 April 2004.
Stewart, D. W., & Shamdasani, P. N. (1990). *Focus groups: Theory and practice*. SAGE.
Swanson, B. G., & Claar, J. B. (1989). The history and development of agricultural extension. In FAO (Ed.), *Agricultural extension: A reference manual* (pp. 1–21). FAO.
USDA. (2003). *Technology Transfer Information Center*. www.nal.usda.gov/ttic/about.htm, Accessed 4 August 2003.
Vanclay, F., & Lawrence, G. (2001). *Farmer rationality and the adoption of environmentally sound practices: A critique of the assumptions of traditional agricultural extension*. http://www.bib.wau.nl/ejae. Accessed 22 January 2001.
Vilela, D., & Bressan, M. (2002). Anais do Projeto Plataforma Tecnológica do Leite - Fase 2 (Proceeding of Dairy Technology Platform Project—Stage 2). Documentos, Embrapa Gado de Leite, Juiz de For a, Brasil.
Walker, D. H., Cowell, S. G., & Johnson, A. K. L. (2001). Integrating research results into decision making about natural resource management at a catchment scale. *Agricultural Systems, 69*(1–2), 85–98.
Yin, R. K. (2003). *Applications of case study research*. Sage.

CHAPTER 7

# The Role of Rural Community in Tackling Complexity of Agricultural Technology and Extension System in China

*Chengdong Liao and Yiyu Liu*

## Introduction

Farmers who cultivate less than two hectares of land and rely on household labours are commonly recognized as a smallholder. In China, smallholder farmers account for nearly ninety percent of total farmers, and nearly eighty per cent of which cultivate less than two-third of a hectare (equal to 10 mu) (Liu & Kong, 2018). Drastic changes are happening in China's agricultural enterprise, for example, the availability of arable land has been reduced owing to land conversion to industrialization, urbanization and creation of protected areas (Chen et al., 2009;

C. Liao (✉) · Y. Liu
Institute for Science, Technology and Society, South China Normal University, Guangzhou, China
e-mail: chengdong.liao@uqconnect.edu.au

Y. Liu
e-mail: liu_yiyu@126.com



C. Baldwin and S. van Bommel (eds.), *Rural Development for Sustainable Social-ecological Systems*,
Palgrave Studies in Environmental Sustainability,
https://doi.org/10.1007/978-3-031-34225-7_7

Yi et al., 2013; Zhang & Chen, 2014); and the food self-sufficiency in the domestic rice market is threatened because of the drastic increase of rice imports (Wang et al., 2018; Yang & Liu, 2019) and price fluctuation (Xiao & Wang, 2017). China's government has been investing in promoting technological innovation in the field of agriculture to secure its food supply and other related issues but failed to balance their negative effects in rural resource and environmental management. For example, many have reported severe metal pollution detected in farmlands due to the rapid industrialization and lax environmental enforcement (Li & Pontes, 2017; Liu & Kong, 2018); massive water pollution was also reported owing to the non-point pollution in major lakes from agricultural production (Cassou et al., 2018; Ongley et al., 2010; Wang, 2006). Smallholder farmers are more vulnerable to those changes and need support to improve their farming.

Agricultural extension, known as the process of agricultural technology delivery to farm practice and application, has an important role in boosting agricultural production, and in the meantime, balancing other development purposes such as co-innovation (Leeuwis, 2004; Röling, 1988; Spielman et al., 2009). However, the system that supports these goals has encountered series of challenges in China. Many countries have changed the modality of agricultural extension from government-dominant to private-sector dominant as a means to reduce the overwhelming expense of public finance on extension and shift extension cost to farmers themselves (Hunt et al., 2012; Swanson, 2008). However, the extension system in China tends to balance both goals, which in fact impedes farmers' access to extension services due to reduction of officers' time and funds spent on extension activities (Hu et al., 2012; Kong, 2009). Furthermore, arguments arise regarding the role of agricultural extension. Cassou et al. (2018) found that farmers in China have been overusing pesticides because they did not follow extension officers' instructions and recommendations. While Sun et al. (2012) and Cai et al. (2022) found that inadequate extension services from both public and private sectors have also contributed, owing to officers' incompetence

C. Liao
School of Agriculture and Food Sciences, The University of Queensland, Brisbane, Australia

of extension work and insufficient knowledge, and retailers deliberately advising farmers to use more chemicals to expand their sales.

Agricultural extension literatures have documented the role of rural communities in extension activities (Feder et al., 2010; Wellard et al., 2013), some borrowed social network and information sharing perspectives (Hoang et al., 2006; Wang et al., 2020). However, their focus remains on technology adoption, rather than broader agricultural and farming systems. Drawing on Professor Helen Ross's expertise in community development (Berkes & Ross, 2013, 2016), we argue that treating community as an analytical level in broader social-ecological systems can provide insights in understanding collective actions and communal goals. Power relations between communities and the extension systems, as well as within communities themselves, such as elite capture and conflict of interest, are also heavily impacting the processes of agricultural extension (Feder et al., 2010).

This chapter aims to examine the role of rural communities in helping smallholder farmers tackle complexities from the agricultural extension system and the broader farming systems. We briefly summarize the shift in focus from technology adoption to systems thinking in the agricultural extension literatures. Then, we present a case study of smallholder rice farming in Guangxi province, China, to understand the role of rural communities in supporting smallholder farmers dealing with variabilities and complexities.

## From the Traditional Technology Adoption to an Innovation System Approach

China has employed a five-level agricultural extension system established from the 1980s, which rigidly complied with the administration system from the national level agricultural extension department to the township level agricultural extension station (very similar to the public-sector SAES in the USA), for spreading technology and supporting agricultural enterprises (Chen et al., 2009; Hu et al., 2012). Over the last century, agricultural extension systems in countries of the European Union, and the USA and Australia, have transformed from public-funded to private-sector dominant, to switch the cost from public finance to farmers themselves or their production organizations, to meet the pluralistic needs of agricultural development (Hu et al., 2012; Sulaiman V & Davis, 2012; Swanson, 2006).

Meanwhile, extension approaches also have transformed from linear technology transfer to approaches that are inclusive, participatory and knowledge-based, and changed scopes from yield-centric to broader environment, value chain and capacity building, with a significantly reduced extension budget from public finance. This is exemplified by the shift from a public-sector agricultural RD&E (research, development and extension) system to the private sector in Australia (Hunt et al., 2012; Paschen et al., 2017; Rivera & Alex, 2004). However, in developing countries such as China, smallholder farmers are still heavily relying on public-funded agricultural extension services. With the government's many attempts to reduce the financial burden, the extension fund has been drastically cut, affecting extension officers' incentives and their motivation is in fact damaged. For example, grassroots extension officers use motorbike to visit villages and farmers but they have no funds for petrol (Friederichsen et al., 2013; Liu & Kong, 2018; Swanson, 2008). The relations between the extension system and smallholder farmers are shown in Fig. 7.1.

The rigid top-down mechanism and the straight linear delivery approach of the system have contributed a rapid dissemination of agricultural innovations across the nation, especially high-yield seeds and agricultural machinery imported from other countries. However, the

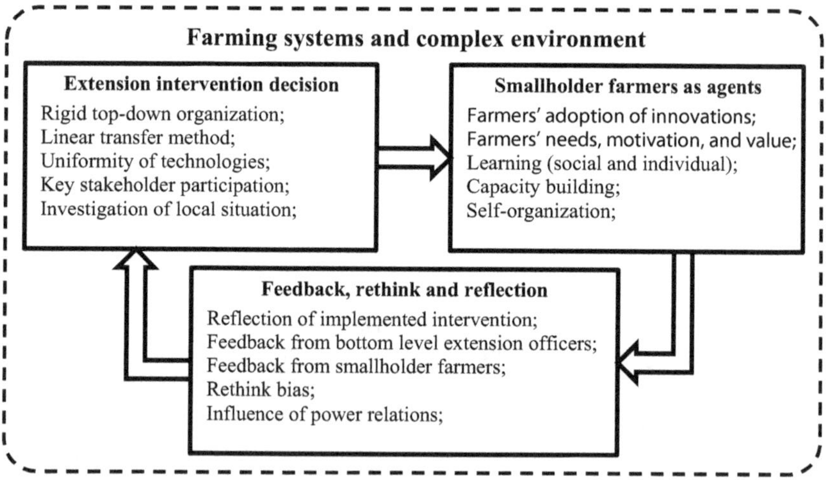


**Fig. 7.1** Loop of agricultural extension system

static system has received growing critiques nowadays for the lack of stakeholder participation, failing to meet pluralistic needs of farmers and use of the one-way delivery approach (Kong, 2009). Jian (2007) pointed out that the top-down extension system was the legacy of the collectivism economy in the 1960s, where the whole nation required quick innovation to boost agricultural development but has shown many contradictions with the needs of farmers nowadays. Earlier studies have pointed out the need for bottom-up mechanisms and farmers' participation in decision-making processes in the extension system (Douthwaite & Hoffecker, 2017; Hu et al., 2007; van de Fliert & Braun, 2002). But their goals remained in boosting technology adoption, which can lead to other consequences such as misuse of natural resources and damage to environment. In the systems language, this is linear thinking, which solves one particular problem that causes another.

In the meantime, growing attention has been to the application of systems thinking to agricultural extension research, prominently the AKIS model by Röling (1988), which specifies role of knowledge and information in agricultural extension, rather than overly emphasizing technology. It also incorporates other approaches that are inclusive and pluralistic in the diffusion processes. Later, Leeuwis (2004), Spielman et al. (2009) and Klerkx et al. (2012) pointed out a need for a holistic understanding of agricultural extension and engineered the Agricultural Innovation System (AIS) model. The AIS views agricultural extension as the process of innovation that generated from interactions among various agents, institutions and environments, it points out that agricultural innovation is broader institutional change, rather than simple technological outcome (Coutts & Roberts, 2011; Klerkx et al., 2012; Leeuwis, 2004; Nettle et al., 2017). Similar opinion is also found in Ampt et al. (2015) on the Australian extension systems, which points out the possibility of the system to embrace the innovation term as agricultural extension gives people the impression of traditional linear technology adoption.

The employment of systems thinking in agricultural extension raises possibilities for theoretical analysis and application compared to the technology adoption tradition. The Complex Adaptive Systems (CAS) theory considers components within the national extension system as agents (Ekboir, 2003), and are capable of self-organising by nature (Spielman et al., 2009). Also, actors in the systems are adaptive to their environment (Klerkx et al., 2010). This suggests a need to understand farmers or community at the micro-level by considering them as a sub-system or hierarchical component of the broader agricultural or farming system, such as community, niche or individual. This is especially important in analysing a rural community's interactions with their upper-level systems, as they proactively act in changes and interventions, rather than passively receive or be intervened upon.

## Complexities in the Agricultural Extension System and the Role of Rural Community

As we revisit the data that collected from the agricultural extension system and smallholder rice farmers, we categorize two main complexities, the organizational and technological complexities, which hinder the effectiveness of the system in meeting farmers' needs. Then, we provide arguments about the incorporation of rural community in tackling those complexities. The data was collected from March to September in 2017, using in-depth interview with 14 officers in the agricultural and extension departments from provincial to township levels, and focus group discussions with smallholder rice farmers in 10 villages (7 to 13 farmers were selected in each village, 95 in total) of two counties that are major rice production sites in Guangxi province. Details of study locations are shown in below Fig. 7.2. The data was then coded and analysed using NVivo 12 platform.

In terms of organizational complexity, firstly, the system is rigidly working top-down and does not allow much feedback from the farmer's side. Evidence was found that, for example, officers at the provincial level reported that they made the decision of promoting non-tillage farm technique to smallholder farmers in accordance with its good yield performance in the neighbouring provinces. Farmers neither participated in the decision-making process nor were consulted about their needs for farming, resulting in reported rejection of this technique. Secondly, the system overlooked feedback from the bottom-level extension officers, especially in political reforms. In our interviews with township level extension officers, many pointed out that their time, funds and incentives for conducting agricultural extension activities were greatly reduced owing to township organizational reform in 2002. The reform has made the extension station in the townships obligated to report to multiple leaderships: township government as the administrative leadership and county level agricultural department as the professional leadership. As a result, extension officers in the township extension station were relocated to other

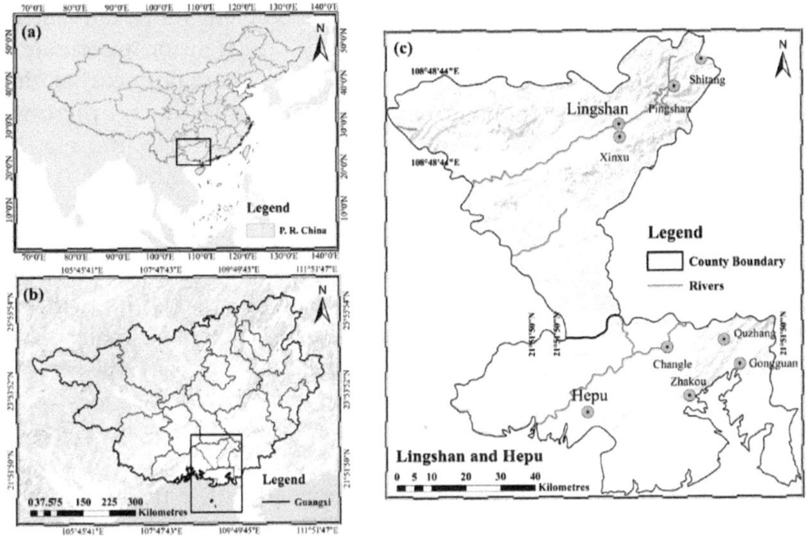


**Fig. 7.2** Study locations (*Source* Created by the first author, using ArcGIS® software by ESRI—"World_Basemap_v2" [basemap]. [1:70,000] in: "World Terrain Base". May 28, 2020. https://services.arcgisonline.com/ArcGIS/rest/services/World_Terrain_Base/MapServer)

non-extension related work duties such as accountancy and village election supervisor; farmers commonly reported that the frequency of officers doing on-farm training and visiting was significantly reduced.

In terms of technological complexity, firstly, technologies promoted by the extension system solved particular farming issues but caused others. Decision-makers reported that adoption of the non-tillage technique can reduce time for land tilling for farmers. However, as the researcher conducted focus group discussions with farmers in several selected villages, farmers commonly reported that their villages did not have conditions to support the employment of the technique such as a stable irrigation water supply. Some reported that the technique caused much more trouble than the traditional techniques, for example, by requiring more use of weedicides. Secondly, social factors played a role in dealing with technological complexity. While extension officers had reduced frequency of visiting villages, farm management information was occasionally absent. This made farm management more difficult for farmers, especially for those who heavily rely on extension information. For example, in a few villages, farmers reported misalignment of time for pesticide spray in their community. During our fieldwork, there was an outbreak of yellow dwarf disease, but with no uniform information, farmers took individual actions across a wide range of timeframe (over a week), which greatly compromised the effect of the spray. Also, a growing number of farmers had gone to town for off-farm work and left acres of farmland unmanaged and settled by pests and rats, making existing farming more difficult. This also has led to more use of pesticides as farmers reported they have used multiple kinds of pesticides than usual.

As we revisit the data, we propose that the conventional rural community of farmers has the possibility of dealing with the complexities in the extension system. Firstly, farmers share their experience in dealing with farm management issues. In villages where the outbreak of yellow dwarf disease emerged, farmers shared their experience, either their own practice or the information that they obtained from extension officers, during our focus group discussions (FGD). For example, during a FGD in Hepu county, a farmer reported that they sprayed a few types of pesticides to counteract the disease, implying that it is transmitted by insects, but it failed to stop the growing of yellow shoots. Then a few farmers shared the information that infected shoots should be removed from the field to stop the spread as they have been advised by officers that the disease is a virus and pesticides would not work. A few farmers also pointed out that pesticides should be applied in the seed breeding stage to prevent the virus being contagious. Some farmers also reported that they would help to inspect the farm of their neighbours who are out in town for off-farm work. Secondly, farmers shared opinions on collective issues. In a few villages, farmers reported that their communities had the opportunity to engage in growing corn and taro by some agricultural businessmen, but eventually they decided to reject it after group meetings. In the taro case, farmers were concerned about not being able to meet the end-production criteria; and for the corn case, farmers concurred that their village is a long distance from the market, which would compromise the quality of the corn. This indicates that farmers, as the main rural community stakeholder, are well aware of the strength of their villages, which are somehow undermined by the agricultural extension system.

Furthermore, there is the possibility to include the township level extension officer, who is a member of the formal system but familiar with local situation and has many interactions with farmers, in the rural community level of the system. Firstly, extension services and training facilitated by extension officers are social learning opportunities for farmers. During our FGDs, when asked about trajectories of sharing farm management information, farmers commonly reported extension officer's on-site technological training. But owning to changes in the organization of the extension system, extension officers have reduced their frequency of visiting villages and thus farmers lose an important occasion to meet specifically for farm management issues. Secondly, extension officers can enhance the strength and capacity building of the rural community. In our field study, we found that farmers are actively exercising agency, e.g. conducting small-scale farm experiment and use of local knowledge, to deal with farming issues. Extension officers can certainly perform the role of coach to validate farmers' ideas, the process of farm experimentation and identification of farm issues. This has been largely pointed out in many farmer-scientist knowledge exchanges and formal and informal knowledge combination literatures (Chambers, 1997; Ingram, 2008; Šūmane et al., 2018). The possibility of considering the township level extension officer as a part of the rural community is shown in Fig. 7.3.

In the overall extension system, the upper level is the rigid top-down system where extension decisions are made, delivered and enforced in a linear way. However, we saw an overlapped role at the township level, in which they are viewed as recipients of technology from upper level and need to obey policies, but also familiar with local situations and have many interactions with farmers in the rural community level. This chapter proposes that at the township level, the extension station or extension officers can be either an intermediary between the agricultural extension system and rural community, or potentially be included as one true member of the rural community level. By raising this point, we argue that the extension system should perform non linear thinking regarding policy-making, to articulate feedback from the bottom level of the system and make changes that would grant more autonomy for local level agricultural extension officers.

## Conclusion

This chapter examines the role of rural communities in helping smallholder farmers manage the limitations in China's agricultural extension system, which impeded smallholder farmers' access to extension services.

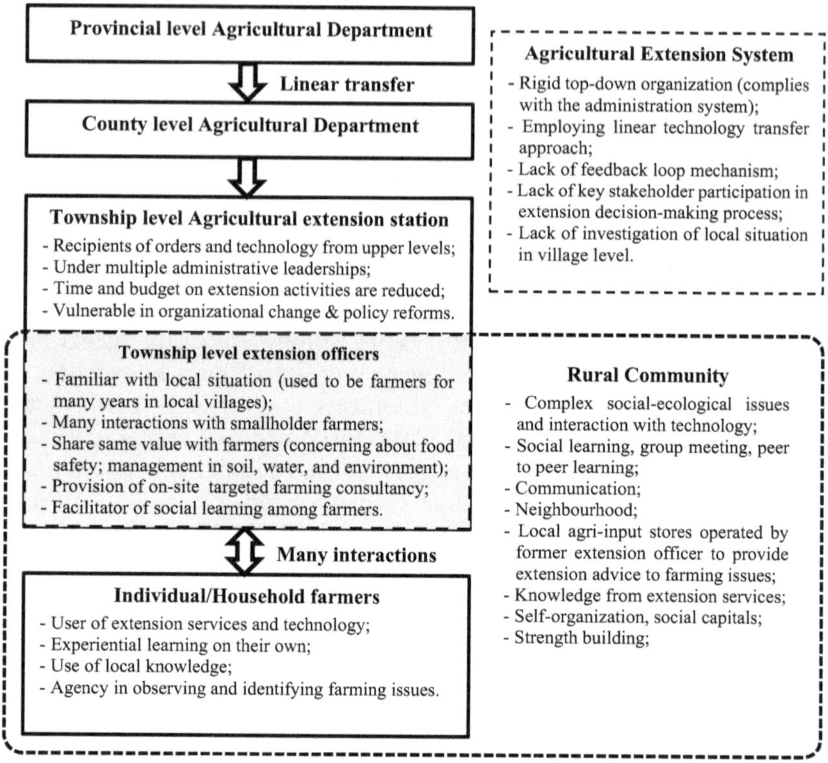


**Fig. 7.3** Hierarchical relations among the extension system, rural community and farmers

Two main limitations were identified in the extension system. Firstly, the system is organizationally top-down and refuses feedback, which causes reduction of extension officers' time and funding in extension activities. Secondly, technologies promoted to villages overlook the local situation, which leads to rejection and further consequences. At the rural community level, in addition to extension services, farmers manage farming issues through social learning, information sharing and exercising agency. Also, extension officers at the township level perform active roles in supporting farmers as they are familiar with local situations and farmers. However, their capacity and autonomy are heavily compromised due to the dysfunctional arrangement of the system. We propose that the system needs to use systems thinking to allow feedback from the township and village levels and build the capacity of extension officers and farmers.

**Acknowledgements** This chapter is based on some reflections of Chengdong Liao's PhD thesis, supervised by Dr Gomathy Palaniappan, Prof Helen Ross and Dr Natalie Jones during 2015-2019, funded by the China Scholarship Council and the University of Queensland International (CSC-UQI) joint scholarship (NO. 201506350079). The collection and analysis of data used for some arguments in this chapter were supported by School of Agriculture and Food Sciences of the University of Queensland. The systems theories and perspectives in this chapter were contributed by Assoc Prof Yiyu Liu, who was senior visiting scholar in Prof Helen Ross's team at the University of Queensland during 2019–2020. We also thank the support from the China Postdoctoral Science Foundation (NO: 2021M701282), the Guangdong Province University Innovation Grant (NO: 2021WTSCX021) and the South China Normal University Early Career Researcher Grant Scheme (NO: 20SK14).

## References

Ampt, P., Cross, R., Ross, H., & Howie, B. (2015). The case for retaining, redefining and reinvigorating extension in agricultural innovation systems. *Rural Extension and Innovation Systems Journal, 11*(1), 157–164. https://doi.org/10.3316/informit.661157849570442

Berkes, F., & Ross, H. (2013). Community resilience: Toward an integrated approach. *Society & Natural Resources, 26*(1), 5–20. https://doi.org/10.1080/08941920.2012.736605

Berkes, F., & Ross, H. (2016). Panarchy and community resilience: Sustainability science and policy implications. *Environmental Science & Policy, 61*, 185–193. https://doi.org/10.1016/j.envsci.2016.04.004

Cai, J., Hu, R., & Hong, Y. (2022). Impact of farmer field schools on agricultural technology extension evidence from greenhouse vegetable farms in China. *Applied Economics, 54*(24), 2727–2736. https://doi.org/10.1080/00036846.2021.1996530

Cassou, E., Jaffee, S. M., & Ru, J. (2018). *The challenge of agricultural pollution*. World Bank Publications.

Chambers, R. (1997). *Whose reality counts: Putting the last first*. Intermediate technology publications.

Chen, X., Zhao, Y., Chen, J., & Luo, D. (2009). *Zhongguo nongcun zhidu bianqian 60 nian (60 Years of Rural Policy Change in China)*. People's Press.

Coutts, J., & Roberts, K. (2011). Theories and approaches of extension: review of extension in capacity building. *Shaping Change: Natural Resource Management, Agriculture and the Role of Extension* (pp. 22–31). Australia-Pacific Extension Network (APEN).

Douthwaite, B., & Hoffecker, E. (2017). Towards a complexity-aware theory of change for participatory research programs working within agricultural innovation systems. *Agricultural Systems, 155*, 88–102. https://doi.org/10.1016/j.agsy.2017.04.002

Ekboir, J. (2003). Why impact analysis should not be used for research evaluation and what the alternatives are. *Agricultural Systems, 78*(2), 166–184. https://doi.org/10.1016/S0308-521X(03)00125-2

Feder, G., Anderson, J. R., Birner, R., & Deininger, K. (2010) 'IDEAS Working Paper Series from RePEc'. St. Louis: Federal Reserve Bank of St Louis.

Friederichsen, R., Minh, T. T., Neef, A., & Hoffmann, V. (2013). Adapting the innovation systems approach to agricultural development in Vietnam: Challenges to the public extension service. *Agriculture and Human Values, 30*(4), 555–568. https://doi.org/10.1007/s10460-013-9433-y

Hoang, L. A., Castella, J.-C., & Novosad, P. (2006). Social networks and information access: Implications for agricultural extension in a rice farming community in northern Vietnam. *Agriculture and Human Values, 23*(4), 513–527. https://doi.org/10.1007/s10460-006-9013-5

Hu, R., Huang, J., & Chen, K. (2012). The public agricultural extension system in China: development and reform. *background paper prepared for the Roundtable Consultation on Agricultural Extension, Beijing, March*, 15–17.

Hu, R., Cao, J., Huang, J., Peng, S., Huang, J., Zhong, X., et al. (2007). Farmer participatory testing of standard and modified site-specific nitrogen management for irrigated rice in China. *Agricultural Systems, 94*(2), 331–340.

Hunt, W., Birch, C., Coutts, J., & Vanclay, F. (2012). The many turnings of agricultural extension in Australia. *The Journal of Agricultural Education and Extension, 18*(1), 9–26.

Ingram, J. (2008). Agronomist–farmer knowledge encounters: An analysis of knowledge exchange in the context of best management practices in England. *Agriculture and Human Values, 25*(3), 405–418. https://doi.org/10.1007/s10460-008-9134-0

Jian, X. (2007). "Yi Nonghu Xuqiu Wei Daoxiang De Nongye Tuiguang Tujing", [New Approach of Agricultural Extension with Farmer's Needs Orientation]. *Science & Technology Progress and Policy, 24*(7), 44–46.

Klerkx, L., van Mierlo, B., & Leeuwis, C. (2012). Evolution of systems approaches to agricultural innovation: concepts, analysis and interventions. In I. Darnhofer, D. Gibbon, & B. Dedieu (Eds.), *Farming systems research into the 21st century: The new dynamic, Springer* (pp. 457–483). Dordrecht: Springer Netherlands. https://doi.org/10.1007/978-94-007-4503-2_20

Klerkx, L., Aarts, N., & Leeuwis, C. (2010). Adaptive management in agricultural innovation systems: The interactions between innovation networks and their environment. *Agricultural Systems, 103*(6), 390–400. https://doi.org/10.1016/j.agsy.2010.03.012

Kong, X. (2009). *Zhongguo Nongye Shehuihua Fuwu: Jiyu Gongji He Xuqiu De Yanjiu, [Agricultural socialized service in China: Based on supply and demand perspective]*. China Renmin University Press.

Leeuwis, C. (2004). *Communication for rural innovation: Rethinking agricultural extension*. Blackwell Science.

Li, L., & Pontes, L. (2017). *World Bank to Support China in Agricultural Land Pollution Control*. World Bank.

Liu, T., & Kong, X. (2018). "Xiaononghu He Xiandai Nongye Fazhan Youji Xianjie: Yiyuan, Shijian Yu Jianyi", [Dynamically connecting smallholder farmers with modern agricultural development: Willingness, practice and implications]. *Rural Economy, 2*, 1–8.

Nettle, R., Klerkx, L., Faure, G., & Koutsouris, A. (2017). Governance dynamics and the quest for coordination in pluralistic agricultural advisory systems. *The Journal of Agricultural Education and Extension, 23*(3), 189–195. https://doi.org/10.1080/1389224X.2017.1320638

Ongley, E. D., Xiaolan, Z., & Tao, Y. (2010). Current status of agricultural and rural non-point source Pollution assessment in China. *Environmental Pollution, 158*(5), 1159–1168. https://doi.org/10.1016/j.envpol.2009.10.047

Paschen, J.-A., Reichelt, N., King, B., Ayre, M., & Nettle, R. (2017). Enrolling advisers in governing privatised agricultural extension in Australia: Challenges and opportunities for the research, development and extension system. *The Journal of Agricultural Education and Extension, 23*(3), 265–282. https://doi.org/10.1080/1389224X.2017.1320642

Rivera, W. M., & Alex, G. (2004). The continuing role of government in pluralistic extension systems. *Journal of International Agricultural and Extension Education, 11*(3), 41–52.

Röling, N. (1988). *Extension science: Information systems in agricultural development*. Cambridge University Press.

Spielman, D. J., Ekboir, J., & Davis, K. (2009). The art and science of innovation systems inquiry: Applications to Sub-Saharan African agriculture. *Technology in Society, 31*(4), 399–405.

Sulaiman V, R., & Davis, K. (2012) 'The "New Extensionist": Roles, Strategies, and Capacities to Strengthen Extension and Advisory Services'. Lindau, Switzerland: Global Forum for Rural Advisory Services.

Šūmane, S., Kunda, I., Knickel, K., Strauss, A., Tisenkopfs, T., Rios, I., d. I., et al. (2018). Local and farmers' knowledge matters! How integrating informal and formal knowledge enhances sustainable and resilient agriculture.

*Journal of Rural Studies, 59*, 232–241. https://doi.org/10.1016/j.jrurstud.2017.01.020

Sun, B., Zhang, L., Yang, L., Zhang, F., Norse, D., & Zhu, Z. (2012). Agricultural non-point source pollution in China: Causes and mitigation measures. *Ambio, 41*(4), 370–379. https://doi.org/10.1007/s13280-012-0249-6

Swanson, B. E. (2008). Global review of good agricultural extension and advisory service practices. Food and Agriculture Organization of the United Nations Rome.

Swanson, B. E. (2006). Extension strategies for poverty alleviation: Lessons from China and India. *The Journal of Agricultural Education and Extension, 12*(4), 285–299. https://doi.org/10.1080/13892240601062488

van de Fliert, E., & Braun, A. R. (2002). Conceptualizing integrative, farmer participatory research for sustainable agriculture: From opportunities to impact. *Agriculture and Human Values, 19*(1), 25–25. https://doi.org/10.1023/A:1015081030682

Wang, G., Lu, Q., & Capareda, S. C. (2020). Social network and extension service in farmers' agricultural technology adoption efficiency. *PLoS ONE, 15*(7), e0235927–e0235927. https://doi.org/10.1371/journal.pone.0235927

Wang, X. (2006). Management of agricultural nonpoint source pollution in China: Current status and challenges. *Water Science and Technology, 53*(2), 1–9. https://doi.org/10.2166/wst.2006.033

Wang, X., Li, Y., Wang, R., & Du, J. (2018). "Woguo Dami Jinkou Jizeng De Xianzhuang, Yuanyin Ji Duice", [Concurrent, causation and implications of the rapid increase of rice Import in China]. *Agricultural Economics, 1*, 118–120.

Wellard, K., Rafanomezana, J., Nyirenda, M., Okotel, M., & Subbey, V. (2013). A review of community extension approaches to innovation for improved livelihoods in Ghana, Uganda and Malawi. *The Journal of Agricultural Education and Extension, 19*(1), 21–35.

Xiao, G., & Wang, Q. (2017). "Liangshi Jinkou Dui Woguo Liangshi Jiage Bodong Yingxiang De Shizheng Yanjiu", [An empirical study of the impact of grain import on grain price fluctuation in China]. *Journal of Social Science of Hunan Normal University, 46*(4), 117–123.

Yang, W., & Liu, Q. (2019). Grain security: Projections for rice supply and demand in China. *Research of Agricultural Modernization, 40*(1), 44–53.

Yi, L., Zhang, Z., Wang, X., Liu, B., Zuo, L., Zhao, X., et al. (2013). "Jin Sanshi Nian Zhongguo Zhuyao Gengdi Houbei Ziyuan De Shikong Bianhua", [Spatial-temporal change of major reserve resources of cultivated land in China in recent 30 years]. *Transactions of the Chinese Society of Agricultural Engineering, 29*(6), 1–12.

Zhang, L., & Chen, F. (2014). "Jiyu Logistic Moxing De Zhongguo Chengzhenhua Yanjin Dui Gengdi Yingxiang Qianjing Yuce Ji Fenxi", [Analysis and forecast on prospect about influence of urbanization gradual progress on cultivated land in China based on Logistic model]. *Transactions of the Chinese Society of Agricultural Engineering, 30*(4), 1–11.

CHAPTER 8

# Reflections on Community Development in the Asia–Pacific: A Researcher's Perspective

*Gomathy Palaniappan*

## How Is Community Development Relevant to Aid Organisations?

Community development practice seeks to empower local communities to identify their strengths and needs and build their capacity and therefore confidence to achieve their own goals (Carlon, 2021). It relies on individuals to participate and collaborate usually in an agreed group process (IACD, 2017). The global reach of the International Association of Community Development (IACD) definition (IACD, 2017) identified eight themes: values into practice; engaging with communities; participative planning; organising for change; learning for change; diversity

G. Palaniappan (✉)
School of Agriculture and Food Sciences, The University of Queensland, St Lucia, QLD, Australia
e-mail: g.palaniappan@uq.edu.au



C. Baldwin and S. van Bommel (eds.), *Rural Development for Sustainable Social-ecological Systems*,
Palgrave Studies in Environmental Sustainability,
https://doi.org/10.1007/978-3-031-34225-7_8

and inclusion; leadership and infrastructure; and developing policy and practice.

Each aid agency has its own purpose to make change: to improve farm productivity, health and well-being; or certain skills, usually through awareness-raising, education, access/transport and provision of materials or equipment. A major challenge is to match the aid available or being offered, to a community's needs and desires, and to make sure that it fits within the broader context of other community activities. This chapter gives examples of cases in the Philippines, Papua New Guinea and Pakistan and raises questions about how a community can identify and access the aid they need, ensure it is shared equitably within the community and have an ongoing benefit.

## The Purpose of My Work in These Communities

I was involved in ACIAR-funded projects which aimed to empower smallholder farmers growing crops, such as rice, vegetables, livestock, pigs based on their needs, to better participate in the markets for greater benefit. It not only required production skills but also marketing and soft skills such as organising and managing group work. Being smallholder farmers, they needed to work collectively. As a community there is hesitation to work together and the projects need to facilitate the process of community engagement and the benefits of working together that would allow participation in markets (e.g. selling in different locations). This would empower the community to negotiate with government about how to improve development activities. Once they start getting attention of governments, they may get better roads, better access to markets and then can explore other opportunities such as tourism. We found that some of the work of aid agencies distracted communities from their goals related to agriculture by not addressing supply chain issues.

## Observations at a Case Study Location in the Philippines

We worked with communities in Cabintan from 2012 in an ACIAR-funded project (Curry et al., 2019). Cabintan is a *barangay* in the city of Ormoc, in the province of Leyte. Its population as determined by the 2020 Census was 2,623. This represented 1.14% of the total population of

Ormoc which was struck by Super Typhoon Haiyan (Yolanda), the Philippines in November 2013, having a direct impact on nearly six million people. The typhoon devastated all the crops and it appears that this has been happening with increasing frequency due to geographic location of villages (see Fig. 8.1). We aimed to support collective action by the producers but found that international aid agencies that arrived in response to the disaster could not organise themselves to work collaboratively. At a local level, aid workers competed for the interests of local bodies.

Each agency had a particular role; one provided funding to help a group of farmers improve their land, but nothing else. Another agency distributed seeds, but not necessarily where the first organisation was working in land preparation. They would have been mutually supportive if they worked together. Without an understanding of the market, it resulted in a big glut of a couple of vegetables in the market, with lower prices and food wastage.

Another agency was concerned with the health of labourers and provided suitable tools. It appointed a couple of nurses in every centre and had a rule of examining health of participating workers. One day as I was talking to a farmer group in a steep field, three nurses contracted by the organisation tested their blood pressure before they walked up the steep slopes, as part of checking their health. The farmers thought this was amusing as all their life they walk up and down on these steep fields, with no adverse effects. They did not understand why funds were spent on this instead of something useful like water infrastructure (e.g. ponds).

## Observations at a Case Study Location in Papua New Guinea

In a similar case, an agency aimed to support families by holding baking classes for communities in Central Province in Papua New Guinea. When queried about whether they attended to learn about nutrition, the local women indicated that they have plenty of local foods that are nutritious, but they were being given things. They indicated that they wanted to come to my agricultural project because they were highly involved in farming and sell in the markets and perceived that it met their priorities: building a house and sending children to school (Fig. 8.2).

Somehow the agency had the idea that women in this rural area had traditional roles of making clothes and cooking, which might have been

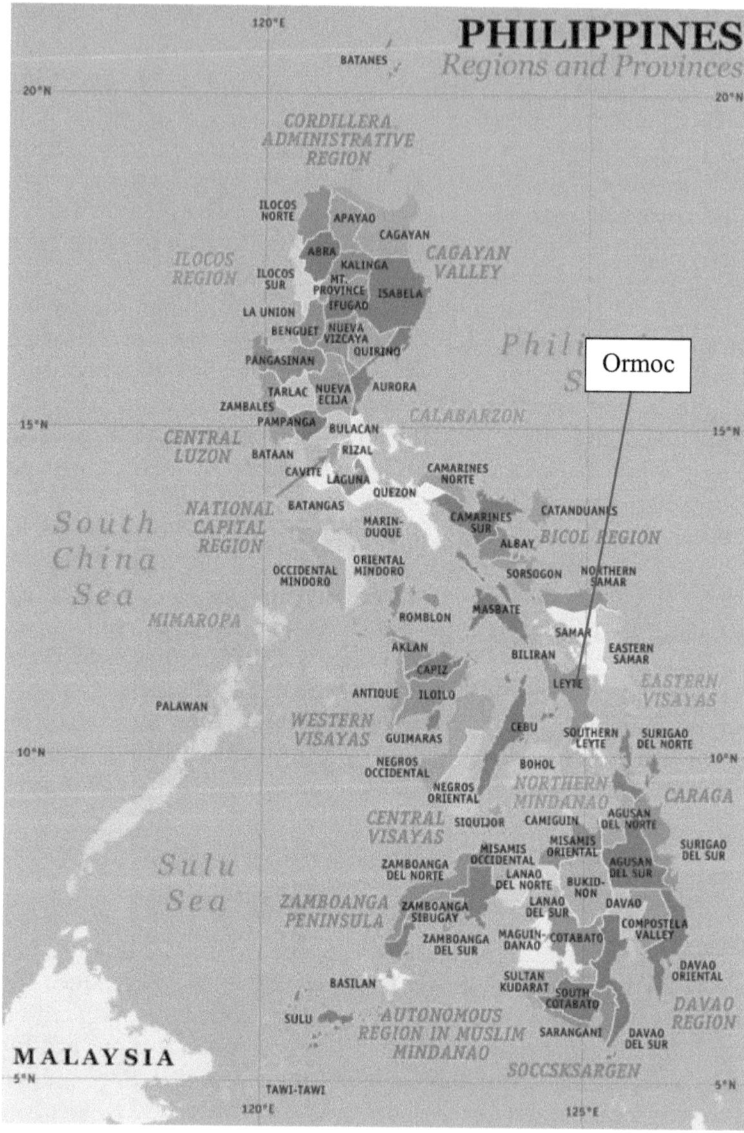


**Fig. 8.1** Location of Ormoc, Philippines (pinterest.com.au/pin/425379127311923574)

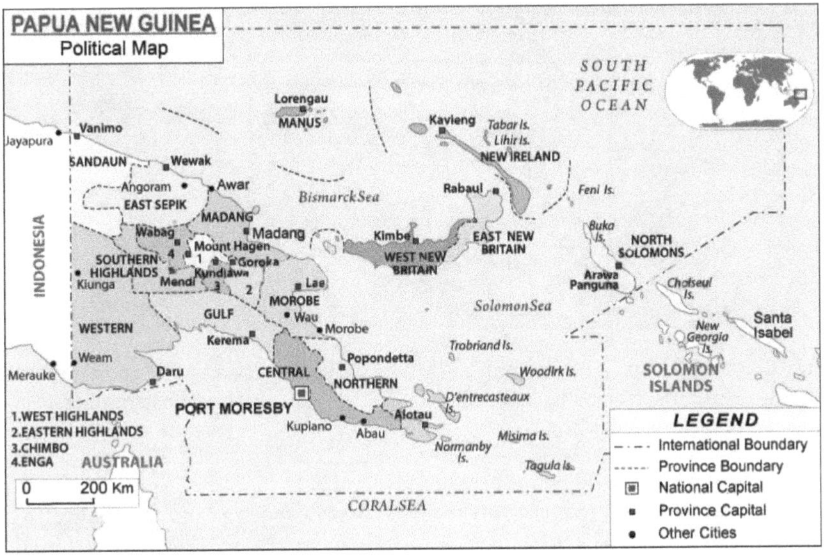


**Fig. 8.2** Location in PNG (maps-papua-new-guinea.com/papua-new-guinea-political-map)

relevant in cities. However, in rural PNG, women have big role in farming and marketing. The agency stereotyped and reinforced a gender role which did not exist. We observed this when we worked in an ACIAR project in PNG in Central Province with small farmer communities (Birch et al., 2014).

## Observations at a Case Study Location in Pakistan

In the village in Sindh province in Pakistan that we worked in an ACIAR project (Bajwa, 2018; Mazhar et al., 2019), an international aid agency provided equipment for drying food such as pineapples and mangoes to add value to their produce (Fig. 8.3). On the surface, it seemed like a good idea and a good way to deal with surplus fruit. They paid women to engage in the project and get trained, thinking it would value-add, but when we arrived a few years later, the project was finished but the dryers had been unused for several years. There had been no link to markets for dried mangoes locally, and if such a product was produced in Pakistan, it

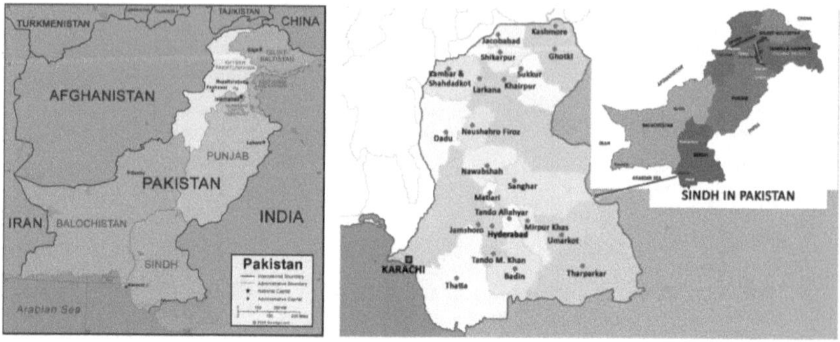


**Fig. 8.3** Location of Sindh province in Pakistan (geology.com/world/pakistan-satellite-image.shtml, destinationpakistanguide.com/sindh-province-of-pakistan/)

would have needed to be at an industrial scale with plans for marketing. They needed someone with skills in identifying the market and following through the supply chain.

It appeared that the farmers were so used to getting something in return for participating in a project (such as wages) that they would not benefit beyond temporary aid. As a community development challenge, it is hard for individuals to transform if they do not see a benefit of being empowered.

## Discussion

On reflecting on these cases, I suggest some of the following lessons be considered.

Any project proposed by an aid agency needs proper consultation at the local level, either with the local authority, farmers group, or tribal leaders. While aid agencies often liaise and get permission at the national level, this level may not understand individual local priorities. International organisations may need to seek assistance from local people to understand their needs.

Aid agencies need to ensure that if the project is aimed at women, a range of women from different socio-economic levels are involved in preliminary discussions, to ensure equitable participation and inclusion.

Since local groups may be dominated by men, it is important that the agency asks for female participation and not just with a leader's wife.

Farmers need to take a stronger position if a project does not meet their priorities, as it can distract them from their higher priorities. If they are not interested in a project, they should encourage the agency to go elsewhere.

In many cases, a good process is initiated, but if there is a natural disaster, as in the case of the Philippines, there may not be the same opportunity for discussion. But learnings from previous interventions from local agencies, government and by other institutions could be taken in account before venturing into aid action.

The private sector often sponsors aid to demonstrate corporate social responsibility but they need to ensure a consultation process and full understanding of the local context.

In some cases, aid agencies have provided incentives for local participation, either wages for attendance, or provision of equipment or materials. While this is often done to reduce time and effort to initiate and maintain engagement, it results in high expectations that this will be the case with all projects. It undermines empowerment which is embedded in the concept of community development, and can undermine projects by other agencies which cannot offer incentives, or which expect that a project will be successful if a local community is self-motivated.

## Conclusion

The analysis presented in the scenarios articulates the limitations of community development approaches practised by international aid organisations. Participative planning does not take into account people with less power who are unable to contribute to the process. Also, community development in practice does not consider the tensions that may exist between development agencies and community of practice. To strengthen the practice of community development, a grassroots approach of empowering the community needs to be agreed and practised by all development agencies.

## References


Bajwa, B. (2018). *Strengthening vegetable value chains in Pakistan for greater community livelihood benefits*. https://www.aciar.gov.au/sites/default/files/project-page-docs/fact_sheet_hort_2016_012_a.pdf

Birch, C. J., Bonney, L., Doyle, R. B., Sparrow, L. A., Chambers, B., Palaniappan, G., Boersma, M., Seta-Waken, P., Dell, M., Nivi, J., Gracie, A. J., Benny, D., Simeon, L., & Wesis, P. (2014). *Increasing vegetable production in Central Province for Port Moresby markets*. https://www.aciar.gov.au/sites/default/files/project-page-docs/final_report_smcn.2008.008.pdf

Carlon, C. (2021). Contesting community development: Grounding definitions in practice contexts. *Development in Practice, 31*(3), 323–333. https://doi.org/10.1080/09614524.2020.1837078

Curry, P., Nicetic, O., & Palaniappan, G. (2019). *Developing vegetable and fruit value chains and integrating them with community development in the southern Philippines—Final report*. The Australian Centre for International Agricultural Research (ACIAR). https://www.aciar.gov.au/publication/technical-publications/developing-vegetable-and-fruit-value-chains-and-integrating-them-community-development

IACD. (2017). *Towards common international standards for community development practice: A draft consultation paper for IACD Members*. Accessed 18 May 2020. www.iacdglobal.org/wp-content/uploads/2017/11/IACD-2017-Draft-Standards-Guidance.pdf

Mazhar, M., Bajwa, B. E., McEvilly, G., Palaniappan, G., & Kazmi, M. R. (2019). Improving vegetable value chains in Pakistan for sustainable livelihoods of farming communities. *Journal of Environmental and Agricultural Sciences, 18*, 1–9. https://www.researchgate.net/publication/332567302_Improving_Vegetable_Value_Chains_in_Pakistan_for_Sustainable_Livelihood_of_Farming_Communities

CHAPTER 9

# Insights from Community Development for Strengthening Landcare in Queensland

*Ken Keith*

## Introduction

Cavaye (2001) identified challenges in moving community development beyond service provision, information dissemination and provision of resources to meet perceived needs. He identified (p. 110) the following dimensions of community that community developers need to engage: act on existing motivation; build enthusiasm and confidence; challenge community attitudes and perceptions; support 'hidden' informal leaders in communities; re-think apparent needs and redefine community assets; gain access to appropriate information and resources; and build relationships with key individuals inside and outside their communities—guidelines for a community development facilitator but also criteria for effectiveness of a community group itself. These principles formed a basis

K. Keith (✉)
Retired Ex School of Natural and Rural Systems Management, The University of Queensland, Brisbane, QLD, Australia
e-mail: hkeith1540@bigpond.com



C. Baldwin and S. van Bommel (eds.), *Rural Development for Sustainable Social-ecological Systems*,
Palgrave Studies in Environmental Sustainability,
https://doi.org/10.1007/978-3-031-34225-7_9

to examine how well Landcare in Queensland has fitted a community development model.

Cavaye and Ross (2019) identified key characteristics of community development:

- simultaneously utilising and enhancing the 'capitals' of communities
- building community cohesion, identity and a sense of place or community
- engagement and empowerment of community members
- collaborative action for collective benefits rather than competition
- uncovering hidden or unrealised assets and capacity in communities
- ongoing improvement through the cumulative effects of discrete initiatives
- action based on values of inclusion, social justice and self-determination.

These characteristics provide a basis to discuss some ways to strengthen community-based natural resource management in Queensland, specifically for Landcare groups.

Before examining the fit between Landcare and community development principles, an outline of Landcare in Queensland and similarities with other Australian states follows.

## Landcare in Queensland and Elsewhere; Development and Evolution

### *The Meaning and Scope of Landcare*

Various meanings have been applied to the term 'Landcare': an ethic or philosophy of stewardship; a grassroots movement encapsulating this philosophy; and a model involving a network of bodies addressing issues of land degradation (Australian Framework for Landcare Reference Group, 2010; Otteson, 2019; Robins, 2018).

Before 'Landcare', there were land carers motivated by a land stewardship ethic, though it was not clearly formulated (Roberts, 1984). Through the growth of the Landcare movement, the stewardship ethic is now well recognised. The Landcare movement is about action by self-selected volunteers concerned about land degradation, working together for the health of the land; mobilising community effort to improve the management of their soil, water, vegetation and other natural resources that are not always directly owned by the group (Campbell, 1991; Cramb, 2004; Ross et al., 2002).

Landcare as a model with its sharing and support mechanisms including groups, networks and government and non-government partnerships has endured a roller-coaster of optimism, reality checks and almost rejection by some policy-makers.

Initially, both governments and the pioneering groups that formed the Landcare movement focussed on erosion, salinity, weeds, feral animals and tree decline. However, from the commencement of the Decade of Landcare (1990), Landcare was seen to include Dunecare, Bushcare and Coastcare, as well as urban Landcare. And nature conservation issues were part of the mix. Queensland's Decade of Landcare Plan principles included:

- Soil, water, vegetation and wildlife resources are interdependent and are all affected by the development and use of land.
- Fauna and flora conservation can be integrated with agricultural and pastoral production and urban land use (Department of Primary Industries Queensland, 1992).

The Queensland Landcare Council identified the goal of Landcare as: to empower the community, both urban and rural, to undertake local action to achieve economic and ecologically sustainable use and management of Queensland's natural resources through a partnership between government and community (Queensland Landcare Council, 1997).

### *History of the Landcare Movement, Queensland*

The Landcare movement began with the launch of Victoria's LandCare program in November 1986 or alternatively with the announcement of the National Landcare Program by Prime Minister Hawke in July 1989. These landmark 'commencements' did not emerge from a vacuum. From the 1950s or 60s, all states can boast community action of some sort to improve land management. But almost invariably, group action was initiated by government agents rather than local leadership. In the 1980s, Western Australia set up a network of community-based committees responsible for prioritising and putting action plans in place.

Meanwhile, in Victoria, several landholder-initiated groups formed and sought government agency support for specific land management issues in the early 1980s. Government fostered the concept in conjunction with the Victorian Farmers' Federation, leading to the launch of Victorian LandCare in November 1986.

In Queensland, government action on land degradation began in the 1940s. Where cooperative action was needed to control erosion, an early approach was to prepare sub-catchment plans and discuss them with landholders. In the 1960s, a next step towards community participation was to form discussion groups that could raise concerns and discuss options before preparation of a plan, making uptake more likely (Pink & Roberts, 1962). Although many landholders cooperated, inevitably some would not invest effort and finances in joint action. After serious erosion events in the prime agricultural land of the Darling Downs and Burnett regions in the early 1970s, the implementation of on-farm soil conservation structures became mandatory. This 'big-stick' approach was partly appeased by a subsidy (Pink & Roberts, 1962). Advisory Group Committees were set up to advise on priorities for treatment and issues arising. But the mandatory programme did not achieve its intentions. As it was wound down between 1983 and 1986, government sought ways for community to take its role in land management rather than see government as responsible for erosion control.

The formation of the Lockyer Watershed Management Committee as a Bicentennial project in 1981, under the inspiration of farmer Fred From, provided an example of what could happen. This was Queensland's first truly grassroots catchment cum Landcare group (without the name). Several subgroups allowed landholders to identify and work on local issues. This became a model for groups to take responsibility for their own issues (Queensland DPI, 1987). Over the next few years, other groups emerged in several rural districts.

The six advisory group committees from the abandoned mandatory programme developed into ten Landcare committees, in primarily cropping districts. At about the same time, the Queensland Cattlemen's Union promoted the formation of Landcare Committees in grazing districts. The Cattlemen's Union proposal for Land Care committees across the nation provided impetus to the Farley (National Farmers' Federation) and Toyne (Australian Conservation Foundation) negotiations with the Federal government for a National Landcare Program (Douglas, 1988).

Thus, the emergence and growth of Landcare in Queensland had a strong 'grassroots' element but also government impetus. Government officers actively encouraged formation of groups, but groups also emerged in unexpected places, generated by local concern or industry impetus. Through 1987 and 1988, Queensland government support initiatives were developed and in quick evolution of names, Soil Conservation District Committees became Land utilisation and Conservation Committees and then Landcare Committees. Government launched Queensland Landcare in March 1989 with 24 recognised committees, some three months before the Hawke Declaration of a Decade of Landcare and $340 million Federal support. Twelve of these were government initiated and others were local community initiatives, some motivated by the Cattlemen's Union promotion.

Queensland government support included model constitutions, insurance cover, newsletters, an administration allowance and availability of an extension officer to assist with technical matters. A state Landcare Coordinator had been appointed and when Federal funds became available; seven Regional Landcare facilitators were appointed to foster formation of groups. Federal money assisted groups with educational and on-ground action projects.

Comparison of early progress with Victoria and New South Wales is possible through presentations at a seminar in February 1990 to launch the educational material working together for Landcare (Chamala & Mortiss, 1990): Victoria 66 accredited Landcare groups and probably three times as many groups in various stages of development (Edgar, R); New South Wales about 60 groups with 20 of those being Dunecare groups (Marston, D); and Queensland nearly 60 Land Care committees in various stages of formation (Begbie, D). Victoria obviously had the lead but all three were comparable at that stage. By June 1990, there were 70 groups in Queensland (see Fig. 9.1). This might imply a huge portion of the state still uncared for, but the map of spatial areas covered by the groups gives a different picture (Berndt, 1990). A difference between Queensland groups and those formed in Victoria needs to be noted. Groups in Queensland tended to be larger than those in Victoria; 'shire' sized groups that looked at issues across a district. Many Victorian groups were smaller single issue groups.

Another significant difference between Queensland and the southern states was the pre-existence of a formal catchment management structure in New South Wales and Victoria, within which Landcare groups

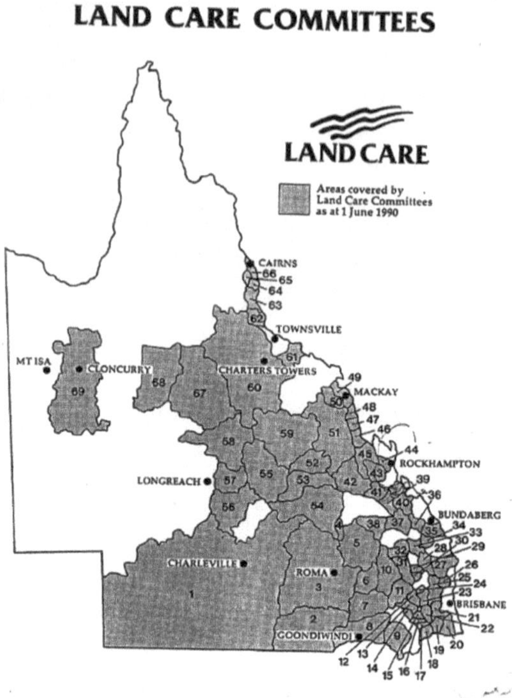


Fig. 9.1 Map of Landcare Committee coverage in Queensland, June 1990

nested. For instance, in NSW, the Total Catchment Management Program established statutory regional catchment committees with members representing state agencies, local government and landholders. Landcare was a subset of this. In Queensland, the need for integration of management across a catchment was 'wedged in' after Landcare group formation was well under way. In 1991, a state government policy for integrated catchment management (ICM) saw catchment coordination committees formed, generally encompassing several Landcare districts. ICM was seen as classic river basin management without regulatory processes, while Landcare was seen as the process whereby ICM actions are put in place

(Bellamy et al., 2002, p. 100). While this was the policy-makers' perception of the two step process, it was rarely seen this way by Landcare groups and often not by catchment committees.

From 1997 to 2002, funding through the National Heritage Trust (NHT1) of $1.25 billion (with a requirement for matching investment by the States) gave capable groups the opportunity to conduct large-scale NRM projects for their community (as shown in the case examples below). However, during this time the Australian government moved towards regional management of funds provided. In Queensland, this meant another tier of NRM management as catchment coordinating committees were not statutory bodies—Regional Management Groups (RMGs) were introduced. This double change in tiers of governance, non-statutory catchment committees and then RMGs, did not occur in NSW and Victoria, which had statutory catchment bodies before introducing Landcare, so groups nested easily into the structure.

The superposition of ICM committees over Landcare groups was challenging at the peak body level as well. Queensland Landcare Council as Queensland's peak NRM body now had a rival; the Queensland Catchment Management Coordinating Committee. This was still an issue when priority actions for the Decade of Landcare Plan were re-developed in 1997: 'Dispel the confusion over Landcare and ICM; Review the functions and roles of the Queensland Landcare Council and Catchment Coordinating Committee' (Queensland Landcare Council, 1997). A joint body, the Landcare and Catchment Management Council, resulted, but with regionalisation much of its function was soon overtaken by a Regional Groups Collective. The Council was dissolved and the Queensland Water and Land Carers (QWaLC) was formed as the peak body for NRM volunteers.

Most analyses of Landcare in Australia have been based on the New South Wales and Victorian experiences (Campbell, 1991; Love, 2012; Robins, 2018; Sobels et al., 2001), without reference to Queensland, which had a slower growth in numbers. Perhaps a focus of government energy on establishing ICM contributed to this slower growth. Resources invested by state governments in southern states saw large numbers of groups formed. In Queensland, it appeared to some commentators that groups had 'self-organised' rather than be part of a state-wide programme commitment to Landcare by the Queensland government. This could be due partly to adoption of a low key nurturing approach in Queensland, rather than highlighting Landcare as a government programme as occurred in NSW and Victoria. From a community development point of view, the Queensland approach had merit, provided support services were maintained.

The next phase of Australian Government funding from 2003, National Heritage Trust Extension (NHT2) and the National Action Plan for Salinity and Water Quality (NAP), formalised the Regional Bodies as recipients of funds to implement agreed Regional Action Plans. The introduction of the Caring for Country programme in 2008, in which funds were quarantined towards specific Federal priorities, and open to competitive tender by a wider group of providers meant that the opportunity for Landcare groups to deal with their priorities largely evaporated (Robins, 2018). The processes installed for market-based tendering for funds to achieve specific targets impacted heavily on the concept of community-based NRM (Curtis et al., 2014; Robins, 2018). In 2013, Caring for Country was replaced by a new National Landcare Program but still largely evidence-based tendering for specific targets, unsuitable to meet the needs of community-based action (Curtis et al., 2014).

By 2022, about 440 community groups with NRM as part of their function were listed on the QWaLC website. Of these, 133 were named Landcare or Catchment or Coastcare groups, although others, including many of the 'Friends' and 'Community action' groups, undertook land care activity. About 85 groups identified as nature conservation and about 20 as environment groups. Over 40 were Fencing Clusters. Fourteen Indigenous organisations were registered.

Sparse population and great distances have made Landcare group activity difficult in much of Queensland's rangelands. The QWaLC website indicates about 25 Landcare groups in Queensland's rangelands, as well as Fencing Cluster groups. Fencing clusters are small groups of contiguous landholders who gain government grants towards dog-proof fencing material. Although fencing to protect stock has potential to improve land condition, not all are motivated by a Landcare ethic in their management practices. A suggestion as to how the social capital inherent in Cluster groups can become a force for the delivery of ecosystem services is offered later.

Opportunities for Indigenous participation in Landcare through co-management have been advanced (Ross et al., 2009), and factors favouring and limiting Indigenous participation have been identified (Hill et al., 2013). Some opportunities are explored in the Ways Forward section.

## Four Case Examples of Queensland Landcare Groups

### *Lockyer Watershed Management Association (LWMA)*

The emergence of the Lockyer group in 1981 and its place in formulating a Landcare model for Queensland has been mentioned. At a Department of Primary Industries seminar (April 1987) to generate staff action towards formation of landholder committees, an invited speaker from a subgroup within LWMA (Richard Scanlan) said in part:

> The next thing was that we decided that interest was building up. We were concerned about seeing our soil getting washed away. We realised it couldn't go on forever and that we would have to do something about it. A meeting was called. You understand that the Lockyer Watershed Management Committee had been operating for a couple of years before this, and our little committee was formed as an organisation under the Lockyer Watershed Committee, which is now an Association. Since that first meeting on 21 March 1984, I am quite proud to say that in that period we have had an average of a little over 20 people at our meetings. ... Towards the end of 1984 we gained funds through Greening Australia; about 40 farmers were involved and its main objects were to eradicate lantana, to establish better pastures, establish leucaena and plant trees... We are also concerned about the natural fauna and flora of the area and we would like to see it maintained in every way. (Queensland DPI, 1987, p. 24)

What a magnificent example to follow, an enthusiastic group looking into local issues within the umbrella of a body taking a catchment perspective. And what a tragic sequel to this story, when government superimposed ICM in 1991, instead of choosing LWMA as the Lockyer catchment committee, a separate body, the Lockyer Resource Management Group (later Lockyer Catchment Coordinating Committee), was installed (presumably to widen the stakeholders). The fallout was significant; a submission to a national catchment management enquiry (HRSCEH, 1999) decried the set-back to Lockyer catchment projects and slated the State government's top-down action. Then when a Regional body was inserted as yet another tier, LWMA was reduced to a Landcare group. It had lost its spirit and within a few years expired. An alternative Lockyer Valley Landcare group has emerged, but is more contained in scope.

### *Waggamba Landcare Group (Based at Goondiwindi)*

A public meeting of Waggamba Shire ratepayers in February 1988 resolved to establish the Waggamba Shire Conservation Committee aiming to advise on conservation within Waggamba Shire to achieve a balanced ecosystem in which soils; the native flora and fauna species are maintained along with the highest sustainable long-term productivity of the land; and to achieve these aims through goodwill and cooperation (Waggamba Conservation Committee, 1988). When Landcare emerged, the name converted to Waggamba Landcare Group. Significantly, it has recently (2020) re-badged as Macintyre Ag Alliance to emphasise the last part—new aims are to work collaboratively with community to achieve positive outcomes in agricultural productivity, enhanced technology and healthy environments, aiming to pass on a more productive and stable environment to the next generation.

The group has undertaken large-scale resource management projects and displayed ability to raise the funds to do so, sometimes from interstate, without relying on the Regional NRM group. Projects from 1989 to 2020 included: soil conservation work, helping create and coordinate sub-catchment groups; Grass Check, measuring changing biodiversity in grass species; coordinated feral animal control; noxious weed control; and fencing off riparian areas. The largest project was for soil conservation work, $470K in 2003–2004.

### *Dalrymple Landcare Committee*

Dalrymple Landcare Committee (Charters Towers) was formed in 1988 as a producer-driven self-help group aimed at improving producer awareness and understanding of land management (Fig. 9.2). It has conducted more than 60 projects, including forage budgeting service, City-Country Days, gully and scalded land demonstrations, weed control, best management practices for wildlife conservation in grazing lands and riparian fencing, the group's largest project, with $1m NHT funds from 1998 to 2002.

Dalrymple Committee has the advantages of an energetic secretary who has been with the group since inception and is also a grazing land management extension specialist; and a state government funded group coordinator—the only one in Queensland (Note that the NSW

**Fig. 9.2** Michael Penna of 'Riverview' Charters Towers explains his irrigated cropping programme to a group of interested townsfolk during the Dalrymple Landcare Committee's annual City-Country Day, 2010

and Victorian State governments invest millions of dollars to fund part time coordinators throughout their states).

### *Cambooya Landcare Group*

This example represents seven still existing groups formed from the Darling Downs Advisory Group Committees (Apart from Cambooya, four). Four are mainly oriented towards sustainable land management; another group majors in nature conservation and the other in tree planting for koala corridors.

The Cambooya group is near Toowoomba, showing some peri-urban characteristics. Its aims are to: provide Landcare related information and education; maintain a resource of publications and tools to assist landholders; enable farmers and landholders to manage their land sustainably; and partner government and other Natural Resource Management Groups. Projects of interest include: enhancing biodiversity values of roadside reserves; feral pig trapping; Lantana Busters (a group of volunteers ridding the district of lantana); smallholder field days on weeds, pasture, basic cropping; and a grassland Reserve project. The group's largest project, Integrated Area-Wide management, ran from 2003 to 2005, receiving a $250K NHT Grant.

## Elements of Community Development Evidenced in the Landcare Movement

Cavaye (2001) identified service provision, information dissemination and provision of resources as standard supports for community development. A first step in exploring the fit between Landcare groups and community development is to consider how well these have been provided.

*Service provision*: Initially, government agents attended meetings to provide technical advice and assist group meetings. With the depletion of government extension services after about 2000, this role fell to RMGs who lacked resources to provide the same type and level of support. Although each Region had a Regional Landcare Facilitator funded by the Federal government, these were often diverted from group support to extension of regional programmes (Keith, 2017).

*Information dissemination*: Newsletters, personal attendance at meetings by extension agents and networking by facilitators provided information dissemination for some years but has declined, partly due to the loss of government agency staff. The internet has become an important vehicle for information and Facebook enables groups to keep in contact with each other's activities.

*Provision of resources*: As mentioned, in early years, the Queensland government provided a small fund for administration and covered insurance. Under the National Landcare Program and NHT1, groups could access substantial project funding. However, over time, support for groups became 'one of instrumentalising communities and their members as on-ground implementation agents of NRM strategies endorsed at higher levels' (Curtis et al., 2014, p. 190). Aspirations for participatory governance expressed by Ross et al. in 'Laying down the ladder' (2002) appear to have reverted to top-down processes with groups on the lower rungs of Arnstein's (1969) ladder of participation hierarchy, which Ross et al. tried to lay down.

A second step is to consider Landcare in context of Cavaye's list of neglected dimensions of community development.

*Act on existing motivation*: Motivation was high in the early 'golden' phase when funds were available for local priorities. However, of 70 Queensland Landcare groups listed in June 1990, just 54 appear in the QWalC list of 2007 and 35 in the 2022 list (It would be interesting to check which attributes of social resilience [Maclean et al., 2014] were lacking in groups that fell and which were catalysts for thriving in the survivors). The number of Landcare plus Catchment plus Coastcare groups (land and water), fell from 206 in 2007 to 133 in 2022 and of those, just 94 had continued since 2007. Even though positive to see new groups, the loss of over 50% of the 2007 cohort is disturbing. On these grounds, it is difficult to claim strong resilience in the Landcare movement. The data suggest that motivation has shifted from land management to nature conservation. Curtis et al. (2014, p. 190) called on policy-makers to recognise the importance of strengthening motivation of community members to contribute to NRM initiatives.

*Build enthusiasm and confidence*: Enthusiasm and confidence were built in early years through development of the Decade of Landcare Plan, through conferences and awards, and through national publicity, as well as availability of funds for local priorities. For many groups, the enthusiasm appears to have waned, shown by the number of collapsed groups. Other groups such as Waggamba/McIntyre have re-energised through a fresh direction.

*Challenge community attitudes and perceptions*: A perception in the 1970s that soil erosion control was a government responsibility shifted as Landcare groups gradually took responsibility for action on their priority issues. One outstanding example of the shift was the call by Jock Douglas, then President of the Cattlemen's Union, to the grazing community to embrace a Landcare ethic (Douglas, 1988). However, in spite of energetic efforts to organise workshops, seminars and field days, Landcare groups have generally not experienced great increases in number of supporters. Spread of the Landcare ethic has not reached the aspirations of the Decade of Landcare Plan.

*Support 'hidden' informal leaders*: Cavaye suggests that leaders need to be inherently good at reconciling or managing conflict over community values and be able to embrace new forms of community involvement, such as coalitions and networking. A chapter on leadership skills in *Working together for Landcare* (Chamala & Mortiss, 1990) emphasises that leadership can be learned; and so 'hidden' leaders who do not see themselves as leaders could develop skills and confidence to add to their underlying eminent qualities. I am not aware of specific training being offered for Landcare leaders, and slightly pessimistic about enthusiasm to undertake it.

*"Re-think" apparent needs and redefine community assets*: The inception of the Landcare movement during the 1980s demonstrated a remarkable re-think by many rural communities—from passive reliance on government to ensure land was protected, to realisation that local communities needed to take the initiative. Jock Douglas catalysed a re-think within the grazing industry, as did Bob Martin, retiring President of the United Graziers Association in 1987, reflecting on the plight of the mulga lands:

> I want to turn now to what, to my mind anyway, is the biggest single problem facing the mulga lands of this continent and especially of Queensland. That is the declining productivity of these red soil regions. I know it is not a popular subject for discussion in some quarters, but it will eventually have to be faced at all levels. The signs are obvious to all who care to look: vast areas of galvanised burr, turkey-bush, sandalwood and other weeds both woody and otherwise; and equally vast areas completely devoid of topsoil. The cause without doubt is overstocking. (Martin, 1987)

Martin moved on from heading a production-oriented organisation to foundation President of the South-West Queensland Rural Conservation Committee.

*Gain access to appropriate information and resources and build relationships with key individuals inside and outside their communities*. When Landcare groups first formed, many organisations saw efficiencies in interacting with groups but that seems to have diminished; perhaps because many groups failed to attract the majority of landholders. The explosion of knowledge available on the Internet over the past twenty years, including Facebook, compensates to some extent for loss of face-to-face interaction. But, for most, to move from awareness to action requires an intervening step of discussion/interaction (Tully, 1967).

### *Challenges for Queensland Landcare as a Catalyst for Community Development*

The Landcare movement fits a community development model, but with deficiencies that hamper its potential contribution to sustainable communities and to cooperative natural resource management. It is a model strongly reliant on bottom-up motivation and initiatives but which also requires substantial support to achieve success in tackling major natural resource management issues. What are some of the weaknesses that need strengthening?

1. *Technical and coordination support.* The withdrawal of technical and facilitation services by state government agencies has not been effectively replaced by Regional NRM groups.
2. *Funding support for projects.* Government's retreat from community implementation to tenders and contracts has hampered community development.

> Prevailing neoliberal approaches ... presume that it is more efficient to purchase outcomes directly from landholders or invest directly in on-ground work than it is to invest in engaging or building human and social capital. Our view is that complementary policy instruments are needed, but there remains a strong imperative for building and engaging social and human capital as a precursor to effecting the changes that lead to improved resource condition (Curtis et al., 2014, p. 193).

3. *Collaboration with tiers of NRM bodies.* Conflict and poor linkages with higher-tier organisations such as catchment management groups and RMGs has been financially and motivationally damaging. The relationship between Landcare groups and catchment management groups has generally been indefinite, often avoided. The relationship between Landcare groups and RMGs has often lacked trust and respect on both sides. Evaluating catchment management in the Murray-Darling Basin, Bellamy et al. (2002, p. 57) noted that roles and responsibilities between catchment or regional organisations and community has lacked clarity and has evolved in an ad hoc manner, leading to tension and conflict, undermining cooperation. When RMGs were installed in the early 2000s, Landcare groups

were wary (We will have a look at the Plan and see what problems it tackles; we need to see that problems in areas of our interest are going to be addressed) (Keith et al., 2006, p. 32). Over time, wariness gave way to a view that support depended on successfully tendering for government priorities. Partnership it was not.

4. *Resource acquisition.* Many groups relied on government funding, neglecting to engage alternative sources such as business and not-for-profits. This has compromised group viability. A participative action management model (Chamala, 1995) stresses the importance of engaging widely with community.
5. *Community engagement.* Whether by overworked or underwhelming leadership or by lack-lustre member commitment, many groups have not successfully engaged their communities. Compton and Beeton (2012), claimed that Landcare maintains the status quo or makes only small incremental change, and that Landcare encourages participation by less innovative landholders. *Call of the Reed Warbler* (Massy, 2020) describes how various regenerative agriculture systems were developed by innovators acting as individuals rather than in cooperative groups.

## Ways Forward for Landcare Groups in Queensland

There is much to be done, even to reach the goals of the Decade of Landcare, set more than three decades ago (minimise the adverse effects of land use on natural resources and associated ecosystems; integrate nature conservation with rural and other forms of land use; and more). Should government heed the calls to invest in community-based action (Curtis et al., 2014), the first two challenges mentioned previously might be alleviated and a surge occur. Regardless, the characteristics of community development outlined by Cavaye and Ross (2019) offer opportunities for groups to move beyond the current status quo.

### *Simultaneously Utilising and Enhancing the 'Capitals' of Communities*

Flora et al. (2018) identify scope for groups to apply the strengths they bring to community to catalyse community benefits for others but also reap returns for themselves. Landcare groups can contribute natural capital, social capital and cultural capital, while perhaps gaining access to human and financial capital.

One way to strengthen local support is to identify how land care fits the values of other community organisations. Based on the strength-of-weak-ties principle (Ashman et al., 1998; Granovetter, 1973), linkages between organisations within a Darling Downs community were mapped, using a simple butcher's paper exercise, to open discussion about opportunities to strengthen links with service clubs, business, church groups, schools, council, producer organisations and environmental organisations (Keith & Ross, 2005). Network ties might enhance NRM participation and also create opportunity for Landcare groups to contribute to wider community needs; for instance organising government run drought or wellbeing seminars or workshops; thus, highlighting links between community development and resilience (Cavaye & Ross, 2019).

Community-based NRM is a cost-effective platform for rural development that extends beyond NRM to provide an important part of the social capital in rural areas, particularly in areas where other institutions have been in decline. In Victoria, there is evidence of Landcare being a platform that enabled governments and NGOs to engage affected landholders effectively during the (bushfire) recovery process (Curtis et al., 2014, p. 191).

### *Building Community Cohesion, Identity and a Sense of Place; Engagement and Empowerment of Community Members*

Landcare groups usually display strong bonding social capital. Although Compton and Beeton (2012) have pointed out that this can lead to stagnation, it is a desirable characteristic when team projects are planned. The Lantana Busters at Cambooya are a strongly bonded group. The riparian fencing by Dalrymple Landcare entailed an extraordinary team effort over several years.

Sometimes bonding can fall short of its potential; collaborative trials could more successfully identify best practices in regenerative agriculture for a district than unconnected efforts by individual farmers.

The term 'agroecology' today refers to the whole food system from production to consumption (Francis et al., 2003; Rosset & Martínez-Torres, 2012). While principles of regenerative agriculture are part of that, agroecology extends into practices and processes that require collective action and co-creation, including recycling, circular economy and even marketing (FAO, 2018). Landcare groups are well-placed to engage in these dimensions as teams in a single Landcare group or across a network of groups on activities such as:

- co-innovative research trials into diversity measures such as cover cropping, intercropping, companion planting or polyculture suited to the district;
- joint approach to adopting integrated pest management systems;
- cooperative action to make compost using 'waste' materials;
- recycling;
- coordinating the marketing of 'Landcare accredited' farm produce.

Often these activities would engage a wider community than landholders; for instance in providing compostable material or marketing locally accredited produce.

### *Collaborative Action for Collective Benefits Rather Than Competition*

One way to generate funds or resources is to form partnerships. For example, the Macintyre Ag Alliance seeks to work collaboratively with like-minded businesses and organisations to achieve progressive agricultural and environmental outcomes.

Partnerships between not-for-profit organisations can take several forms: memoranda of understanding, service level agreements and shared responsibility agreements. In preparation for shared responsibility, prospective partners need to identify issues and objectives of common interest, clarify roles and relationships and establish a climate that develops responsiveness and achieves shared understanding (Felkins, 2002). Guidelines on what is needed to maintain long-term commitment to a collaborative partnership were provided by Felkins et al. (1993). These include: mutual trust through continuing interaction; respect for individual rights and dignity; open communication; confronting issues in a direct and respectful manner; and sharing the benefits of collective performance.

One form of collaboration much requiring repair is the relationship between the various tiers of a nested community-based NRM structure (that is, linking social capital). Within the constraints imposed by government funding policy, what does it take to lay down the participation ladder to achieve more collaborative relationships so that true partnership with shared responsibility agreements is enjoyed? Is it possible for RMGs and Landcare groups to come together to workshop the above principles and construct new working relationships?

### *Uncovering Hidden or Unrealised Assets and Capacity in Communities*

Indigenous land management practices, effective for millennia, have been largely unnoticed or neglected until relatively recently but are now providing models of cooperative management as well as practices suited to management of Australia's peculiar landscape. Four examples show different facets of the potential.

Ross et al. (2009) highlighted progress made in co-management of Indigenous Protected Areas, with social and environmental benefits. Improved partnerships and communications enabled Indigenous and non-Indigenous collaborators to share ways of achieving species and landscape conservation. Can this be applied to lands other than designated protected lands—for instance, Landcare groups collaborating with Indigenous land managers to enhance biodiversity on both public and private land? Through early research into First Nations anthropology and a deep appreciation of principles of community-based NRM, Professor Ross has injected valuable insights into the opportunities.

The adaptive governance skills of the Girringun Aboriginal Corporation in North Queensland show how socio-ecological initiatives can be implemented without direct land ownership (Maclean et al., 2013). 'Girringun engages proactively and highly effectively with other regional environmental governance organisations to develop innovative co-operative management arrangements' (p. 98). Girringun practises 'caring for country' through specific initiatives that improve the health of country, including its people. It partners with research organisations for the co-production of knowledge. 'Its adaptive governance approaches include power-sharing, partnership development and visionary and innovative leadership that practices reconciliation between Aboriginal and non-Aboriginal Australians' (p. 100). Landcare groups could do well to follow this management style.

The Cape York Weeds and Feral Animals Inc. uses Indigenous familiarity with country to manage damage caused by European grazing management methods. Operated mainly by Indigenous volunteers and part time staff, it offers on-ground weed and feral animal management and training in property maintenance (https://capeyorknrm.com.au/).

A fortuitous meet-up by Victor Steffensen from Kuranda, Djabugay country, with two Kuku-Thaypan Elders, Tommy George and George Musgrave of Laura, led to the recording of a wealth of traditional knowledge, in particular cultural burning, through video and a computer database (Steffensen, 2020). Documentation, which led to Honorary Doctorate awards through James Cook University for the two Elders, categorised and linked topics such as plants, story places and country type (Hill et al., 2013), providing community development benefits for environmental management and culture. In recent years, Steffensen has shared this knowledge through cultural burning workshops organised by Landcare groups in southern Queensland, see Fig. 9.3.

**Fig. 9.3** Cultural burning demonstration by Victor Steffensen, Condamine Headwaters Landcare, 2017

### *An Ongoing Process of Improvement, Often Involving the Cumulative Effects of Discrete Initiatives, Projects or Actions*

Language about soil condition has changed from preventing further deterioration to improving 'soil health'—improving structure, carbon content and the biotic component. Australia's National Soil Strategy with a 20-year programme from 2021 to 2041 offers Landcare groups a chance to synchronise this core business with a government priority.

Many goals of the strategy engage research and coordination beyond the local level but groups have a place in co-innovation processes to achieve these goals. Some goals (DAWE, 2021) that Landcare groups can and should seek to contribute towards include:

- group action to mitigate the risks of land management practices on the broader environment and community;
- participation in development of effective approaches to promote the adoption of best management practices to increase soil organic carbon and its sequestration; and
- participation in trials to develop better understanding of the relationship of soil condition and trend to land management practices, soil organisms and climate for different soil classes in different landscapes.

The programme aims to increase the frequency and quality of soil testing by land managers and to provide access to this data in a secure national platform. There would be considerable benefit if Landcare group members do this in a coordinated way to develop suitable group extension efforts to other landholders in their district.

Similarly, it will be advantageous for groups of farmers seeking recompense for improved biodiversity and carbon farming to work in groups to reduce costs of technical advice and audits. Landcare groups could coordinate contiguous groups of farmers interested in this form of income diversification to the benefit of the landscape. This could also apply to Fencing Cluster groups. Queensland's Land Restoration Fund (Evans & Butler, 2019) is one accessible source of such income. It supports landholders, farmers and First Nations peoples to generate new, regular income streams through carbon farming projects while providing valuable co-benefits such as healthier waterways, increased habitat for threatened species and more resilient landscapes.

## Conclusion

The aim was to explore how well Landcare fitted various dimensions of community development as it applies to Australia and to recommend ways to overcome deficiencies. Landcare is an outstanding case of community development for natural resource management and healthy thriving communities. Landcare groups engage community development through their own projects to enhance environmental health, through the social capital they contribute to community functioning and through facilitating innovative processes that can regenerate natural systems and human livelihoods.

Hindrances to community development for Landcare include lack of technical and coordination support, lack of funds for significant projects, collaboration within the natural resource management (NRM) structure, resource acquisition and community engagement. Suggested ways forward, with or without increased government support, include: enhancing community capital through better linkage with other organisations; using strong bonding capital to undertake significant team projects, including agro-ecological activities; learning from and collaborating with First Nations organisations; and forming efficient collective teams to enter the National Soil Strategy and carbon farming ventures.

Reflecting on the transition in Queensland's Landcare over three decades, the richer enthusiasm for biodiversity and nature conservation, expressed throughout both rural and urban communities, is noteworthy. On the other hand, disbandment of many land and water groups means that expectations when Landcare was set in place are not met. Could recognition of Landcare's role in community development generate the support that will re-establish powerful outcomes that are possible through community-based action?

Finally, as I reflect on background reading about community-based natural resource management, community development, collaborative partnerships and First Nations land management, the nimble mind of Professor Helen Ross seems to appear everywhere: exploring, clarifying and conjecturing on what is and what might be.

## References


Arnstein, S. R. (1969). A ladder of citizen participation. *Journal of the American Institute of Planners, 35*(4), 216–224.

Ashman, D., Brown, L. D., & Zwick, E. (1998). The strength of strong and weak ties: Building social capital for the formation of governance of civil society resource organizations. *Nonprofit Management & Leadership, 9*(2), 153–171.

Australian Framework for Landcare Reference Group. (2010). *Australian Framework for Landcare.* Sydney. https://www.agriculture.gov.au/agriculture-land/farm-food-drought/natural-resources/landcare/the-australian-landcare-framework

Bellamy, J., Ross, H., Ewing, S., & Meppem, T. (2002, January). *Integrated catchment management: Learning from the Australian experience for the Murray-Darling Basin.* CSIRO Sustainable Ecosystems.

Berndt, R. D. (1990). *Landcare Focal Program, Annual report 1989–90.* D. o. L. M. Q. D. o. P. Industries. Canberra.

Campbell, A. (1991). *Landcare: Testing times.* National Landcare Facilitator 2nd Annual report. National Soil Conservation Program.

Cavaye, J. (2001). Rural community development: New challenges and enduring dilemmas. *Journal of Regional Analysis & Policy, 31*(1), 109–124.

Cavaye, J., & Ross, H. (2019). Community resilience and community development: What mutual opportunities arise from interactions between the two concepts? *Community Development (Columbus, Ohio), 50*(2), 181–200. https://doi.org/10.1080/15575330.2019.1572634

Chamala, S. (1995). Overview of participative action approaches in Australian land and water management. In S. Chamala & K. Keith (Eds.), *Participative approaches for landcare, perspectives, policies, programs* (pp. 5–42). Australian Academic Press.

Chamala, S., & Mortiss, P. D. (Eds.). (1990). *Working together for Landcare: Group management skills and strategies.* Australian Academic Press.

Compton, E., & Beeton, R. J. S. (2012). An accidental outcome: Social capital and its implications for Landcare and the "status quo." *Journal of Rural Studies, 28*(2), 149–160. https://doi.org/10.1016/j.jrurstud.2011.12.004

Cramb, R. A. (2004, February). Social capital and soil conservation: Evidence from the Philippines. *48th Annual Conference Australian Agricultural & Resource Economics Society*, 10–13.

Curtis, A., Ross, H., Marshall, G. R., Baldwin, C., Cavaye, J., Freeman, C., et al. (2014). The great experiment with devolved NRM governance: Lessons from community engagement in Australia and New Zealand since the 1980s. *Australasian Journal of Environmental Management, 21*(2), 175–199. https://doi.org/10.1080/14486563.2014.935747

DAWE. (2021, April). *National Soil Strategy*. D. o. A. W. a. t. Environment. Commonwealth of Australia.

Department of Primary Industries Queensland. (1992). *Decade of Landcare Plan, Queensland*. Queensland Government.

Douglas, J. (1988, March 22). *A proposal to establish land care programmes for Australian Agriculture: A discussion paper issued by the Cattlemen's Union*. Cattlemen's Union.

Evans, M., & Butler, D. (2019, July). *Update in the Land Restoration Fund*. Rangelands Dialogue. Royal Society of Queensland.

FAO. (2018). *Scaling up agroecology to achieve the sustainable development goals*. Proceedings of the second FAO International Symposium. Food and Agriculture Organisation of the United Nations.

Felkins, P. (2002). *Community at work: Creating and celebrating community in organizational life*. Hampton Press.

Felkins, P., Chakiris, B. J., & Chakiris, K. N. (1993). *Change management: A model for effective organisational performance*. Quality Resources.

Flora, C. B., Flora, J. L., & Gasteyer, S. P. (2018). *Rural communities: Legacy and change*. Routledge.

Francis, C., Lieblein, G., Gliessman, S., Breland, T. A., Creamer, N., Harwood, R., et al. (2003). Agroecology: The ecology of food systems. *Journal of Sustainable Agriculture, 22*(3), 99–118.

Granovetter, M. S. (1973). The strength of weak ties. *American Journal of Sociology, 78*(6), 1360–1380.

Hill, R., Pert, P. I., Davies, J., Walsh, F. J., & Falco-Mammone, F. (2013). *Indigenous land management in Australia: Extent, scope, diversity, barriers and success factors*. Citeseer.

HRSCEH. (1999). *Lockyer Landcare submission to Inquiry into Catchment Management August 1999*. House of Representatives Standing Committee on Environment and Heritage, Australia (HRSCEH).

Keith, K. (2017). Thriving or just surviving: Social capital and resilience of long-term Landcare groups in Queensland. *Rural Extension and Innovation Systems Journal, 13*(2), 11–21.

Keith, K., & Ross, H. (2005). *Understanding Pittsworth: Social networks and natural systems*. The University of Queensland.

Keith, K., Ross, H., & Gorringe, S. (2006). *Regional partnership agreements on prioritised investment strategies for the Burnett Mary Region*. The University of Queensland.

Love, C. (2012). *Evolution of Landcare in Australia: In the context of Australian Government natural resource management policy and programs*. Australian Landcare Council, Australian Government Department of Agriculture,

Fisheries and Forestry. https://www.agriculture.gov.au/agriculture-land/farm-food-drought/natural-resources/landcare/publications/evolution-of-landcare-in-australia

Maclean, K., Cuthill, M., & Ross, H. (2014). Six attributes of social resilience. *Journal of Environmental Planning and Management, 57*(1), 144–156. https://doi.org/10.1080/09640568.2013.763774

Maclean, K., Ross, H., Cuthill, M., & Rist, P. (2013). Healthy country, healthy people: An Australian Aboriginal organisation's adaptive governance to enhance its social–ecological system. *Geoforum, 45*, 94–105. https://doi.org/10.1016/j.geoforum.2012.10.005

Martin, R. (1987, June). *The President's farewell*. The Mulga Line.

Massy, C. (2020). *Call of the Reed Warbler—A new agriculture a new earth*. University of Queensland Press.

Otteson, H. (2019). *The value of Landcare: A literature review*. National Landcare Network.

Pink, H., & Roberts, M. (1962, August). *The group approach to soil conservation as an extension medium*. Australian Agricultural Extension Conference. Hawkesbury Ag. College, NSW.

Queensland DPI. (1987, April 23). *The contribution of landuser committees to achieving soil conservation*. Queensland Department of Primary Industries Conference and Workshop Series. QC87009 National Library of Australia, Toowoomba.

Queensland Landcare Council. (1997). *Queensland decade of Landcare plan review priority actions 1997–2000*. Queensland Government.

Roberts, B. (1984). *Land ethics—A necessary addition to Australian values*. Land Degradation Conference. Australian National University, Canberra.

Robins, L. (2018). More than 30 years of 'Landcare' in Australia: Five phases of development from 'childhood' to 'mid-life' (crisis or renewal?). *Australasian Journal of Environmental Management, 25*(4), 385–397. https://doi.org/10.1080/14486563.2018.1487342

Ross, H., Buchy, M., & Proctor, W. (2002). Laying down the ladder: A typology of public participation in Australian natural resource management. *Australian Journal of Environmental Management, 9*(4), 205–217. https://doi.org/10.1080/14486563.2002.10648561

Ross, H., Grant, C., Robinson, C. J., Izurieta, A., Smyth, D., & Rist, P. (2009). Co-management and Indigenous protected areas in Australia: Achievements and ways forward. *Australasian Journal of Environmental Management, 16*(4), 242–252. https://doi.org/10.1080/14486563.2009.9725240

Rosset, P. M., & Martínez-Torres, M. E. (2012). Rural social movements and agroecology: Context, theory, and process. *Ecology and Society, 17*(3), 17–17. https://doi.org/10.5751/ES-05000-170317

Sobels, J., Curtis, A., & Lockie, S. (2001). The role of Landcare group networks in rural Australia: Exploring the contribution of social capital. *Journal of Rural Studies, 17*(3), 265–276. https://doi.org/10.1016/S0743-0167(01)00003-1

Steffensen, V. (2020). *Fire country: How Indigenous fire management could help save Australia*. CSIRO Publishing.

Tully, J. (1967). *Agricultural Extension Research—Another view*. Paper presented at the Australian Agricultural Extension Conference, Hobart, Tasmania.

Waggamba Conservation Committee. (1988, February 12). *Constitution of the Waggamba Shire Conservation Consultative Committee adopted at a public meeting of Waggamba Shire ratepayers*.

PART II

# Community-Based Natural Resource Management, Conservation and Co-Management

CHAPTER 10

# Introduction to Section 2: Community-Based Natural Resource Management, Conservation and Co-management

*Valerie A. Brown*

During a lifetime of research, Professor Helen Ross worked in partnership with communities seeking sustainable management of their social and environmental resources. Every community had responsibility for co-management by the many interested parties, whether they accepted the responsibility or not. A major challenge for such inquiries was how to bring all the interests together and to support individual contributions. The challenge is great enough to be called a wicked problem, a problem arising from within the society that has produced it, and for which there is no established solution (Rittel & Webber, 1973).

V. A. Brown (✉)
Fenner School, Australian National University, Canberra, ACT, Australia
e-mail: Val.Brown@anu.edu.au



C. Baldwin and S. van Bommel (eds.), *Rural Development for Sustainable Social-ecological Systems*,
Palgrave Studies in Environmental Sustainability,
https://doi.org/10.1007/978-3-031-34225-7_10

Since wicked problems arise within the community in which they are found, any solution brings with it changes in that community. Section 2 of 'Putting the community first' contains six case studies of co-management linking diverse interests, most in partnership with Professor Ross. Each case study identifies a different avenue for working together and links different interests under different conditions. The problems they face include hearing all the voices, water management, engaging with Indigenous knowledge, empowerment of women, co-managing coastal resources, and plastic waste through local eyes. In common to them all was the need to listen to a wide range of voices and use a mode of collaboration able to bring the interests together.

An example of the first challenge is the need to hear all the voices. Early in Professor Ross's career, she worked with former leading public servant and member of the Council for Aboriginal Affairs, 'Nugget' Coombs (Coombs, 1978). Dr Coombs was a forceful exponent of recognition of Indigenous rights to their land and governance of country, and that Indigenous people should be resourced to choose their own directions for development. Dr Coombs frequently told of the close of an inquiry, where after multiple participants had given a wide range of evidence, he asked why they were all male (Ross, 1989–1995). He was told that the women were waiting outside. Having moved the inquiry outside, the women spoke to Dr. Coombs in their own voice; they danced what they had to say. When Prof. Ross worked with Dr Coombs on the East Kimberley Impact Assessment Project from 1986 to 1989 (Ross, 1991), women were highly active in all events and research.

The voices in co-management may be dancing, writing a report, and sending a text message. However different they may be, there is still need to respect the many voices of all interested parties even, or especially, if they have been rejected by the mainstream. Co-management requires the full range of individual voices of the many different interests, including the community concerned. The frequent use of 'stakeholders' for the voices can be misleading, since it conveys a sense of economic ownership. The term 'interests' is more accurate, and more inclusive, since it includes not only the current interests but those not actually present, such as dissidents, ancestors, and the unborn.

Second, by definition, the co-management of socio-ecological issues has to bring the interests together, not only to meet, but to work together across boundaries. This is impeded by the Western tradition of compartmentalising knowledge. Social and environmental explanations are routinely separated, as they are in education, organisation, and research. There are marked differences in accessing different constructions of knowledge: according to the source of the evidence. If the source was an individual, it was dismissed as anecdote. If it was from members of the community, it would be rejected as gossip. If from experts, it would be called facts and given dominance over the others. Evidence from an organisation would be assumed to be biased. A holistic thinker, drawing on all these forms of evidence, would be considered pretentious or impractical (Brown, 2010). When those voices try to speak to each other, they are more likely to be in competition rather than collaboration, as the six case studies demonstrate.

In spite of the frequent dissonance in shared management, there have been many accounts of communities' success in achieving co-management of social and natural resources. Nobel Prize winner Eleanor Ostrom drew on numerous case studies from several countries in her work on the governance of common-pool resources. The standard expressions of the time were choices between community management, private ownership, and state control.

Ostrom established the esssential role of community co-management of socio-ecological resources in arriving at sustainability and community well-being. She also stressed the role of the individuals who make up each of the interest groups. She recognised that, as can be noted in the case studies of this section, new principles for co-management speak with a different voice. For instance, she coined the phrase 'governing common-pool resources' to replace 'management of competing interests'. She also turned the then dogma of the tragedy of the commons (the inevitability of a greedy community degrading their resources) on its head (the increased value of the shared resources when all interests care for them [Ostrom, 2015]).

Another success story: in 1989, an Australian country-wide community-managed programme, Landcare, brought together community members, farmers, government, and conservation advocates in grassroots cooperation, replacing competing conservation and farming interests. Describing local environmental management systems as 'efficient management by separate stakeholders' compared to Ostrom's 'common pool resources governed by collective responsibility of the community'. Over the next 50 years, Landcare became a major country-wide programme governed by over 6,300 volunteers. Unfortunately, Landcare eventually lost momentum (Lockie, 1995) but continues today largely due to community support and ingenuity.

Australian leaders have used their influence to bring together competing 'stakeholders' in natural resource management. Examples are Prime Minister Hawke's round tables for Ecologially Sustainable Development in 1985 and Prime Minister Rudd's 2020 National Summit which included sustainability as a national goal. In both these initiatives, participants included local government, policy development, advocacy, business, and research as a common-pool resource, with varying success (Brown, 2008).

So far, the introduction to Section 2 of 'Putting the Community First' has explored the critical role of multiple voices and ways of arriving at collective thinking systems in co-management of wicked problems. Section 2 also contains six case studies of co-management in which the authors make use of these ideas. This chapter summarises each of the six chapters in relation to their responses to wicked problems of threatened socio-ecological resources. In the examples of co-management, such as Ostroms' governing of the commons, the method for bringing interests together does not take the form of a hierarchy, a continuum, or a consensus. Rather, Ostrom treats differences among many voices and the widest range of contributors to co-management as a rich resource.

In Chapter 11, the authors speak in the voice of a wicked problem and use collaboration in bringing the interests together in the context of water management. They found that the collaboration was not static, rather changed over time. Since many interests also change over time, co-management is an evolving process.

In Chapter 12, the emphasis is on the crucial contribution of Indigenous peoples working in concert with Western scientists. The shared contribution joins two voices with different understandings of the role of conservation. The shared context provided a template for socially just resource management: it included Indigenous Protected Areas, conservancy and land use agreements, and climate adaptation strategies.

In Chapter 13, the twin essential voices of community-based tourism—financial security and exploitation of the culture—are grounded in the Western tradition of specialisation and opposing interests. A fortunate outcome comes from a different direction. Cultural traditions unite the older women in cooperation in preparing and serving local foods, leading to a thriving tourist trade.

Chapter 14 develops criteria that allow all parties to learn from co-management of natural resources in Vietnam. The criteria establish a broad context and a form, a voice for evaluating the processes of co-management and advances in sustainability. The tool for establishing coastal management plans was funded by the World Bank. The plans needed to include the details of the supply chain and a governance model which is based on integration of ideas and cooperation among all those involved.

Chapter 15 confronts the wicked problem of plastic waste on the beaches of Indonesia. The language is that of the local people. The conclusion is that there is little hope of a co-management system enlisting the major interests that pursue their own agendas.

The case studies of community-based co-management of socio-ecological issues give an overview of how it might function in practice. They included voices that needed support, and forms of collaboration which match the interests of the participants. Co-management is not a simple process. The linking systems that connected the participants in co-management ranged across empowerment, partnerships, cooperation, collaboration, non-controlling guidence, clear criteria, and listening. Each of these is discusssed in more detail in the chapters of Section 2.

In conclusion, the Section 2 of '*Putting Communities First*' offers a benchmark for the individual who joins in a co-management process. Reading through the six chapters, the following elements of 'proper practice' of a co-manager of socio-ecological resources emerged as being:

- Committed to a future that is sustainable, just, and harmonious;
- Open-minded to new and different voices;
- Prepared to cross knowledge boundaries and break new ground;
- Thinking collaboratively and constructively; and
- Considering the Commons as a common-pool resource.

**Acknowledgements** I wish to thank Professor Helen Ross for her warm friendship and skilled professional support over seven years at the Centre for Resource and Environmental Studies.

Thanks to Dr John Harris for yet again contributing his limitless patience and editorial skills.

## References

Brown, V. A. (2008). *Leonardo's vision: A guide to collective thinking and action.* Brill.

Brown, V. A. (2010). Collective inquiry and its wicked problems. In V. A. Brown, J. A. Harris, & J. Y. Russell (Eds.), *Tackling wicked problems through the transdisciplinary imagination.* Earthscan.

Coombs, H. C. (1978). *Kulinma: Listening to aboriginal Australians.* Australian National University Press.

Lockie, S. (1995). Beyond a 'good thing': Political interests and the meaning of Landcare. *Rural Society, 5*(2–3), 3–12.

Ostrom, E. (2015). *Governing the commons.* Cambridge University Press.

Rittel, H. W. J., & Webber, M. M. (1973). Dilemmas in a general theory of planning. *Policy Sciences, 4*(2), 155–169. https://doi.org/10.1007/BF01405730

Ross, H. (1989–1995). *Personal communication while working at the Centre for Resource and Environmental Studies.*

Ross, H. (1991). The east Kimberley impact assessment project. *Interdisciplinary Science Reviews, 16*(4), 313–322.

CHAPTER 11

# SEQ Healthy Waterways Partnership: Reflections on an Innovative and Evolving Collaboration

*Jennifer Bellamy and Brian Head*

## Introduction: Collaborative Challenges

This chapter builds on the extensive research that examines how collaboration for innovation in policies and programmes can improve natural resource management (NRM). Since the 1980s there has been a global research focus on assessing NRM regional governance outcomes (e.g. Brown et al., 2015; Westley et al., 2011). In Australia, decade-long

J. Bellamy (✉)
School of Agriculture and Food Sciences, University of Queensland, Brisbane, QLD, Australia
e-mail: jenny.bellamy@uq.edu.au

B. Head
School of Political Science and International Studies, University of Queensland, Brisbane, QLD, Australia
e-mail: brian.head@uq.edu.au



C. Baldwin and S. van Bommel (eds.), *Rural Development for Sustainable Social-ecological Systems*,
Palgrave Studies in Environmental Sustainability,
https://doi.org/10.1007/978-3-031-34225-7_11

droughts, increasing frequency of major flooding and fires, accelerating urbanisation, and growing recognition of climate change and related environmental impacts have refocused policy attention on collaborative processes for addressing complex issues.

Within river basins and catchments, including both rural and urban communities, the policy focus has moved beyond water scarcity alone, to wider challenges of improving water management and governance under conditions of on-going social and environmental change (Bellamy, 2007; Bowmer, 2014; Dunn et al., 2018; Head, 2014). In this context, innovative forms of collaborative governance have become the instruments of choice across all levels of government in Australia, owing to the widespread desire to incorporate diverse interests into NRM and water policy, planning and practice, and particularly for wicked problems with competing interests at a regional, river basin, or catchment scale (Bellamy, 2007; Bellamy et al., 2017; Head et al., 2016).

Collaborative innovations in NRM and water management in Australia have been pursued since the 1990s by a diversity of actors (leaders, stakeholders and institutions) seeking to manage conflict over divergent interpretations of problems and solutions. Collaborative structures and processes have been developed and evolved, either as government-mandated strategic interventions, or as emergent self-organising approaches managed by stakeholders. These governance innovations have taken different forms at the regional scale, whether utilising catchment or watershed boundaries, or administrative regional areas (Bellamy et al., 2012, 2017; Head, 2022; Head et al., 2016). Collaborative innovations frequently encounter challenges arising from interconnected issues and from diverse sources of knowledge and experience across multiple governance levels and scales (Fischer et al., 2015; Folke et al., 2016; Patterson et al., 2015; Preiser et al., 2018). Collaboration is also a cornerstone of research initiatives within the sustainability sciences, with the formation of collaborative networks of researchers and stakeholders conducting joint research aiming to influence societal and policy change (Chambers et al., 2022; Folke et al., 2016; Preiser et al., 2018).

In Australia, the complex and commonly contested 'wicked' challenges that underlie NRM practice, policy and governance have largely been addressed as single-issues in NRM policy and investment (such as rural production, water supply, nature conservation), rather than more holistically as complex social-ecological systems for sustainability (Bellamy, 2007; Bellamy et al., 2012; Bowmer, 2014; Fischer et al., 2015). Research on social-ecological systems (SES) increasingly emphasises that social systems and ecological systems are not just 'connected' but truly intertwined and jointly co-evolving across levels and scales (Fischer et al., 2015; Folke et al., 2016). These SES display different dynamics at a single scale, across scales or at multiple scales and in moving between scales (e.g. local, regional and national) (Ansell & Torfing, 2015; Walker, 2019). In this context, collaboration and innovation in NRM and water governance is very demanding. It requires high levels of interaction and dialogue, knowledge exchange, co-production and integration, joint planning, and cross-scale interaction that engage diverse actors including governments (all levels), business, farmers and landholder interests, community stakeholders, and scientific and related social researchers (Bellamy et al., 2012, 2017; Bowmer, 2014; Patterson et al., 2015; Ross et al., 2016).

## The Wicked Problem of Water Pollution

Wicked problems in NRM and water management are typically complex, uncertain, conflictual, and unpredictable. Both the nature of the problem and the preferred solutions are contested by various stakeholders (Head, 2022; Head & Alford, 2015). Moreover, wicked problems cannot be solved in a piecemeal way as any intervention will likely result in unintended consequences (Preiser et al., 2018). Thus, science-derived solutions alone may not be able to 'solve' the underlying and evolving problem in a definitive way (Dunn et al., 2018; Head et al., 2016). Sustainability approaches for social and economic development exhibit all the classic characteristics of 'wicked' problems—poorly defined requirements, unclear boundaries, and contested causal factors which no single agency or knowledge discipline can address (Bellamy, 2007; Brown et al., 2015; Westley et al., 2011). Head and Alford (2015) identify three key factors characterising wicked problems: *social pluralism/diversity* (i.e. multiple interests and values of stakeholders); *institutional complexity* (i.e. the context of interorganisational cooperation and multilevel or cross-scale governance); and *scientific uncertainty* (i.e. complexity, fragmentation, and gaps in reliable knowledge). Where sustainability is the focus, wicked problems are commonly nested in complex intertwined social and ecological systems, co-evolving, and being shaped by one another (Folke et al., 2016). Sustainability systems and challenges interplay in complex ways through relations, interactions and feedback across spatial, temporal and functional scales and involve conflicting values and perceptions, and considerable uncertainty about their nature, as well as about management and governance options (Folke et al., 2016; Walker, 2019).

Tackling wicked NRM problems, such as water pollution, are complex and highly challenging, but it is also 'an evolving art' that requires innovative and collaborative approaches (Australian Public Service Commission, 2007; Bellamy et al., 2012). The challenge of water quality management is not just about managing individual factors or stressors, but also the interaction of various stressors that affect the utility of the water resource or the status of river/estuarine health at any given time (Bowmer, 2014). For example, seagrasses are critical components of marine and brackish aquatic ecosystems all along the east coast of Australia, including the Great Barrier Reef. Their loss is attributed to the interaction of pollutant discharge from coastal urban centres in conjunction with changing hydrological patterns of catchment systems (Bowmer, 2014; Bunn et al., 2010). Globally, there are many areas with high levels of sewage contamination (Wear et al., 2021). These contamination hotspots frequently occur close to coral reefs, salt marshes, and fish-rich river systems, thus raising major concerns over the health of both human and natural systems. Such challenges have led to urgent calls for more innovation through cross-sector collaboration (Wear et al., 2021).

Globally, the overall availability of 'natural' nutrient sources of nitrogen (N) and phosphorus (P) has been declining in past decades due to massive land transformations, while by contrast 'anthropogenic' sources of nutrients (through agriculture, sewage treatment, and aquaculture) are becoming increasingly dominant (Beusen et al., 2022). For example, the global proliferation of harmful algal blooms has been attributed to increasing diffuse-source pollution loads in freshwaters and their export to marine waters. Point sources have also become a significant source of freshwater and marine pollution, mostly due to increasing urban populations with sewerage systems (Beusen et al., 2022; Wear et al., 2021). Collaboration offers one way of recognising the complexity of such wicked problems and engaging the multiplicity of actors affecting the 'wickedness' of a problem.

## Collaboration and Innovation

Collaboration is defined variously in the social innovation and the governance literature as:

> a process through which parties who see a situation very differently can constructively explore those differences and seek solutions that go beyond their own limited vision of what is possible .... a process of joint decision-making among key stakeholders of a problem domain about the future of that domain. (Gray, 1989, pp. 5–11)

> a "high intensity" mode of interaction .... that nurtures mutual interdependence and joint action while preserving the autonomy of collaborating parties. (Ansell & Gash, 2018, p. 16)

In the context of multi-stakeholder collaboration, 'collaborative platforms' have emerged as important for sustaining emergent self-organising approaches in wicked problem situations (Ansell & Gash, 2018; Folke et al., 2016). They are described variously as:

> combining diverse types of knowledge for understanding complex social-ecological systems are emerging that try to learn from history. (Folke et al., 2016)

> the [governance] platform concept emphasizes how its architecture is deployed to leverage, catalyse, and harness distributed social action—as opposed to steer, control or hold it accountable. (Ansell & Miura, 2020, p. 264)

> [platforms] do not mandate collaboration or collaborative governance, but rather catalyze and facilitate voluntary efforts .... A collaborative platform is an organization or program with dedicated competences, institutions, and resources for facilitating the creation, adaptation and success of multiple or ongoing collaborative projects or networks. (Ansell & Gash, 2018, p. 20)

Regional collaborations to address a wicked problem are not simply a matter of relevant parties deciding to work together, as Gray's definition above may imply (Ross et al., 2016). Nor is innovation a known destination at which one arrives but rather an ongoing process essential to resilient and sustainable societies (Westley, 2017). Collaborative and innovative approaches in wicked problem situations require the engagement of diverse actors (people, institutions), among whom power and agency is widely dispersed across multiple scales. Part of the solution to the problem involves these actors fostering sustained behavioural change within the context of on-going social and environmental change (Dunn et al., 2018; Fischer et al., 2015; Westley, 2017; Westley et al., 2016). Folke et al. (2016) argue that wicked problems require a focus on managing and governing for emergence and continuous learning through 'collaborative platforms' that generate diverse types of knowledge for understanding complex social-ecological systems (SES). These platforms require social and ecological understanding at many scales through continuous learning, gaining experience, and building adaptive capacity for dealing with moving targets like social-ecological systems and their intertwined dynamics (Folke et al., 2016).

## The SEQ Case Study

Using an historical lens covering two decades, 1993–2012, we examine the evolution of collaboration and innovation through a single case study of an emergent self-organised collaborative innovation, the Healthy Waterways Partnership (HWP) in Southeast Queensland (Fig. 11.1). SEQ was one of four case study regions analysed by this paper's authors, collaboratively with Professor Helen Ross, in a joint ARC-funded research project on 'Collaboration as a wicked problem in natural resource management' (Bellamy et al., 2012, 2017; Head et al., 2016; Ross et al., 2016).

The HWP focused on the wicked problem of pollution (point source and diffuse source) and water quality management for healthy regional waterways across land and marine boundaries. SEQ covers an area of nearly 23 000 km$^2$ (see Fig. 11.1; Bunn et al. 2007, 2010). SEQ is a sub-tropical region that includes 19 major catchments, the capital city of Brisbane, 19 local councils and the coastal waters, estuaries, and islands, stretching from the tourism centres of the Sunshine Coast in the north to Moreton Bay and the Gold Coast in the south; together with their inland freshwater catchments west to the Great Dividing Range. With a current population of more than 3 million people, SEQ has been one of the fastest growing regions in Australia. It has experienced rapid urbanisation with associated land use changes and pressures from population growth and climate change (Abal et al., 2001; Head, 2014; Lloyd et al.,

2000). The agricultural areas of the region also contribute significantly to the local and regional economy.

The SEQ region is a multi-use environment providing residential, rural, industrial, and recreational land with water supply, conservation, biodiversity and natural amenities, as well as a very important spiritual place for indigenous Australians (Bunn et al., 2010; McPhee, 2017). It includes a marine park listed as a wetland of international significance under the RAMSAR agreement for the protection of wetland habitats and migratory birds (Lloyd et al., 2000; McPhee, 2017). Moreton Bay and its seagrass areas and wetlands are of high conservation value for dugongs, turtles, dolphins, whales, and migratory wading birds. They also support significant commercial and recreational fisheries and host the major port of Brisbane. Healthy wetlands are important for reducing floods, producing clean water and food, improving water quality, and providing habitat for animals, birds, and plants, as well as coastal protection against erosion (Healthy Land & Water, 2021; McPhee, 2017).

Since European settlement the ecological assets of SEQ have been greatly changed. Salt marsh and intertidal mudflats are an integral component of the Moreton Bay wetlands, but less than half the pre-European wetlands remain (HLW, 2021). These wetlands are now listed as a vulnerable ecological community under Australia's Environmental Protection and Biodiversity Conservation Act. The SEQ catchment hydrology has also been substantially altered since the nineteenth century. Construction of dams and weirs as well as changes in land-use, vegetation cover, and dredging for the major seaport at the Brisbane River mouth, have substantially altered runoff impacts from heavy rainfall events in SEQ, including 'flashier' stream flows (Bunn et al., 2010; Head, 2014). Climate change appears to intensify these events. For example, the rapid urbanisation of SEQ involved significant changes in land use, increased land clearing, the building of new dams and weirs to meet changing urban and rural water supply needs, and the dredging of a major seaport at the Brisbane River mouth (Bunn et al., 2007; McPhee, 2017; Dennison & Abal., 1999). The outcome of these changes has been declining water quality, changes to the volume and frequency of water flows into Morton Bay marine environment the loss of aquatic biodiversity (Bunn et al., 2007; Head, 2014; Head et al., 2016; Ross et al., 2016; McPhee, 2017). Major water crises (droughts and floods) often linked to climate change seem to intensify these effects (Bunn et al., 2007; Head, 2014; McPhee, 2017).

## Methods and Analysis

Against a background of on-going social and environmental change in SEQ, this paper examines the wicked problem of water quality and pollution in SEQ riverine, estuarine, and coastal/terrestrial waterways through examining the HWP. In this paper, we reflect on the HWP, an innovative and emergent self-organising collaboration, as it evolved over two decades in SEQ. We focus on the historical emergence and evolution of collaborative forms, and the opportunities, challenges, and tensions encountered over two decades from 1993 to 2012. We draw on the authors' (including Prof. Helen Ross as a collaborating researcher) own professional experiences with the HWP initiative, as well as their collaborative regional case study research on the HWP and related documentation (Bellamy et al., 2012, 2017; Head et al., 2016; Ross et al., 2016). We overview the various attempts at collaboration and how and why they arose, how the focus evolved, and how the collaborative model was adapted to the changing dynamics of the environmental, social, and political situation of the WP. Finally, we reflect on the emergence and evolution of long-term collaborative innovation process and identify some paradoxes and tensions that emerged in the process.

The HWP case study synthesises diverse data sources, including examination of organisational and government documentation, academic research, and 17 semi-structured interviews with key informants undertaken in 2011. Informants were either directly involved in the HWP collaboration, or in partnering or observer organisations. We also draw on the practical experience of each of the three researchers beyond their joint work in the ARC research project, given their various long-term connections and associations with HWP's Scientific Expert Panel (SEP), or with HWP stakeholder communication activities. The interviews, lasting one to two hours, were undertaken by the three collaborating researchers either face-to-face or by phone during 2011. They examined historical and institutional themes, inviting participants to review their personal and organisational histories and experiences of involvement in NRM activities and the evolving collaborations in the SEQ region. Informants were asked about their diverse scientific, organisational, and community roles with respect to the underlying wicked problems, the contributions of various actors, the HWP collaboration itself, and NRM more generally. They were invited to elaborate on the key changes they had experienced during their years of involvement. The interviews were analysed in terms of these key themes, and their experiences were mapped against the documentary materials concerning the evolution of the HWP collaboration.

The larger joint research activity with Prof. Helen Ross involved four separate case studies, one of which being the SEQ HWP (Head et al., 2016). In each case, the research synthesis examined how and why collaborative approaches emerged in response to the wicked problems of water pollution; the processes for initiating and enabling the evolution of self-organising collaborative initiatives; and the challenges faced in addressing the NRM wicked problems. The research explored a long-term perspective (over two decades) in order to unpack the patterns of change and adaptation in the collaboration and how emerging tensions and conflicts were managed. Each case study thus covered the development of regional collaborative processes and the evolutionary dynamics of these arrangements, its links with science and professional expertise; and the effects on regional initiatives and organisational change arising from overlapping government programmes which eventually came to operate in a similar space.

## Emergence and Evolution of SEQ HWP, 1993–2012

The SEQ Healthy Waterways Partnership (HWP) has been a highly innovative and emergent self-organised 'not for profit' collaboration between the Queensland Government, local councils, community organisations, and industry and research groups. HWP's focus was to improve the ecological and human health of SEQ waterways through effective catchment management strategies, community education, and programmes monitoring ecosystem health. Drawing on our previous work (Bellamy et al., 2012, 2017; Head et al., 2016; Ross et al., 2016) as well as other relevant published literature, in this chapter we map the emergence and evolution of the HWP collaborative innovation over two decades from early 1993 until the absorption of the HWP processes into the broader government-mandated regional NRM process after 2012.

The SEQ region (Fig. 11.1) is a unique and complex inter-linked terrestrial-marine system, which since the 1980s has consistently been one of the fastest growing urban regions in Australia. The coastal receiving waters of Moreton Bay support a large number of species recognised as having global natural conservation significance. Government and community concerns emerged strongly in the 1990s over the potential impact of rapid urban growth on the ecological health of SEQ's diverse marine and freshwater waterways and catchments and the health of its human population (Abal et al., 2001; Dennison & Abal, 1999; Lloyd et al., 2000). This complex set of water quality and environmental problems challenged scientists, government administrators, environmentalists, and industry and concerned citizens alike.

Until the 1990s, management and planning for water quantity and water quality in SEQ proceeded through parallel but disconnected paths (Head, 2014). Water quantity management focused on building water infrastructure (i.e. dams and weirs for water supply for urban and agricultural purposes and for flood mitigation). Water quality management focused on treating algal blooms in Moreton Bay, river dredging in the lower Brisbane River estuaries, and sediment deposits at the river mouth for extending the port of Brisbane (Abal et al., 2001; Bunn et al., 2010; McPhee, 2017). Moreover, the city of Brisbane and surrounding councils continued to channel untreated sewage and stormwater into the receiving waters of the lower Brisbane River estuaries and Moreton Bay (Werbeloff et al., 2016).

Concerns grew rapidly about the declining water quality situation and the on-going impacts on the ecological health of the lower Brisbane River and Moreton Bay areas. The ecological impacts included loss of seagrass, local outbreaks of algal blooms, sediment accumulation, loss of species diversity, and harm to human health (e.g. contact dermatitis) (Abal et al., 2001). The need for improved wastewater management practices became increasingly evident in the 1990s. Local governments in SEQ had an important role because during those decades they jointly or separately owned and operated storages and water supply and wastewater systems in SEQ (Abal et al., 2001; Head, 2014).

In the absence of any existing Queensland standards for water quality management, tensions arose when a coalition of scientists and local governments was formed to oppose a Queensland Government proposal to apply New South Wales standards for regulating water quality in Queensland. In response to these tensions, the Brisbane City Council (BCC) joined with the Queensland Department of Environment in 1993 to establish the Brisbane River Management Group (BRMG) as a community, government, and industry initiative to better coordinate the management of waterway health within the lower Brisbane River area. In 1993, the BRMG initiated the Brisbane River and Moreton Bay Wastewater Management Study or BRMBWM (Abal et al., 2001; Lloyd

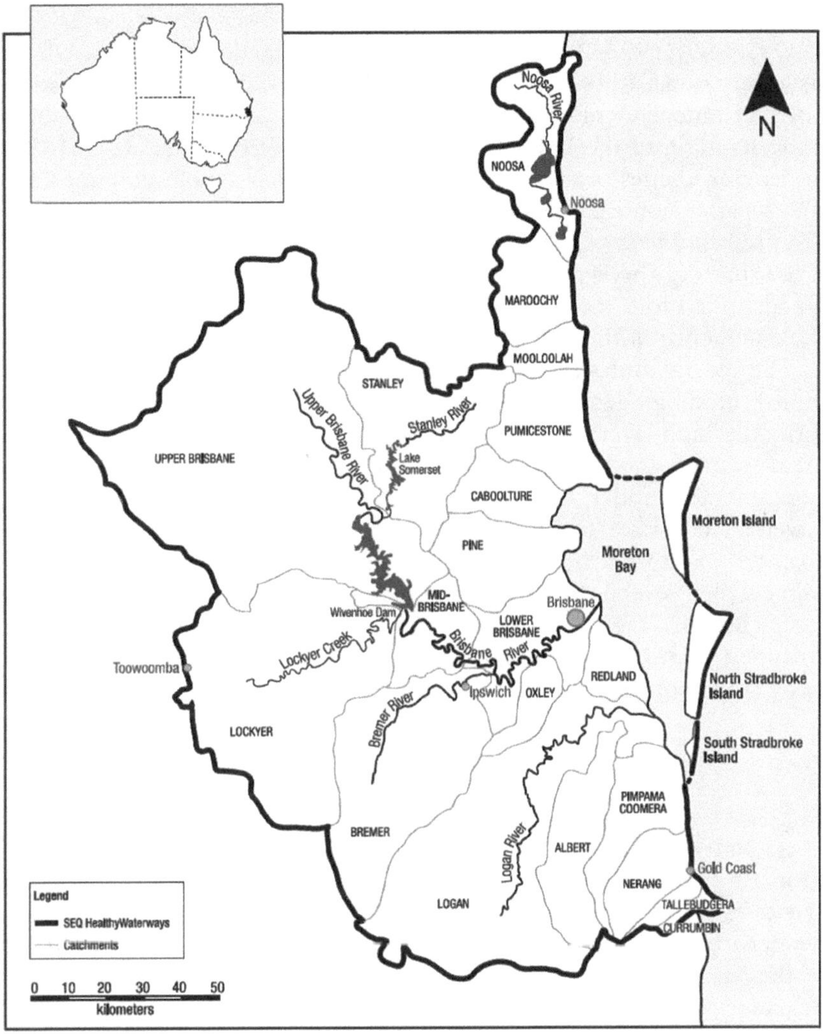


**Fig. 11.1** SEQ healthy waterways region and catchment areas

et al., 2000). This 'Moreton Bay Study' later became known as 'The SEQ Study' (1993–2001). It comprised a research partnership among six local councils bordering Moreton Bay, the state government, and a body of independent experts and dedicated to improving the scientific understanding of the Bay (Bunn et al., 2007; Werbeloff et al., 2016). It undertook a series of scientific studies of point source pollution in parallel to a separate consultative process involving a highly diverse group of key SEQ stakeholders.

In charting the evolution of these developments from 1993 to 2012, we identified four inter-linked phases of collaboration and innovation in the emergence of the HWP (see Fig.11.2).

The federal and state governments, which had been supplying much of the funding, determined that the NRM regional structures should form the high-level frames for ongoing water and land-based improvement strategies around Australia; and from 2013, the SEQ organisations were merged under the name 'Healthy Land & Water' with a new governance board; those later developments are not analysed in this chapter. Sustaining the HWP collaboration as a self-organising partnership entailed several shifts in direction and proved challenging in the long term (Bellamy et al., 2012; Head et al., 2016; Ross et al., 2016). In analysing each of the four phases, we note a series of key factors and challenges, along with some legacies.

### *Phase 1: 1993–1998: SEQ Regional Water Quality Strategy*

Focussing on the urban areas of Moreton Bay and the lower Brisbane River estuaries, Phase 1 collaboration involved two collaborative processes running in parallel—one scientific and one consultative. These two processes were critical success factors leading to the development of the SEQ Regional Water Quality Strategy (SEQWQS) which in turn was later to shape the emergence of the Healthy Waterways Partnership (HWP) in 2002 (Abal et al., 2001; Lloyd et al., 2000; Werbeloff et al., 2016).

Firstly, 'the Moreton Bay Study' 1993–2001 (later known as 'the SEQ Study') was operationalised as a series of background scientific studies of point source pollution. It was conducted by a diversity of scientists not only from local research and academic institutions, but also from two federal government research agencies, two state government agencies, and

| Phase 1: 1993-1998 | Phase 2: 1999-2001 | Phase 3: 2002-2007 | Phase 4: 2008-2012 |
|---|---|---|---|
| **Moreton Bay Water Quality Management Strategy (MBWQMS)** | **SEQ Regional Water Quality Management Strategy (SEQRWQMS)** | **SEQ Healthy Waterways Strategy (SEQHWS)** | **A multi-stakeholder collaborative platform for SEQ catchment health** |
| Focus: Moreton Bay & coastal estuaries only Single issue: urban point source water quality decline<br><br>Scoping research studies find (a) Sewage Treatment Plants (STPs) major source of urban point source pollution; (b) diffuse-source N as key source of algal blooms in Moreton Bay.<br><br>Moreton Bay Water Quality Management Strategy (MBWQMS) released Sept. 1998.<br><br>MBWQMS underpinned by: (a) credibility of the science knowledge of the "Moreton Bay study (later 'the SEQ study'); (b) consultative processes engaging farmers, local government and community; (c) strong scientific & political leadership & local government advocacy. | Expanded regional coverage to all SEQ freshwater catchments & all pollution sources (diffuse, point)<br><br>2001 – 'The SEQ Study' and BRMG formed the Moreton Bay Water Quality Partnership (MBWQP) - an inclusive stakeholder-based collaboration.<br><br>Catchment hydrology altered substantially from dams and weirs interception, & changes in land use, vegetation and runoff.<br><br>2001 – SEQ Regional Water Quality Management Strategy (SEQRWQMS).<br><br>Local council planning for upgrades to sewage treatment plants (STPs) begin<br><br>Design of regional estuarine/marine Ecosystem Health Monitoring Program (EHMP) and DSS tool (EMSS) | Focus: SEQ region; Integrated adaptive management of water quality in Moreton Bay, estuaries, catchments<br><br>MBWCP renamed SEQ Healthy Waterways Partnership (HWP), an inclusive stakeholder-based collaboration<br><br>Establishment of a Scientific Expert Panel (SEP); major investments in STP upgrades; Annual Report Cards<br><br>2005 'Healthy Catchments-Healthy Waterways' initiative; focus on sustainable loads for pollutants<br><br>Millennial drought shifts focus from water quality to quantity and removed many responsibilities for water management away from local govt.<br><br>Release SEQ Healthy Waterways Strategy (2007-2012) (SEQHWS) | Focus: SEQ region Major shift in structure to a 'network' that members had to buy into, i.e. a collaborative platform for catchment health in SEQ.<br><br>SEQHWS based on 12 integrated action plans focussed on water quality commences.<br><br>Whole-of-Water Cycle Management Initiatives:<br>(a) 'Healthy Country' - diffuse sources<br>(b) 'Water by Design' & WSUD - urban pollution<br><br>2011-13: Major floods following Millennial drought: shifts focus to water quantity as dominant SEQ water management issue and HWP with a focus on technology transfer of issues relating to water quality.<br><br>Amalgamation of SEQ Local Councils leads to reduction from 19 to 10 entities |
| 6 Local councils plus state and federal governments | 19 Local councils plus state and federal governments | 19 Local councils plus state and federal governments Mandated NRM & Water initiatives | 10 Local councils plus state and federal governments Mandated NRM & Water initiatives |

**Fig. 11.2** Emergence and evolution of the HWP collaborative innovation in SEQ, 1993–2012

three consultancy groups. Initially, the focus was on urban point source pollution of the lower catchment and estuarine areas of the Moreton Bay region (Dennison & Abal, 1999; Lloyd et al., 2000). The major point sources identified were urban sewage treatment plants (STPs) in the lower Brisbane River and western Moreton Bay, with nitrogen (N) identified as a pollutant of particular concern linked to algal blooms (Dennison & Abal, 1999). Later in Phase 2, the focus was expanded to diffuse pollution from stormwater as well as the freshwater and tidal ecosystems in the region (Bunn et al., 2007; Werbeloff et al., 2016). The study identified a major plume of organic-rich sediments at the mouth of the Brisbane River (from upstream diffuse source pollution) as causing continual resuspension of fine sediments into Moreton Bay. The plume was considered responsible for declines in shallow seagrass habitats and declines in the populations of dugong, turtles, and other biota in Moreton Bay and estuaries (Bunn et al., 2007; Dennison & Abal, 1999; McPhee, 2017).

Secondly, a parallel consultative process to 'the Moreton Bay Study' operated through engaging three diverse group of key SEQ stakeholders: (a) an independent *Scientific Advisory Group (SAG)* involving representatives of local Universities, state and federal agencies, water companies, and 6 local Councils to peer review research; (b) a *Technical Advisory Group* (TAG) of representatives from State and local Government groups; and (c) an *Industry and Community Advisory process* with non-government representatives including industry, fishing, conservation, catchment management, Landcare, and Indigenous groups (Bunn et al., 2007; Lloyd et al., 2000).

Four key legacy factors in the above process underpinned the subsequent publishing of the SEQ Regional Water Quality Strategy (SEQWQS) in 2001 and the emergence of the HWP in 2002. These factors were critical for shaping HWP's future structure and related collaborative processes and had critical implications for diverse stakeholder engagement.

The first significant factor was the development and release of the Moreton Bay Waterways Water Quality Management Strategy (MBWQMS) in September 1998 involving six local governments adjacent to Moreton Bay. Critically, its development was supported by scientific experts, industry, and community interests through the consultative process provided through the independent Scientific Advisory Group (SAG) of diverse scientific interests, a Technical Advisory Group (TAG) of State and local government groups, and a process of extensive consultation with industry and the wider community (Bunn et al., 2007, 2010; Head, 2014; Werbeloff et al., 2016).

The second factor was that the collaborative model created for the MBWQMS development was extended to the whole of SEQ, including 19 local governments (SEQRWQMS 2001). This was fundamental to the design and evolution of the HWP collaborative process established in 2002. The development of the SEQRWQMS led to the formation of the Moreton Bay Waterways and Catchments Partnership (MBWCP) in 2002, which later evolved into the SEQ Healthy Waterways Partnership (the HWP) in 2005.

The third factor relates to the collaboration processes being underpinned by highly credible scientific and policy research, coupled with strong scientific and political leadership and advocacy including four charismatic leaders/gate-keepers (Bunn et al., 2010; Head, 2014; Lloyd et al., 2000; Werbeloff et al., 2016).

The fourth legacy factor of Phase 1 is that policy development was connected closely with on ground delivery. As one interviewee argued: 'If you want change, you've got to get the necessary connections working. That means that it's got to be negotiated at the right spatial scale for on ground delivery' (Interview A).

Overall, Phase 1 can be seen as a critical and timely turning point regarding water quality management in SEQ catchments and the waterways of Moreton Bay. It not only facilitated the on-going emergence of a process of improved scientific understanding of the water quality (terrestrial and marine) in SEQ; it also critically brought together all relevant stakeholders in a collaborative way to systematically and holistically investigate the needs of the region's waterways and catchments (Head et al., 2016; Ross et al., 2016; Werbeloff et al., 2016).

### *Phase 2: 1999–2001: Moreton Bay Waterways and Catchments Partnership*

From 1999 to 2001, it was estimated that 60% of the sediment in Moreton Bay originated from 30% of specific soil types found predominately in the south-western catchments of SEQ (Abal et al., 2005; Bunn et al., 2007, 2010). This led to a shift in focus to include the diffuse sources of sediments, nutrients, and toxicants found in Moreton Bay and estuaries resulting from both human/anthropogenic actions and natural processes (Bunn et al., 2010). At this time, the regional geographical scope was also expanded to include freshwater catchments from Noosa in the north down to the Queensland/NSW border in the south, and inland to the west to the Great Dividing Range. Funding was project-based and involved 19 Local Councils, as well as the Queensland State Government and Federal Governments (Fig. 1.1).

Initiated in 2000, a regional estuarine/marine and freshwater Ecosystem Health Monitoring Program (EHMP) was designed to provide a regional assessment of the health for each of SEQ's major catchments, river estuaries, and Moreton Bay zones. This involved the measurement of 16 parameters, including measures of fish and invertebrate biodiversity, ecosystem processes, and water quality (Bunn et al., 2007). Under the EHMP banner, several initiatives were implemented to maintain the visibility of and engagement with waterway health (Abal et al., 2005). For example, research and monitoring results were synthesised and communicated through the annual *Report Card*, which was established to communicate the health of the region's waterways. The Report Cards graded each waterway on a scale from A+ to F which were released annually by local politicians in a public, televised event (for the current version, see Healthy Land & Water, 2021).

In July 2001, 'the SEQ Study' group joined with the Brisbane River Management Group (BRMG) to form the Moreton Bay Waterways and Catchments Partnership (MBWCP). The MBWCP involved an alliance of 19 local councils, 2 state agencies, environmental groups, and research scientists. Subsequently in 2002, as 'the SEQ Study' was completed, the MBWCP was re-designated as the SEQ Healthy Waterways Partnership (SEQHWP), constituting a highly visible single entity and a mechanism for ongoing engagement of diverse stakeholders (Head et al., 2016; Ross et al., 2016; Werbeloff et al., 2016).

### *Phase 3: 2002–2007: SEQ Healthy Waterways Partnership*

This phase, marked initially by the formal renaming of the MBWCP as the SEQ Healthy Waterways Partnership (HWP), provides a fine example of an inclusive stakeholder-based collaboration. It was underpinned initially through the establishment of a HWP Scientific Expert Panel (SEP), major investments by local councils in sewage treatment plant (STP) upgrades, an annual water quality reporting programme by catchment. By 2005, major investments had been made in 25 sewage treatment plant upgrades, to reduce point source wastewater pollution in rivers and estuaries (Head et al., 2016). Ultimately, a major milestone of this phase was the release of the SEQ Healthy Waterways Strategy (2007–2012), i.e. the SEQHWS (Healthy Waterways Partnership, 2007).

A new focus emerged for the collaborative HWP centred on an integrated adaptive management approach for Moreton Bay, estuaries and inland waterways, and catchments of SEQ (Abal et al., 2001). The HWP adaptive management approach was informed by scientific advice from the newly established SEP. One example was the Ecosystem Management Monitoring Program (EHMP) which was fully implemented in 2002–2003 (Bunn et al., 2010). It included 16 performance indicators, to be measured twice per year at 120 stream sites across SEQ. The implementation of this thorough water quality monitoring was communicated widely through an annual HWP *Report Card* at a catchment level across the whole of SEQ (Abal et al., 2005; Bunn et al., 2007). Public presentations about the findings of these annual reports on the health of waterways were made to politicians and senior policy-makers and at specific public forums. The report cards not only served to raise public awareness, but also revitalised political support for HWP (Bunn et al., 2007; Head et al., 2016). Another important development, from 2002 through to 2008, was subregional meetings and forums that enabled local government, industry and NGO stakeholders to network and discuss emerging issues, and to connect with the best available science in a timely manner. The water quality improvement plan for Moreton Bay attracted federal funding and supported by the state government and 19 local councils. At the same time, federal and state agreements were also being negotiated to promote regional NRM plans.

The HWP became a key driving force behind Brisbane's urban stormwater quality improvement programme. In 2005, a capacity building programme, *Water by Design*, was established to build industry expertise in designing, constructing, and maintaining Water Sensitive Urban Design (WSUD) assets (Head, 2014; Werbeloff et al, 2016). By involving all relevant stakeholders, HWP fostered a collective commitment from all parties about the need for action in relation to waterway health, enabling relatively swift agreement on management actions to be implemented (Head, 2014; Werbeloff et al., 2016). Under the HWP banner, new technologies and approaches for WSUD were developed and trialled to build confidence in these 'soft' engineering approaches to delivering improved waterway health. HWP, however, lacked any formal authority to translate its findings into policy or practice and relied on informal strategies of ongoing engagement, persuasion, and communication to create a culture of valuing waterway health to bring about change on the ground (Werbeloff et al., 2016).

The emergent sustainability agendas of the first decade of the HWP collaboration were increasingly disrupted during Phase 3 and Phase 4 by the long-lasting Australian 'Millennium Drought' that extended from around 2000 to 2010. As it continued to deepen over several years, it progressively impacted on the environmental and natural resource policy priorities of government in SEQ. In particular, the drought generated fears of water shortage in the mid-to-late 2000s. The response of the Queensland government was to shift the focus of water management away from water quality towards water security, investing in water purification plants and constructing the SEQ 'water grid' to improve water supply throughout SEQ (Head, 2014; Head et al., 2016).

### *Phase 4: 2008–2012: A Collaborative Platform for Catchment Health*

With the increasing importance of government-mandated regional NRM structures and processes during the first decade of the new century, it became inevitable that the HWP initiatives would find it difficult to retain their prominent and distinct identity.

In 2009, the emphasis shifted towards 'whole-of-water' cycle management, which included actions to prevent water scarcity under conditions of future climate change. Two initiatives, the 'Healthy Country' and the continuing 'Water by Design' programmes, underpinned a more integrated approach. The HWP concern with water quality in the western catchments was further developed in the *Healthy Country* project, which focused on reducing sediment loading in the upper catchments (Logan, Bremer and Lockyer).

In the shadow of the ongoing Millennium Drought (2000–2010), major changes also occurred in the structure of local government in Queensland in 2008. Through amalgamations, the number of relevant local councils in SEQ was reduced from 19 to ten. Subsequently, there was a hiatus in local government interest and financial support for the healthy waterways agendas. The SEQ HWP, therefore, faced significant changes and reduced capacity to implement its Strategy. In effect, it was faced with transitioning from an innovative and evolving collaboration with a clear Strategy, focused on addressing the wicked problem of water quality and pollution in SEQ catchments, towards becoming a new multi-stakeholder collaborative platform addressing 'catchment health', and with a more traditional technology transfer role. This transition is implicit in the range of choices outlined by Ansell and Miura (2020) regarding possible directions for collaborative platforms:

> interaction, production and innovation leverage constitute generative mechanisms for platforms and their evolution .... Some platforms become innovative and collaborative, while others slide back into a more traditional technology-transfer role. (Ansell & Miura, 2020, p. 264)

As Werbeloff et al. (2016) argue, the declining interest in stormwater quality and 'water sensitive cities' was reflected in a diminution of structural support for WSUD. A sector-wide restructure had significantly overhauled organisational roles and responsibilities in SEQ. Subsequently, there was a loss of both institutional knowledge and expertise in relation to whole-of-water-cycle initiatives and industry champions for this approach.

In addition, there was no Healthy Waterways Strategy developed that extended beyond 2012. The disruptions continued for the HWP collaborative partners, with a switch in crises from drought to floods. Shortly after the breaking of the drought in 2009–2010, SEQ was hit with devastating floods in 2011–2012, leading to a heightened governmental and community interest in flood protection. Consequently, there was a reduced appetite for ongoing investment in the water quality agenda, making it difficult to secure funding for stormwater quality improvement and other initiatives.

Extensive restructuring ensued, such that priorities were reshaped, and collaborative relationships reconstructed. The establishment of SEQ Catchments, as the regional NRM body for SEQ, created ambiguity and potential competition, leading to an initial accommodation between the HWP focus on water quality, and the regional NRM body focus on land-based action. Ultimately, a loose integration of HWP into the broader regional NRM processes was negotiated from 2013, followed by a formal organisational merger which became known as 'Healthy Land & Water' (2021).

## Challenges for Emergent Collaborative Processes

The evolution and impact of collaborations is highly variable and depends on many factors. Preiser et al. (2018) argue that collaboration depends strongly on the structure of social networks and the flow of information within them. The capacity to maintain close communication and interaction is crucial. An appropriate mix of diverse participants allows for a range of useful interactions, and over time, there are inevitably some changes in the structures and functions of the system components. The patterns are marked by adaptation, evolution, and transformation across time and space.

Innovation in complex and uncertain collaborative situations requires understanding the challenge from a systems perspective, recognising the importance of innovative ideas and the significant presence of paradox, conflict, and cross-scale dynamics (Waardenburg et al., 2020; Westley, 2017). The term 'paradox' refers to conflicting demands, opposing perspectives, and sometimes illogical findings (Lewis, 2000). Paradoxes in NRM are often complex and can create conditions where it is difficult to resolve conflicts between values, goals, and actions (Browne & Bishop, 2011). Collaborative frameworks may highlight paradoxes and tensions that facilitate options for innovation and adaptation over longer time frames. In the complex dynamics of social innovation, Westley (2017) argues that paradoxes and tensions can be catalysts for evolution and that the focus of organisational innovation is not to eliminate conflicts, but rather to manage conflicts and tensions in a productive way.

Engle et al. (2011) point out that integrated cross-sectoral approaches may be more 'legitimate and accountable' than top-down command approaches, but the scope for innovation also depends on encouraging the 'flexible, experimental, and self-organizing' aspects of adaptive management. In addition to promoting innovation, other governance challenges include the selection of appropriate spatial scales and timeframes. For example, environmental strategies are often poorly matched with the spatial scale of environmental problems (hence the modern emphasis on catchment scale); this is exacerbated by the arbitrary nature of administrative-jurisdictional boundaries, which seldom match the scale of effective solutions (Ansell & Torfing, 2015). Within water policy, other tensions arise concerning how to integrate competing priorities, such as water quantity and water quality, or rural and urban interests (Head, 2014).

Sustaining the HWP collaboration as an emergent self-organising collaborative innovation in the long term proved very difficult. Our broader comparative study of regional NRM collaborations in Australia (see Bellamy et al., 2012, 2017; Head et al., 2016; Ross et al., 2016) found that emergent challenges included: managing the intersection of diverse competing interests and mandates; navigating difficult issues despite uncertain knowledge; maintaining focus and effort over long periods; building local and regional momentum and continuity while adjusting to policy changes at other levels of governance; recognising opportunities for collaboration and adaptive change; and generating a collaborative platform for linking science-policy-community leadership. Other researchers argue that social innovations can emerge out of dynamic tensions between agency (vision) and the balance of opportunities and constraints in the system (Walker, 2019).

The tensions underlying the interplay of competition and collaboration have proved particularly challenging for the HWP in practice. For example, one interviewee commented that:

> There's always a strong tension between the state and local governments, and [this] depends on the personalities of the day and the issues of the day. Sometimes that tension can come together in a very dynamic and substantive way, and you get good co-investment and co-leadership..... Other times....the same tension has become very destructive of relationships and also destructive of some of the good things that were happening around coordination and co-investment. (Interview A)

Regional NRM collaborations develop within a web of organisations, obligations, and policy contexts. Members of the collaboration have dual loyalties, because they retain their 'home base' priorities and responsibilities while simultaneously seeking to increase their capacities and create mutual interests (Bellamy et al., 2012, 2017; Head et al., 2016; Ross et al., 2016).

As both a government-mandated strategic intervention or as an emergent self-organising stakeholder network for addressing wicked NRM problems, collaborative forms have provided major contributions for reframing problems, and for building cross-sectoral bridges in multi-level regional NRM arrangements. Collaborative forms also help to link science, policy, and community arenas in policy, planning, and practice (Bellamy et al., 2012, 2017). Significant changes were made in SEQ. Once established, collaborative forums did make diverse contributions.

Although HWP in its original evolving form has been superseded by new initiatives, it is clear that during its first decade and beyond; it was able to provide an evolving self-organised collaborative platform that facilitated and guided a number of innovative and collaborative projects, including the science-policy response to the algal blooms crisis and sediment deposition in Moreton Bay, the Ecosystem Health Monitoring Program (EHMP) linked to the annual Report Card by catchment, and the facilitation of wetlands restoration and water-sensitive urban design (WSUD) (Healthy Land & Water, 2021; Werbeloff et al., 2016).

In conclusion, the HWP as an emergent self-organised collaboration successfully evolved in SEQ over two decades pursued by a diversity of actors (leaders, stakeholders, and institutions) who sought to improve scientific knowledge relating to water quality in SEQ and to manage associated conflict in the face of on-going social, institutional, and environmental change. Prof. Helen Ross made an invaluable contribution to the HWP collaborative process through both research and practice.

## References


Abal, E. G., Bunn, S. E., Dennison, W., Collier, C., Curtis, L., & Moore, K. (2005). *Healthy waterways, healthy catchments: making the connection in South East Queensland, Australia*. Moreton Bay Waterways and Catchments Partnership.

Abal, E. G., Dennison, W. C., & Greenfield, P. F. (2001). Managing the Brisbane River and Moreton bay: An integrated research/management program to reduce impacts on an Australian estuary. *Water Science and Technology, 43*(9), 57–70. https://doi.org/10.2166/wst.2001.0508

Ansell, C., & Gash, A. (2018). Collaborative platforms as a governance strategy. *Journal of Public Administration Research and Theory, 28*(1), 16–32. https://doi.org/10.1093/jopart/mux030

Ansell, C., & Miura, S. (2020). Can the power of platforms be harnessed for governance? *Public Administration (London), 98*(1), 261–276. https://doi.org/10.1111/padm.12636

Ansell, C., & Torfing, J. (2015). How does collaborative governance scale? *Policy and Politics, 43*(3), 315–329. https://doi.org/10.1332/030557315X14353344872935

Australian Public Service Commission. (2007). *Tackling wicked problems: A public policy perspective*. Australian Public Service Commission.

Bellamy, J. (2007). Adaptive governance: The challenge for regional natural resource management. In A. J. Brown & J. A. Bellamy (Eds.), *Federalism and regionalism in Australia: New approaches, new institutions* (pp. 95–117). ANU E-Press.

Bellamy, J., Head, B., & Ross, H. (2012, February 6–8). *Blurring boundaries and building bridges: Challenges of managing across borders in the Lake Eyre Basin*. Paper presented at the Proceedings of 6th Australian Stream Management Conference—Managing for Extremes.

Bellamy, J., Head, B. W., & Ross, H. (2017). Crises and institutional change: Emergence of cross-border water governance in lake Eyre Basin, Australia. *Society & Natural Resources, 30*(4), 404–420. https://doi.org/10.1080/08941920.2016.1272729

Beusen, A. H. W., Doelman, J. C., Van Beek, L. P. H., Van Puijenbroek, P. J. T. M., Mogollón, J. M., Van Grinsven, H. J. M., et al. (2022). Exploring river nitrogen and phosphorus loading and export to global coastal waters in the shared socio-economic pathways. *Global Environmental Change, 72*, 1. https://doi.org/10.1016/j.gloenvcha.2021.102426

Bowmer, K. H. (2014). Water resources in Australia: Deliberation on options for protection and management. *Australasian Journal of Environmental Management, 21*(3), 228–240. https://doi.org/10.1080/14486563.2014.913269

Brown, R. R., Deletic, A., & Wong, T. H. F. (2015). Interdisciplinarity: How to catalyse collaboration. *Nature, 525*(7569), 315–317. https://doi.org/10.1038/525315a

Browne, A. L., & Bishop, B. J. (2011). Chasing our tails: Psychological, institutional and societal paradoxes in natural resource management, sustainability, and climate change in Australia. *American Journal of Community Psychology, 47*(3–4), 354–361. https://doi.org/10.1007/s10464-010-9390-1

Bunn, S. E., Abal, E. G., Greenfield, P. F., & Tarte, D. M. (2007). Making the connection between healthy waterways and healthy catchments: South East Queensland, Australia. *Water Science & Technology: Water Supply, 7*(2), 93–100. https://doi.org/10.2166/ws.2007.044

Bunn, S. E., Abal, E. G., Smith, M. J., Choy, S. C., Fellows, C. S., Harch, B. D., et al. (2010). Integration of science and monitoring of river ecosystem health to guide investments in catchment protection and rehabilitation. *Freshwater Biology, 55*(1), 223–240. https://doi.org/10.1111/j.1365-2427.2009.02375.x

Chambers, J. M., Wyborn, C., Klenk, N. L., Ryan, M., Serban, A., Bennett, N. J., et al. (2022). Co-productive agility and four collaborative pathways to sustainability transformations. *Global Environmental Change, 72*, 102422. https://doi.org/10.1016/j.gloenvcha.2021.102422

Dennison, W., & Abal, E. G. (1999). *Moreton Bay study: A scientific basis for the healthy waterways campaign*. South East Queensland Regional Water Quality Management Strategy.

Dunn, G., Bos, J. J., & Brown, R. R. (2018). Mediating the science-policy interface: Insights from the urban water sector in Melbourne, Australia. *Environmental Science & Policy, 82*, 143–150. https://doi.org/10.1016/j.envsci.2018.02.001

Engle, N. L., Johns, O. R., Lemos, M. C., & Nelson, D. R. (2011). Integrated and adaptive management of water resources: Tensions, legacies, and the next best thing. *Ecology and Society, 16*(1), 19. https://doi.org/10.5751/ES-03934-160119

Fischer, J., Gardner, T. A., Bennett, E. M., Balvanera, P., Biggs, R., Carpenter, S., et al. (2015). Advancing sustainability through mainstreaming a social–ecological systems perspective. *Current Opinion in Environmental Sustainability, 14*, 144–149. https://doi.org/10.1016/j.cosust.2015.06.002

Folke, C., Biggs, R., Norström, A. V., Reyers, B., & Rockström, J. (2016). Social-ecological resilience and biosphere-based sustainability science. *Ecology and Society, 21*(3), 41. https://doi.org/10.5751/ES-08748-210341

Gray, B. (1989). *Collaborating: Finding common ground for multiparty problems*. Jossey-Bass.

Head, B. W. (2014). Managing urban water crises: Adaptive policy responses to drought and flood in Southeast Queensland, Australia. *Ecology and Society, 19*(2), 1–14. https://doi.org/10.5751/ES-06414-190233

Head, B. W. (2022). *Wicked problems in public policy: Understanding and responding to complex challenges*. Springer International Publishing AG.

Head, B. W., & Alford, J. (2015). Wicked problems: Implications for public policy and management. *Administration & Society, 47*(6), 711–739. https://doi.org/10.1177/0095399713481601

Head, B. W., Ross, H., & Bellamy, J. (2016). Managing wicked natural resource problems: The collaborative challenge at regional scales in Australia. *Landscape and Urban Planning, 154*, 81–92. https://doi.org/10.1016/j.landurbplan.2016.03.019

Healthy Land & Water. (2021). *Report card 2021*. Brisbane. https://hlw.org.au/report-card/

Healthy Waterways Partnership. (2007). *South East Queensland healthy waterways strategy 2007–2012*. Healthy Waterways Partnership.

Lewis, M. W. (2000). Exploring paradox: Toward a more comprehensive guide. *The Academy of Management Review, 25*(4), 760–776. https://doi.org/10.2307/259204

Lloyd, T., McMahon, P., Gibbes, B., & Collier, C. (2000, March 6–9). South East Queensland regional water quality management strategy. *Coast to Coast 2000 Conference* (pp. 42–47).

McPhee, D. P. (2017). *Environmental history and ecology of Moreton Bay*. CSIRO Publishing.

Patterson, J. J., Smith, C., & Bellamy, J. (2015). Enabling and enacting 'practical action' in catchments: Responding to the 'wicked problem' of nonpoint source pollution in coastal subtropical Australia. *Environmental Management, 55*(2), 479–495. https://doi.org/10.1007/s00267-014-0409-5

Preiser, R., Biggs, R., De Vos, A., & Folke, C. (2018). Social-ecological systems as complex adaptive systems: Organizing principles for advancing research methods and approaches. *Ecology and Society, 23*(4), 46. https://doi.org/10.5751/ES-10558-230446

Ross, H., Bellamy, J., & Head, B. (2016). *The challenges of collaboration in environmental governance* (pp. 175–196). Edward Elgar Publishing. https://doi.org/10.4337/9781785360411.00017.

Waardenburg, M., Groenleer, M., de Jong, J., & Keijser, B. (2020). Paradoxes of collaborative governance: Investigating the real-life dynamics of multi-agency collaborations using a quasi-experimental action-research approach. *Public Management Review, 22*(3), 386–407. https://doi.org/10.1080/14719037.2019.1599056

Walker, B. (2019). *Finding resilience: Change and uncertainty in nature and society*. CSIRO Publishing.

Wear, S. L., Acuña, V., McDonald, R., & Font, C. (2021). Sewage pollution, declining ecosystem health, and cross-sector collaboration. *Biological Conservation, 255*, 109010. https://doi.org/10.1016/j.biocon.2021.109010

Werbeloff, L., Brown, R. R., & Loorbach, D. (2016). Pathways of system transformation: Strategic agency to support regime change. *Environmental Science & Policy, 66*, 119–128. https://doi.org/10.1016/j.envsci.2016.08.010

Westley, F. (2017). *The evolution of social innovation* (pp. 239–256). Edward Elgar. https://doi.org/10.4337/9781786431158.00018.

Westley, F., Olsson, P., Folke, C., Homer-Dixon, T., Vredenburg, H., Loorbach, D., et al. (2011). Tipping toward sustainability: Emerging pathways of transformation. *Ambio, 40*(7), 762–780. https://doi.org/10.1007/s13280-011-0186-9

Westley, F. R., McGowan, K. A., Antadze, N., Blacklock, J., & Tjornbo, O. (2016). How game changers catalyzed, disrupted, and incentivized social innovation: Three historical cases of nature conservation, assimilation, and women's rights. *Ecology and Society, 21*(4), 13. https://doi.org/10.5751/ES-08811-210413

CHAPTER 12

# Building Collaborative Models for Change: Engaging Indigenous People and Country into in Socially just Conservation

*Melissa Nursey-Bray*

## Introduction

In Australia, the terrible twins of climate change and biodiversity loss have coiled in on themselves and with each other to become behemoths that amplify the impacts of all other problems—soil loss, water quality, feral weed and animal management to name a few.

Building sustainable social ecological systems in rural areas is an ongoing challenge, beset by reducing populations, impacts of climate change and changing policy environments. The Indigenous peoples of Australia are disproportionately affected by these challenges. Despite decades of attempted reform, Indigenous peoples across Australia remain amongst the most socio-economically disadvantaged, and still suffer from

M. Nursey-Bray (✉)
School of Social Sciences, University of Adelaide, Adelaide, SA, Australia
e-mail: melissa.nursey-bray@adelaide.edu.au



C. Baldwin and S. van Bommel (eds.), *Rural Development for Sustainable Social-ecological Systems*,
Palgrave Studies in Environmental Sustainability,
https://doi.org/10.1007/978-3-031-34225-7_12

a range of health and other issues. Their land and seas are also being increasingly affected by climate change—middens are being washed away by sea level rise, and climate changes are disrupting seasonal harvesting practice and eroding sacred sites. As such, Indigenous knowledge itself is under threat.

Yet, despite these challenges, Indigenous Australians have led the world in building conservation regimes that address social justice, and care for their Country via deployment of Indigenous knowledge in partnership with scientific knowledge (Garnett et al., 2018). These models include co-management, Indigenous Protected Areas, various conservancy and land use agreements and climate adaptation strategies, including innovative fire burning programs. In so doing, Indigenous Australians have asserted their agency and voice into wider governance regimes.

This work is especially significant as its focus on social justice highlights the importance of how to reconcile the tension between people, economy and environment that confronts us all. In this chapter, I first introduce the different models of Indigenous environmental management in Australia. The challenges in implementation are then discussed, highlighting the tensions in the role of Indigenous knowledge and used of Country. The chapter concludes with consideration of how knowledge partnerships can be built to progress socially just conservation into the future. I conclude that in moving forward we need to establish modes of knowledge co-existence and governance to find ways to resolve the inherent conflict between social justice and conservation imperatives and establish significant and much needed templates for confronting a world that is experiencing unprecedented change.

## Indigenous Models of Management

In Australia, there has been an ongoing history of formalised First Nation engagement with Western resource management, beginning in the late 1980s with the establishment of programs in conservation and land management in the Batchelor Institute and the Cairns TAFE. At this time, Indigenous Rangers, employed via the CDEP program, enrolled in certificates to build their skills and knowledge in resource management. In caring for their Country, these original initiatives have led to the creation of what is now one of the most diverse suite of Indigenous-led initiatives in the world. These include Working on Country Ranger Programs, Indigenous Protected Areas, Sea Country, Climate Change and Healthy Country planning, Traditional Use Marine Resource Agreements, co-management and joint management initiatives and Indigenous Alliances.

## Working on Country—Indigenous Rangers

Indigenous Ranger projects were first funded as a nation-wide program in 2007 through the Working on Country program. Since then, the program has created more than 2,100 full-time, part-time and casual jobs across over 900 Ranger Units. Collectively, they provide a workforce in land and sea management that not only enables Indigenous peoples to be meaningfully employed on their Country but to manage 20 per cent of Australia's land and sea Country (Allam, 2021).

Indigenous Ranger work programs are very diverse and enable the application of traditional knowledge combined with conservation training to protect and manage land, sea and culture. Activities include bushfire mitigation, protection of threatened species and biosecurity compliance. Indigenous Ranger groups also develop partnerships with research, education, philanthropic and commercial organisations to share skills and knowledge, engage with schools and generate additional income and jobs in the environmental, biosecurity, heritage and other sectors. Indigenous Ranger work also enhances Indigenous well-being.

## Indigenous Protected Areas

Indigenous Protected Areas (IPAs) are areas of land and sea managed by Indigenous groups as protected areas for biodiversity conservation through voluntary agreements with the Australian Government. IPAs are an essential component of Australia's National Reserve System, which is the network of formally recognised parks, reserves and protected areas across Australia. There are currently 78 dedicated IPAs over 74 million hectares. These account for more than 46 per cent of the National Reserve System. IPAs provide multiple benefits as they ensure both environmental as well as cultural benefits by facilitating Indigenous management of their Country and recognition of their cultural values (Rist et al., 2019). They have created jobs for Indigenous peoples and assisted in building economic security and livelihoods (Godden & Cowell, 2016). Rangers employed within IPAs undertake interpretation, monitoring and surveillance, research, cultural history, bush tucker and language projects (Ross et al., 2009).

## Sea Country Planning

This initiative, originating as part of Australia's Oceans Policy, was designed to "involve Aboriginal and Torres Strait Islander peoples in the use, conservation and management of Australia's marine jurisdiction". As part of this, Indigenous people were engaged to build Sea Country Plans, which sought to help Indigenous people negotiate with other marine managers and users to develop policies and institutional arrangements that are respectful of Indigenous peoples' rights, interests and responsibilities in sea Country. Consistent with the idea of socially just conservation, a Sea Country Plan is designed to (i) improve Indigenous participation in marine planning and management, (ii) address a range of cultural, ecological and economic issues and enable Indigenous people to identify opportunities to derive greater social and economic benefit from the management of sea Country, (iii) facilitate Indigenous participation in Sea Country management at appropriate geographical and cultural scales, and (iv) help others develop greater understanding of Indigenous peoples' Sea Country interests and responsibilities. Sea Country Planning is now well established (Bock et al., 2021).

## Healthy Country Planning

Allied with Sea Country Planning have been initiatives under the *Healthy Country Planning*, which leads groups through a series of five steps including (i) deciding what the plan is about, (ii) making the plan, (iii) doing and monitoring the work, (iv) deciding if the plan is working and (v) communicating the results. There are multiple Healthy Country Plans across Australia that provide the basis for partnerships between Indigenous people and others in order to build biodiversity management and protection (Carr et al., 2017; Moorcroft et al., 2012).

## Traditional Use of Marine Resources Agreements (TUMRA)

Indigenous harvesting of charismatic and sometimes endangered species is often a contentious issue. To address this, one major partnership has been via the creation of the Indigenous Land and Sea Country Partnerships Program, which comprises a $20 million investment in Traditional Owner management of the Reef (Zurba, 2009). An essential part of this program is the active support of what are known as Traditional Use of Marine Resources Agreements (TUMRAs). Traditional Use of Marine Resources Agreements are community-based plans for management of traditional resources which are accredited in legislation and embed agreements between the Great Barrier Reef Traditional Owner groups and the Australian and Queensland governments about how to manage traditional use activities on their Sea Country (Havemann et al., 2005).

## Climate Initiatives

The last decade has also seen a rise in climate-focused initiatives (Nursey-Bray & Palmer, 2018). These include development of climate adaptation plans such as with the Arabana in South Australia and the Torres Strait at Australia's very north-eastern tip. Fire for carbon management is also being trialled. For example, the Karlantijpa North Kurrawarra Nyura Mala Aboriginal Corporation, has been allocated funds for carbon credits under the Australian Government's Emissions Reduction Fund. In this program, the Corporation secures payments for sequestering 24,100 tonnes of carbon by undertaking savanna burning early in the season. Elders are working with youth and in many other contexts to build knowledge transmission programs to build adaptive capacity to climate change. In the Northern Territory, the Miriwoong people of Keep River are developing a climate adaptation tool using traditional ecological knowledge based on the weather and observed environmental change. Involving extensive community engagement, this project resulted in an interactive seasonal calendar, which enables the Miriwoong people to discuss ongoing and future weather patterns and help ensure younger generations can adapt to those changes (Nursey-Bray et al., 2019).

## Co-Management

Co-management agreements among Indigenous people, state agencies and other stakeholders offer substantial promise as a way of dealing with natural resource conflicts in a participatory and equitable manner. Co-management emphasises participatory and collaborative management and is also ideally placed to have the policies and mechanisms in situ for dispute management. It is a model that can embed power sharing mechanisms and allow platforms for Indigenous voices to be heard and problems to be resolved—it then becomes an ongoing process. Other important elements within co-management include (i) shared responsibility; (ii) balanced power regimes; (iii) cooperation; (iv) participation; (v) discussion; (vi) education and sharing of information; (vii) communication; (viii) consensus; (ix) flexibility; and (x), the use of traditional ecological knowledge and scientific knowledge (Nursey-Bray & Rist, 2009).

There are many examples of co-management in Australia. Girringun Aboriginal Corporation located in Cardwell in Queensland has been running a highly successful co-management program since the early 2000s, which has undergone many iterations, and continues to exemplify how Indigenous people can work equitably and in partnership with scientists and policymakers (Zurba et al., 2012). In South Australia, a number of national parks have been dedicated under co-management arrangements and are constituted by a partnership between the state government and Aboriginal groups in ways that combine traditional knowledge with contemporary park management. Such initiatives also seek to advance reconciliation processes. There are now 12 co-management agreements in place over 35 of South Australia's parks and reserves, covering 13.5 million hectares or 64% of the reserve system which equates to 13% of the land area of South Australia.

## Joint Management

Joint management is another form of co-management, and early on was trialled in high-profile parks, such as Uluru and Kakadu. In the Northern Territory, for example, the Parks and Wildlife Commission of the Northern Territory (Parks and Wildlife) manages 85 areas as parks and reserves, and of them, 33 are subject to formal joint management with Aboriginal traditional owners. Joint management in these cases is largely the result of negotiated settlement of land and/or native title claims.

As with co-management, joint management arrangements are predicated on principles of equity, empowerment, economic sustenance, protection of cultural values and the integration of traditional with Western scientific knowledge (Carter et al., 2022). Garig Gunak Barlu National Park (Cobourg Marine Park), Nitmiluk National Park, Tnorala (Gosse Bluff) Conservation Reserve, Barranyi (North Island) National Park and Djukbinj National Park are just a few of the current examples of jointly managed parks.

## Indigenous-led Alliances

Increasingly, Indigenous voices in environmental management are being navigated via Indigenous-led alliances and organisations. An early and highly successful example of this type of initiative is the North Australian Indigenous Land and Sea Management Alliance Ltd (NAILSMA) which is an Indigenous-led not-for-profit company operating across north Australia. NAILSMA has over twenty years' experience in supporting Indigenous groups across Northern Australia to build their skills, care for their Country and build partnerships with scientists, government and managers. NAILSMA (nailsma.org.au) has led initiatives in water policy and planning, regional marine planning, Carbon, Indigenous Livelihoods and Indigenous youth leadership. In so doing it has assisted in the development and then delivery of $30 million worth of projects across northern Australia. Another alliance—the Indigenous Desert Alliance states it: "plays a vital role in securing the future health of the desert and its people by strengthening the existing *connections* between desert people, empowering desert people to look after their Country and by *connecting* the desert story with the story of *our country*" (indigenousdesertalliance.com).

Balkanu Aboriginal Economic Corporation, based in Cairns, is another example of an Indigenous organisation with members from across the Cape York and focuses on supporting the Aboriginal people of Cape York to improve and build their economic and social capacity while preserving their heritage and culture. Again, the focus is not just on culture, or economy, or environment, but an amalgam of all three so that social just conservation and socio-ecological outcomes can be achieved. Other partnerships are galvanised between Indigenous groups and industry. For example, in 2011, a number of native title claim groups in the Pilbara signed agreements with Rio Tinto Iron Ore which negotiated a range of economic and other benefits to the native title claim groups and included an income stream from mining on their lands, training and job opportunities, access to contracts for services for Rio Tinto and support for environmental and heritage activities.

In 2018, the South Australian First Nations Sea-Country Research (SA FNSCR) Alliance was established. This alliance is an Indigenous-led advisory group comprised of South Australian Sea Country Traditional Owners as well as SA Government and University researchers/academics and seeks to build a bridge between First Nation peoples and researchers to foster culturally collaborative Sea Country Research, Development and Extension (RD&E) initiatives.

## Indigenous Land Use Agreements (ILUAs)

ILUAs are another model of Indigenous resource management and are voluntary agreements between native title parties and other people of bodies where there are mutual agreements made about the use and management of specific land and/or seas/waters (Campbell & Hunt, 2013). For example, an ILUA may be declared on a pastoral property and prescribe the rights, use, access and protection of different parties to that piece of land, or it may focus on a relationship between and Indigenous and mining groups (Doohan et al., 2012). For example, there is an ILUA in Western Australia (WA), called the Yawaru Agreements, which is between the WA Government, the Yawuru people and others, and covers an area of over 973 square kilometres. These agreements resolve heritage issues, release the State from any liability for compensation and provide for $56 million in monetary benefits for the Yawuru for (i) capacity building, (ii) preservation of culture and heritage, (iii) economic development, (iv) housing and (v) joint management of a proposed conservation estate. Another example is that of the Indigenous land use agreements between the Bundjalung people of Byron Bay and the NSW Government.

## Challenges in Implementation

This range of initiatives showcase the diversity of Indigenous cultures and their range in terms of resource management. Collectively, they now make a compelling contribution to Australia's environmental policy and management platforms. Yet, these programs continue to face challenges in their implementation. Indigenous voices are still not always heard, and

Indigenous people not accommodated into environmental governance regimes as a matter of practice. Indigenous peoples, despite 60,000 years of inhabiting the continent, are persistently left out of policy discussions. For example, despite a resurgence of interest in Indigenous fire knowledge and being holders of distinct spatial legal rights, Indigenous people in south-eastern Australia, affected by the 2019–2020 fires, were left out and marginalised in subsequent policymaking forums (Williamson et al., 2020).

Another illustrative example is the fire that destroyed 50% of K'Gari (Fraser Island), where the lack of First Nations engagement prompted a recommendation from the commission of inquiry into it to build better cooperation with the Butchulla people. Programs designed to benefit biodiversity conservation and Indigenous peoples result in inequitable outcomes that marginalise Indigenous interests. In practice, moreover, rhetoric about social justice and conservation often confront each other as cultural and economic differences come into focus. There is inherent conflict in the demand to maintain cultural identities and systems, while simultaneously seeking to maintain livelihoods, and economically progress out of poverty. As Vucetich et al. (2018) note: "In some communities' poverty precludes freedom to pursue a healthy, meaningful life for a large portion of humans. This conservation-inhibiting poverty is importantly a consequence of extreme inequality in the distribution of wealth, both within and among nations". Working out how to work together remains a delicate dance. There are many reasons why this is the case, but for the rest of the chapter, I explore the idea that the driving force behind this ongoing implementation challenge is (i) the tension between knowledge systems and (ii) the way that the idea of Indigenous Country is used and abused.

## Tensions in Knowledge Management Inhibit Socially Just Conservation

Effective and socially just conservation management is necessarily underpinned by knowledge and how it is deployed. Indigenous knowledge systems and Western scientific paradigms dominate policy. While deploying Indigenous knowledge into environmental policy arenas is necessary, the challenge is how to also support Indigenous knowledge sovereignty, intellectual property and maintenance in processes of policy and decision-making (Bardsley et al., 2019; Nursey-Bray & Palmer, 2018;

Nursey-Bray et al., 2019). Indigenous peoples in Australia and elsewhere do not wish to give their knowledge away, a factor complicated by the fact Indigenous knowledge is itself under pressure and threatened not just as a result of environmental/climate change, but as a result of colonisation (Cockerill, 2018; Nursey-Bray et al., 2020; Whyte, 2020). Knowledge loss and repair/revitalisation are as important to Indigenous peoples, as is permitting its incorporation into management of problems like climate change and biodiversity loss. The nuances inherent in how knowledge is treated, maintained, used and owned are a litmus test that reflect how different peoples and cultures can work with and engage with each other.

In this context, existing scholarship has focused on how science becomes policy and the strengths and weaknesses of the process (Diver, 2017; Jasanoff, 2004; Matuk, 2020). Important work by Robinson and Raven (2017), Robinson et al. (2018) highlights the importance of recognising customary law to help prevent bio-piracy and connect State and Indigenous law in relation to Indigenous knowledge and species management. Engaging with Indigenous knowledge in the Anthropocene offers insights for stewardship (Hill et al., 2020b). Studies on knowledge *integration* articulate how Indigenous knowledge could be included at the knowledge interface (Hill et al., 2020b; Nadasdy, 1999). But knowledge integration is a fraught issue, involving conflict over knowledge, interests and values (Bohensky & Maru, 2011; Diver, 2017), often creating binaries where Indigenous knowledge and Western science are pitted against each other, and where Indigenous knowledge is not respected nor seen as equal (Evering, 2012). Traditional approaches to integration are also critiqued for ignoring the role of power relations and for prioritising certain kinds of knowledge over others, leading to knowledge products that serve science and the state over the interests of Indigenous communities (Nadasdy, 1999). Attempting to 'integrate' both paradigms creates management tension, as each knowledge system operates entirely differently (Nursey-Bray & Arabana Corporation, 2015).

Knowledge *co-production* has been one alternative mode by which some of the tensions inherent in the idea of integration have been addressed. Co-production is a process that builds knowledge in ways that transcend disciplinary boundaries and includes multiple knowledge realms, to then be jointly produced, and agreed to, the result of partnerships between societal actors, such as scientists and Indigenous peoples. Co-produced knowledge at the knowledge interface challenges entrenched norms of knowledge, including the sciences and increases its legitimacy, relevance and usability (Jasanoff, 2004). While many users still do not accept that co-produced knowledge has utility, successful co-production approaches yield insights into how context affects outputs and outcomes, identify how drivers and barriers to success differ, and challenge the hegemony of particular ways of knowing.

Knowledge *co-existence* provides a mode by which different forms of knowledge can co-exist at the knowledge interface (Nursey-Bray & Arabana Aboriginal Corporation, 2016). Howitt (2020) argues for knowledge co-existence as a means of acknowledging colonial invasion and the enabling practices of mutual recognition, collaborative building of consent, and appreciation of cultural continuity. This will also facilitate options for co-governance, as Smyth et al. (2016) discuss in their paper on Sea Country planning. Indigenous scholar Whyte (2013, 2019) argues that such relational qualities are crucial for cross-societal coordination, and that trust, strong standards of consent and genuine expectations of reciprocity will act to build conservation outcomes. Much of this thinking has merit in progressing ideas around how knowledges—of all kinds—can be used to strengthen and build socially just conservation management.

## Predominance of Simplistic Appreciation of the Idea of Country

Another core element underpinning all discourse around Indigenous models of management, and how Indigenous knowledge can be deployed is how the idea of Country is used—and abused—to progress environmental policy agendas. Indigenous Country is an all-encompassing concept: it is not just the physical geographical location for particular tribal groups, but it also is the bedrock of cultural identity, the source of knowledge and lore, and language. One's 'Country' is the place to which you belong and which you are part of. As Deborah Bird Rose puts it (7):

> Country in Aboriginal English is not only a common noun but also a proper noun. People talk about Country in the same way that they would talk about a person: they speak to Country, sing to Country, visit Country, worry about Country, feel sorry for Country, and long for Country. People say that Country knows, hears, smells, takes notice, takes care, is sorry or happy. Country is not a generalised or undifferentiated type of place, such as one might indicate with terms like 'spending a day in the country' or 'going up the country'. Rather, Country is a living entity with a yesterday,

> today and tomorrow, with a consciousness, and a will toward life. Because of this richness, Country is home, and peace; nourishment for body, mind, and spirit; heart's ease.

The hundreds of First Nations that comprise Australia are bounded by hundreds of individual Countries. The idea of Country is extraordinarily powerful. It has been the cornerstone of multiple models and funding programs, and for Indigenous Rangers and those living in remote communities, Country remains tangible, and demanding, and as Bock et al. (2021, p. 4) reflect "Country is alive, it understands [but] Living Country is not really understood or applied in mainstream policy or programmes or reflected in western science. Nor is science the only way to know or manage Country".

However, as a notion, it is often misunderstood by non-Indigenous peoples, who as they strive to be culturally understanding, appropriate the idea as a romantic and static affiliation with land and sea that equates to their own love for nature. Yet the idea of Country is underpinned by a world view that is fundamentally different from non-Indigenous peoples. Table 12.1, based on Bock et al.'s (2021) conceptualisation of these differences, shows that a simple equating of Country to love of nature is naïve.

In developing joint solutions to the problems facing us, and which are occurring at unprecedented speed, we need to engage with the essential

**Table 12.1** Different world view interpretations of Country

| *Indigenous world view* | *Non-Indigenous world view* |
|---|---|
| Spiritually Oriented | Scientific/sceptical |
| There are many truths | Focus on 'one truth', usually science |
| Everything and everyone is connected | Individual focus |
| Land is sacred | Land is available for use and exploitation |
| Time is non-linear and cyclical | Time is linear |
| Feeling comfortable is about having good relationships | Feeling comfortable is about achieving goals |
| Humans are not the most important thing in the world | Humans are the most important thing in the world |
| Wealth can be amassed for the good of the community | Wealth is amassed for individual gain (Bock et al., 2021, p. 4) |

conflict between cultural world views, and in this case, what the environment actually signifies. It would be useful to broaden Western conceptions of the environment to incorporate the world view underpinning Country without imposing what is essentially a bundle of romantic expectations.

Further, it is important to recognise that physically, Country, like everything else has changed with colonial invasion. Indigenous Countries have sometimes irrevocably changed as colonial and global forces including mining, logging, pastoralism and urban development, have left traditional land and seas scarred by bulldozers, invaded by feral animals and weeds and been used by others for their own economic advantage. Access to Country has been stymied, and some associated knowledge and languages inhibited and often lost or reduced.

Hence, while understanding and acknowledging Country as a reflection of world views *is* crucial, acknowledging the losses it has sustained is also necessary. Doing so opens the door to discussions around how to repair and revitalise Country, how to engage with Country and how to manage it.

However, many Indigenous peoples find it hard to have their voice heard in environmental governance: they are often rendered invisible. This is because many Indigenous peoples, dispossessed of their Country and forced to leave to live in urban areas, missions or to work, now identify historically with many other areas—other people's Country. Given the reality that over 80% of Indigenous peoples in Australia now live in cities, this creates a two-edged sword. Firstly, many Indigenous peoples will not 'speak' for any Country other than their own, and so multiple Indigenous peoples, living in cities, will not speak at all. For example, members of the Arabana people, whose Country is in central South Australia but who live in Darwin in the Northern Territory, will not speak about climate change impacts in the region as it is not 'their Country', yet they have lived there for generations due to being moved on from their own Country in the nineteenth century. Thus, in identifying as Indigenous, the opportunities to contribute as Australian citizens in the broadest sense are curtailed when living in urban areas or outside of one's Country.

Further, policymakers and others do not validate Indigenous people's desire to participate in decision-making about the Country they do live in, at least rarely without also critiquing it or trying to control that participation. Acknowledging 'Country' is *de rigeur* and an important starting point for meetings, and Indigenous peoples may be consulted about the language name for an urban park or an Indigenous walking trail in town. However, Country, in being deployed in these limited ways as the predominant modus for engagement between cultures, is essentially limiting in relation to all the other issues of keen interest to Indigenous peoples that are *living in cities*, this type of engagement is minimal. In being taught about the importance of Country and acknowledging it, policymakers and others—ironically—may run the risk of overlooking Indigenous peoples' perspectives—especially the majority who live in cities—and therefore not reap the benefits of engaging them on a range of other issues of concern. A related challenge is that policymakers are also wary of and do not know how to engage with Indigenous peoples and groups who live in cities but are not the traditional owners. They fear making cultural mistakes.

It is important then to reflect on how the idea of Country has changed for those living in cities—can urban spaces become types of Indigenous Country? How has the idea of Country evolved over time and amidst the sunder caused by colonisation? How are Indigenous peoples who live in cities supported to have a voice over their actual Country, even if they don't live in it? What role can they play? Again, the tensions inherent in the idea of Country and the lived reality of how Country is experienced today, has ramifications for how we construct and understand how to respond to key challenges like climate change and biodiversity loss. With the increasing likelihood that most of us will live in an Australian city by 2050, this is an issue that faces us all and warrants further exploration. Indigenous knowledge, in this context, if understood as a *process* rather than a *corpus* of facts—an evolving process of generation, reparation and revival—offers potential to inform how we all more sustainably and justly live in cities.

## Conclusion

We are living in a world beset by environmental and cultural crises. All future challenges are unprecedented across all cultures due to their intensity and fast-moving change. Hence, the ongoing demarcation of spaces and knowledge only exacerbates and embeds ongoing conflict between groups that need to be working together and wastes what precious time is left to address the impacts of climate change and biodiversity loss. We are in this together and need to unite to move forward. Yet we also need to work out how to confront these challenges in socially just ways. The insistence by Indigenous peoples that economic benefit, employment and sustenance be at the core of conservation management frameworks, is driven by the legacy of colonial invasion and contemporary socio-economic disadvantage, and acts as a mirror to the challenge the entire world faces—how to reconcile economy and environment. It also reflects the need to consider equity in decision-making and acknowledge that social injustice and hence power inequalities, inhibit the implementation of even the best management program. The additional tensions created by what is an increasing urban reality for us all also mean that 'socially just conservation' as an idea needs to be imagined as an urban as well as 'out of the city' construct, as we move towards green and sustainable cities that are functioning ecological systems.

The move towards knowledge co-existence is in the right direction and the current range of Indigenous-led conservation enterprises inspiring. Yet there remains the question of how to reconcile the reality of the concurrent presence and application of multiple knowledge systems in contested domains. What are the processes that govern how knowledge is used, and how are these processes woven into policy to progress biodiversity, adaptation and social justice outcomes? How can the inevitable conflicts that arise from the demands of social justice with conservation be reconciled? Indigenous voices still need to be structurally embedded in governance systems, whether for Country *in situ*, or for decision-making within cities. There is a need to further interrogate the impact of the nexus between Indigenous and scientific systems for policy, to generate new understandings about how to build co-managed and socially just conservation enterprises across all physical geographies, and specifically, how to build practices of knowledge repair, renewal, loss, appropriation and maintenance. The result will enable all of us to build our resilience to the challenges ahead (Berkes & Ross, 2013; Ross & Berkes, 2014).

## References

Allam, L (2021) Country needs People: Indigenous rangers safeguarding Australia's rich landscape receive a well-earned boost. *The Guardian*. https://www.theguardian.com/australia-news/2021/aug/01/country-needs-people-indigenous-rangers-safeguarding-australias-rich-landscape-receive-a-well-earned-boost. Accessed 1 March 2021.

Bardsley, D. K., Prowse, T. A. A., & Siegfriedt, C. (2019). Seeking knowledge of traditional Indigenous burning practices to inform regional bushfire management. *Local Environment, 24*(8), 727–745.

Berkes, F., & Ross, H. (2013). Community resilience: Toward an integrated approach. *Society & Natural Resources, 26*(1), 5–20.

Bock, E., Hudson, L., Isaac, J., Vernes, T., Muir, B., Whap, T., & Fell, D. (2021). Safeguarding our sacred islands: Traditional Owner-led Sea Country governance, planning and management in Australia. *Pacific Conservation Biology*. https://doi.org/10.1071/PC21013

Bohensky, E. L., & Maru, Y. (2011). Indigenous Knowledge, Science, and Resilience: What Have We Learned from a Decade of International Literature on "Integration"? *Ecology and Society, 16*(4), 6–.

Campbell, D., & Hunt, J. E. (2013). Achieving broader benefits from Indigenous land use agreements: Community development in Central Australia. *Community Development Journal, 48*(2), 197–214.

Carr, B., Fitzsimons, J., Holland, N., Berkinshaw, T., Bradby, K., Cowell, S., & Weisenberger, F. (2017). Capitalising on conservation knowledge: Using conservation action planning, healthy country planning and the open standards in Australia. *Ecological Management & Restoration, 18*(3), 176–189.

Carter, R., Atkinson, G., Burchill, M., Phillips, R., Humann, D., Mahoney, J., & Wong, N. (2022). Djaara cultural authority drives inclusion of their knowledge and culture in a Joint Management Plan for parks. *Ecological Management & Restoration, 23*(S1), 117–128.

Cockerill, J. (2018). Hauntology on Country, Overland, 22 March 2018. https://overland.org.au/2018/03/hauntologyon-country/

Diver, S. (2017). Negotiating Indigenous knowledge at the science-policy interface: Insights from the Xáxli'p Community Forest. *Environmental Science & Policy, 73*(C), 1–11.

Doohan, K., Langton, M., & Mazel, O. (2012). *From paternalism to partnership: The good neighbour agreement and the argyle diamond mine indigenous land use agreement in Western Australia* (pp. 231–250). Routledge.

Evering, B. (2012). Relationships between knowledge(s): implications for "knowledge integration". *Journal of Environmental Studies and Sciences*, 2(4), 357–368.

Garnett, S. T., Burgess, N. D., Fa, J. E., et al. (2018). A spatial overview of the global importance of Indigenous lands for conservation. *Nature Sustainability, 1*, 369–374.

Godden, L., & Cowell, S. (2016). Conservation planning and Indigenous governance in Australia's Indigenous Protected Areas: Conservation planning in Indigenous Protected Areas. *Restoration Ecology, 24*(6), 692–697.

Harvey, C. (2019). Charting knowledge co-production pathways in climate and development. *Environmental Policy and Governance, 29*(2), 107–117.

Havemann, P., Thiriet, D., Marsh, H., & Jones, C. (2005). Traditional use of marine resources agreements and dugong hunting in the Great Barrier Reef World Heritage Area. *Environmental and Planning Law Journal, 22*(4), 258–280.

Hill, R. (2011). Towards equity in indigenous co-management of protected areas: Cultural planning by Miriuwung-Gajerrong people in the Kimberley, Western Australia. *Geographical Research, 49*(1), 72–85.

Hill, R., et al. (2020a). Knowledge co-production for Indigenous adaptation pathways: Transform post-colonial articulation complexes to empower local decision-making. *Global Environmental Change, 65*, 102161.

Hill, R., et al. (2020b). Working with Indigenous, local and scientific knowledge in assessments of nature and nature's linkages with people. *Current Opinion in Environmental Sustainability, 43*, 8–20.

Howitt, R. (2020). Unsettling the taken (for granted). *Progress in Human Geography, 44*(2), 193–215.

Jasanoff, S. (2004). *States of knowledge: The co-production of science and the social order, states of knowledge*. Routledge.

Johnson, J., & Murton, B. (2007). Re/placing native science: Indigenous voices in contemporary constructions of nature. *Geographical Research, 45*(2), 121–129.

Matuk, T. (2020). Allying knowledge integration and co-production for knowledge legitimacy and usability: The Amazonian SISA policy and the Kaxinawá Indigenous people case. *Environmental Science & Policy, 112*, 1–9.

Moorcroft, H., Ignjic, E., Cowell, S., Goonack, J., Mangolomara, S., Oobagooma, J., & Waina, N. (2012). Conservation planning in a cross-cultural context: The Wunambal Gaambera healthy country project in the Kimberley, Western Australia. *Ecological Management & Restoration, 13*(1), 16–25.

Nadasdy, P. (1999). The politics of Tek: Power and the "integration" of knowledge. *Arctic Anthropology, 36*(1/2), 1–18.

Nursey-Bray, M., & Arabana Aboriginal Corporation (2015). Cultural indicators, country and culture: the Arabana, change and water. *The Rangeland Journal*, 37(6), 555–569.

Nursey-Bray, M., & Corporation, A. (2015). Cultural indicators, country and culture: The Arabana, change and water. *The Rangeland Journal, 37*(6), 555–569.

Nursey-Bray, M., & Palmer, R. (2018). Country, climate change adaptation and colonisation: Insights from an Indigenous adaptation planning process, Australia. *Heliyon, 4*(3), 1–12.

Nursey-Bray, M., & Rist, P. (2009). Co-management and protected area management: Achieving effective management of a contested site, lessons from the Great Barrier Reef World Heritage Area (GBRWHA). *Marine Policy, 33*(1), 118–212.

Nursey-Bray, M., Marsh, H., & Ross, H. (2010). Exploring discourses in environmental decision-making: An Indigenous hunting case study. *Society & Natural Resources, 23*(4), 366–382.

Nursey-Bray, M., Palmer, R., Smith, T., & Rist, P. (2019). Old ways for new days: Australian Indigenous peoples and climate change. *Local Environment, 24*(5), 473–486.

Nursey-Bray, M., Palmer, R., Stuart, A., Arbon, V., & Rigney, L. E. (2020). Scale, colonization and adapting to climate change: Insights from the Arabana people. *Geoforum, 114*, 138–150.

Office of the Inspector-General Emergency Management. (2021). *K'gari (Fraser Island) Bushfire Review*. Queensland Government, Brisbane.

Rist, P., Rassip, W., Yunupingu, D., Wearne, J., Gould, J., Dulfer-Hyams, M., Bock, E., & Smyth, D. (2019). Indigenous Protected Areas in Sea Country: Indigenous-driven collaborative marine protected areas in Australia. *Aquatic Conservation Marine Freshwater Ecosystems, 29*, 138–151.

Robinson, D., & Raven, M. (2017). Identifying and Preventing Biopiracy in Australia: Patent landscapes and legal geographies for plants with Indigenous Australian uses. *Australian Geographer*, *48*(3), 311–331.

Robinson, D., Raven, M., Tari, T., Hickey, F., & Kalfatak, D. (2018). Kava, kastom and indigenous knowledge: Next steps under teh Nagoya Protocol. *Journal of South Pacific Law, 2018*, 77–93.

Ross, H., & Berkes, F. (2014). Research approaches for understanding, enhancing, and monitoring community resilience. *Society & Natural Resources, 27*(8), 787–804.

Ross, H., Grant, C., Robinson, C., Izurieta, A., Smyth, D., & Rist, P. (2009). Co-management and Indigenous Protected Areas in Australia: Achievements and ways forward. *Australasian Journal of Environmental Management, 16*(4), 242–252.

Smyth, D., Gould, J., Ayre, M., Bock, E., & Dulfer-Hyams, M. (2016). Indigenous-driven co-governance of sea country through collaborative planning and Indigenous Protected Areas. *Indigenous Law Bulletin, 8*, 15–20.

Vucetich, J. A., Burnham, D., Macdonald, E. A., Bruskotter, J. T., Marchini, S., Zimmermann, A., & Macdonald, D. W. (2018). Just conservation: What is it and should we pursue it? *Biological Conservation, 221*, 23–33.

Whyte, K. (2013). On the role of traditional ecological knowledge as a collaborative concept: A philosophical study. *Ecological Processes, 2*(1), 1–12.

Whyte, K. (2019). Too late for indigenous climate justice: Ecological and relational tipping points. *Wiley Interdisciplinary Reviews Climate Change, 11*, 1.

Whyte, K. (2020). Too late for indigenous climate justice: Ecological and relational tipping points. *Wiley Interdisciplinary Reviews: Climate Change, 11*(1), e603–n/a.

Williamson, B., Markham, F., & Weir, J. (2020). Aboriginal peoples and the response to the 2019–2020 bushfires, Working Paper No. 134/2020. Centre for Aboriginal Economic Policy Research, Australian National University. https://doi.org/10.25911/5e7882623186c

Wright, A., Yap, M., Jones, R., Richardson, A., Davis, V., & Lovett, R. (2021). Examining the associations between indigenous rangers, culture and wellbeing in Australia, 2018–2020. *International Journal of Environmental Research and Public Health, 18*(6), 3053. https://doi.org/10.3390/ijerph18063053

Zurba, M. (2009). Bringing local synthesis into governance and management systems: The Girringun TUMRA case in Northern Queensland, Australia. *Journal of the Royal Society of New Zealand, 39*(4), 179–182.

Zurba, M., Ross, H., Izurieta, A., Rist, P., Bock, E., & Berkes, F. (2012). Building co-management as a process: Problem solving through partnerships in Aboriginal Country, Australia. *Environmental Management (new York), 49*(6), 1130–1142.

CHAPTER 13

# Women's Empowerment in a Community-Based Tourism Project: A Case Study from the 'Spice Islands' of North Maluku, Indonesia

*Roswita M. Aboe, Harriot Beazley, Price Livingstone, Michelle Prasad, William Ramsay, Kris Syamsudin, and R. W. Carter*

R. M. Aboe
Khairun University, Ternate, Indonesia

H. Beazley (✉) · P. Livingstone · M. Prasad · K. Syamsudin
School of Law and Society, University of the Sunshine Coast, Maroochydore, QLD, Australia
e-mail: hbeazley@usc.edu.au

W. Ramsay
Centre for Transformative Innovation, Swinburne University of Technology, Hawthorn, VIC, Australia



C. Baldwin and S. van Bommel (eds.), *Rural Development for Sustainable Social-ecological Systems*,
Palgrave Studies in Environmental Sustainability,
https://doi.org/10.1007/978-3-031-34225-7_13

## Introduction

Ternate proverb:

> Let us be tolerant like nutmeg and mace,
> thrive together and fall together,
> based on love and affection. (Prasad & Syamsudin, 2022, p. 34)

Community-based tourism (CBT) is often perceived as an appropriate model for improving the well-being of communities in remote regions of developing countries. Within these discussions, natural, cultural and livelihood assets are seen as tourist attractors, with communities only needing to develop commercial products from these assets to achieve tourism success. As with all small businesses, however, most CBTs fail within five years because of internal and external factors; some peculiar to tourism and some inherent in many CBT initiatives. Those that remain viable are frequently accompanied by asset exploitation and unwanted cultural and societal change, or they necessitate ongoing external financial and technical support.

This chapter focuses on the remarkable and sustained success of one CBT initiative in Ternate, Indonesia: the *Cengkeh Afo and Gamalama Spices* (CAGS) project. The community-based project is the result of a serendipitous convergence of the community's desire to engage with ecotourism, their unique global identity created by the sixteenth-century colonial trade with the 'Spice Islands', and their highly sought-after assets in the form of endemic and internationally valued tree species (nutmeg, clove and mace). The CAGS community tourism project is a gastronomic experience, owned and operated by the village women. The commercial product is locally produced cuisine cooked by local women with the famous spices from the spice forest, utilising traditional cooking processes. The food is consumed in a natural setting, served by village youth on traditional plates and cutlery, made with locally grown bamboo. The men

R. W. Carter
Sustainability Research Centre, University of the Sunshine Coast, Maroochydore, QLD, Australia
e-mail: bcarter@usc.edu.au

and youth also offer tourists ecotours through the ancient spice forest plantation, in the foothills of the active volcano, Mt Gamalama.

In this chapter, we describe how participatory workshops and interviews with village stakeholders and external supporters of the CAGS project revealed enhanced cultural pride of those involved in the project, as well as a rise in status of women within the village. The case study also recorded considerable economic benefits to those involved in the project, and to the wider Ternate community.

## Community-Based Tourism

In the past 60 years, development practice has grown significantly in scope and context, from an economic growth perspective to one that seeks a balance with cultural, social and environmental outcomes (Sen, 1999; Radcliffe & Laurie, 2006). This contemporary approach has resulted in community-led projects focused on rights based participatory methodologies, where it is accepted that best development practice stems from facilitating communities to help themselves (Chambers, 1994a, 2006; Westoby & Shevellar, 2018). Community development has been celebrated as a successful way to address inequality through the stimulation of sustainable development initiatives within low socio-economic populations (Ansari et al., 2012; Westoby & Shevellar, 2018).

As community development has gained traction in development practice, CBT has also emerged as an alternative form of tourism, where tourism products are created by small-scale businesses, with interactions between visitors and the host community (Pearce, 1992). Initially aligning with ecotourism, the concept has broadened to include an array of tourism products based on environmental and cultural settings, and the intangible cultural heritage (ICT) of the host community (Khanom, et al., 2019; Kim et al., 2019). ICT includes oral traditions, performing arts, dance, rituals, foods and cuisine, which are regarded by communities as being part of their cultural heritage (Kim et al., 2019; UNESCO, 2003). By 2016, ASEAN (2016, p. 2) had defined CBT as:

> (T)ourism activity, community owned and operated, and managed or coordinated at the community level that contributes to community the well-being of communities through supporting sustainable livelihoods and protecting valued socio-cultural traditions and natural and cultural heritage resources.

Tourist experiences of CBT initiatives range from homestays, festivals, guided food or nature tours and adventure activities, where cultural assets, resources and local heritage are showcased to tourists (Acharya & Halpenny, 2013; Giampiccoli & Kalis, 2012). They are characterised by four main principles: local community empowerment and ownership, conservation of tourism resources, community social and economic growth, and quality visitor experiences (Hiwasaki, 2006). Under a CBT model, profits are shared among community members and used to further preserve cultural and natural assets (Karacaoğlu & Birdir, 2017). Importantly, and in line with community development principles, CBT encourages a bottom-up approach to development. Decision-making is grounded in consensus building and the equitable sharing of the benefits.

Inherent in the establishment and operation of a CBT initiative, therefore, is investment by the community, as well as external support and guidance. Social capital is essential (see Kline et al., 2019; Zielinski et al., 2020; Musavengane & Kloppers, 2020) to ensure collaboration and the bringing together of human capital in terms of knowledge and skills needed to develop a tourism product. The 5-capitals framework (social, human, financial, constructed (produced) and natural) for community development projects (see Sohnrey & Hegemann, 2021) has proven to be useful in identifying needed inputs by both the community and external investors. Similar to social enterprises, revenue created by the product is reinvested back into the community to reinforce and build human, social and cultural capital (Ansari et al., 2012). Such reinvestment strengthens and incentivises the practice and celebration of cultural heritage and identity through reinforcement of traditional cultural activities. In this way, the CBT has been recognised as important for maintaining intangible cultural heritage, particularly within rural communities, and for its potential to increase a community's sense of cultural and social value. The CBT process also has the potential to empower marginalised members in a community, including women.

## Women's Empowerment Through CBT

The empowerment of women through sustainable community development is increasingly recognised as a way to address global poverty and to empower poor communities (Arintoko et al., 2020). Further, the United National World Tourism Organization (UNWTO, 2020, p. 20) has stated that tourism has 'the potential to empower women in developing regions', particularly through community-based tourism (McCall & Mearns, 2021). Recent research focused on the impact of CBTs on women in South Africa demonstrated how such initiatives can play a critical role in reaching the UN 2030 Agenda for Sustainable Development, especially Sustainable Development Goal (SDG) 5, associated with gender equality and women's empowerment (McCall & Mearns, 2021, p. 157). Other relevant studies in Cambodia and Indonesia have shown how community-based tourism has brought socio-economic benefits to local women by empowering them economically, socially, psychologically and politically (Artinoko et al., 2020; Nara & Irawan, 2020). The empowerment of women through community-based tourism also impacts the community as a whole as they strengthen the communities in which they live (McCall & Mearns, 2021). The positive experiences created through CBT activities can therefore result in the creation of new roles for women within a community, thus enhancing relationships between individuals and groups. Accordingly, although the goal of CBT is focused on increasing financial capital, the process also builds human and social capital which assist in alleviating poverty through diversified and more resilient livelihoods.

## Indonesian Context

With more than 17,000 islands in Indonesia, there are numerous opportunities for sustainable tourism growth. Indonesia's continuous economic growth over the past two decades has seen it move to a middle-low socio-economic nation, with a significant rise in both international and domestic tourists. Prior to the COVID-19 pandemic, the Indonesian tourism industry was increasing rapidly, with Windarti (2016) arguing that the growth had produced an important boost to the economy, while simultaneously supporting environmental preservation and benefiting the social sphere through job creation. However, others have observed how the tourism boom resulted in unregulated tourism development, causing an increase in crime, and significant cultural, social and environmental degradation in some parts of the archipelago (Andriansyah et al., 2019).

The Indonesian government has made considerable efforts in recent years to ensure that the nation develops into a sustainable tourism destination that spurs widespread regional growth and enhances the well-being of all Indonesians (OECD, 2020). In remote Indonesian islands, communities are seeking to enter the tourism market but often lack the necessary skills, knowledge, business experience and investment capital. A rich Indonesian history and culture offers a unique tourism product if it can be marketed, packaged and delivered appropriately. The mobilisation of community assets could be a crucial strategy in addressing income inequality, cultural awareness and increase livelihood diversification (Arintoko et al., 2020; Kodir, 2018).

### *Ternate, North Maluku*

Ternate Island is part of the Maluku Islands (the Moluccas), in the Molucca Sea in Eastern Indonesia (Fig. 13.1). At the centre of Ternate Island is the 1715 m high volcano Mt Gamalama. The island's rich volcanic soils and warm, moist climate have resulted in sequential colonisation by plants, exemplified by nutmeg (*Myristica fragrans*) and clove (*Syzygium aromaticum*). Commonly referred to as the "Spice Islands", the two provinces of North Maluku and Maluku were at the heart of the colonial spice trade from the sixteenth century to the early eighteenth century, due to their exclusive resources of cloves, mace and nutmeg (Krohn, 1993).

For more than two and a half centuries, the wealth generated by the spice trade to the Middle East and Europe enabled Ternate's sultans to expand their influence in the region. In the mid-1400s, the spice (and silk) flow to Europe was blocked by the Turkish Ottoman Empire. This precipitated the sixteenth-century era of European Colonialism, driven by the enormous profits to be gained through exploitation of the Spice Islands. Throughout the 1500s, successive Sultans negotiated uneasy alliances with the Portuguese, Spanish, Dutch and British, enabling continued colonial occupation of Ternate. While the spices continued to flow to Europe, colonial control was punctuated by fierce battles between the Portuguese, Spanish, Dutch, and British, often supported or led by the sultans of either Ternate or their rival Tidore, initiating a 200-year monopoly over the spice trade by the VOC (*Vereenigde Oostindische Compagnie*: Dutch East India Company) and establishment of clove plantations on the foothills of Mt Gamalama. The Spanish co-existed on Ternate with the VOC for over 50 years, eventually withdrawing in the 1660s (Fig. 13.2). The power void was filled by the British East India Company, resulting in more forts being built throughout the archipelago. The bloody battle for control of the islands ended in 1801, when the

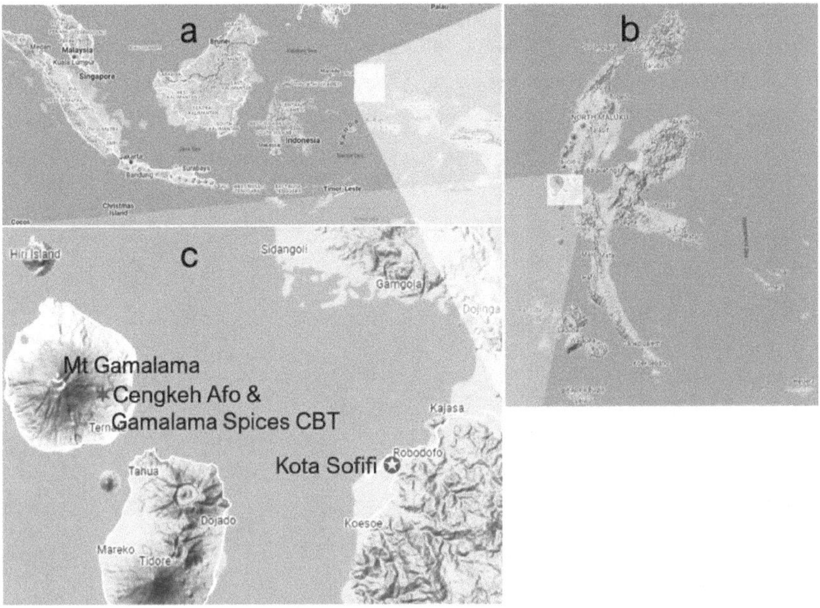


**Fig. 13.1** Location of the Cengkeh Afo & Gamalama Spices (CAGS) CBT. a. North Maluku Province in Indonesia; b. Ternate Island and town in North Maluku Province; c. CAGS on slopes of Mt Gamalama, Ternate (from Google Maps)

British violently took Ternate from the Dutch. At the same time, plantations were established in Mauritius, Penang and Zanzibar from clove and nutmeg seedlings smuggled out of the Moluccas. Subsequently, Ternate was no longer the centre of the spice trade and by the mid-nineteenth century, the Moluccas had become a backwater for European traders.

In 1858, during his collecting expedition to the 'Malay Archipelago', British naturalist Alfred Russel Wallace wrote to Charles Darwin from Ternate to share his theoretical reflections in what has become known as the 'Ternate Essay'. Wallace concluded that species diversity was the result of survival of "the best fitted". The essay prompted Darwin to publish his and Wallace's ideas about natural selection and evolution jointly in London in 1858. In 1869, Wallace's book *The Malay Archipelago* was published, with a chapter on Ternate. Today, the Dutch clove plantations still remain on the slopes of Mt Gamalama. The cloves are used in

**Fig. 13.2** Fort Oranje, part of Ternate's colonial and spice cultural heritage. The fort was once the centre for the Dutch trade in spices from the Maluku islands

the distinctive, local fragrant cuisine; a fusion of traditional and colonial influences, and a powerful symbol of Ternate's cultural heritage.

## Case Study Location

The focus of the remainder of this chapter is an exploration of the factors that led to the success of a predominantly female-led CBT initiative in Ternate, North Maluku, Indonesia: the *Cengkeh Afo & Gamalama Spices* (CAGS) project. Positioned on the eastern slope of Mount Gamalama, CAGS is owned and managed by a small community, adjacent to the Dutch clove plantations, which include some of the world's oldest clove trees. As described, the Ternate community has a rich cultural tradition, influenced by their environmental setting and traditional livelihoods, and expressed in their cuisine, music and dance. These attributes have led to the creation of a CBT project, designed through the empowerment of the community to produce an authentic dining experience, relying solely on the tourism assets of local produce and labour.

## Cengkeh Afo & Gamalama Spice (CAGS) Project

The *Cengkeh Afo & Gamalama Spice (*CAGS) project was conceptualised by Professor Bill Carter, David Shirley and Kris Syamsudin in Ternate as part of an Australia Awards Short Course. The concept was supported through crowdfunding with no government funding involved, which exemplifies the community's extraordinary strength and perseverance. The name *Cengkeh Afo* refers to the world's oldest clove tree; *Gamalama* is the name of the active volcanic mountain, where the village and restaurant are located. 'Gamalama Spice' refers to the abundance of spices that can be found in the forest around the village and volcano.

Before the CAGS project, the spice islands were only ever written about, or presented to tourists, from a historical perspective. No one had considered how indigenous people in Ternate continue to use local herbs and spices in their everyday lives for cooking and medicinal purposes. The combination of ingredients used by the *Cengkeh Afo* community is exclusive to the village and this uniqueness is what has inspired community commitment to developing the CBT project. It was understood that such an initiative would preserve local traditions and knowledge about the spices that grow in the foothills of Mt Gamalama, and help to educate visitors and the younger generation about the importance of the traditional food and cultural heritage (Fig. 13.3).

The CAGS project invites tourists to experience elements of the traditional way of life of the North Maluku people. The project's tourism activities emphasise an ecological-cultural experience for visitors. The main attraction is the exotic food that is prepared and cooked using traditional techniques with bamboo. The spices and ingredients are locally sourced from the forest, with many local community members working closely with the restaurant. The exquisite taste of the food is said to come from the fact that the dishes are cooked in the hollow stems of bamboo, referred to as batik bamboo (*Bambu tutul)* (Fig. 13.4).

As a community-based initiative, the CAGS project ensures that profits go back into preserving the community's traditional culture and customary practices. The community has carefully created a system that benefits every community member, including women, youth, the elderly

**Fig. 13.3** CAGS community members shelling nutmeg. Traditional methods of harvesting and preparing spices have been passed down for generations

and the disabled. Earnings are distributed equitably among community members, and the tourism activities are primarily overseen by women. The CAGS restaurant is managed and operated by twelve local women who offer a culinary-cultural experience by cooking dishes in the bamboo tubes which are then carried from the kitchen in traditional woven bags (*Saloi*) (Fig. 13.5), and served to guests in handcrafted bamboo huts (*Sabua*) under the rainforest canopy. The youth from the village work as waitresses, waiters and guides for tours of the forest. The men offer visitors cultural performances of their ancestral heritage, including traditional music, dance and *moro-moro* poetry. They also sell local spices and provide spice education programmes and ecotours for visitors.

**Fig. 13.4** The CAGS restaurant's traditional cuisine cooked in bamboo tubes

## Methods

This CBT case study was an international collaboration between staff and students of Khairun University in Ternate, and staff and Masters of International Development students from the University of the Sunshine Coast in Australia. To identify the factors of success for a sustainable CBT project, a deep understanding of the community's power dynamics, market feasibility, community-specific cultural identities and gender roles is required (Ashley & Garland, 1994; Reed, 1997). Many studies of CBT use participatory rural appraisal (PRA) as a form of analysis to critically engage with CBT projects (Chambers, 1994a, 2006; Nair & Hamzah, 2015). As a result, a combination of PRA methods, focus group discussions and one-on-one interviews were conducted to gather social, cultural and economic data about the creation and impact of the CBT in the village. These data were used to evaluate the reasons for the success of the CBT project. Participants were assured that any information shared would be anonymous, and any personal details were confidential. At the beginning of each activity, it was stressed that all activities were for the sole purpose of evaluating the success of CAGS.

**Fig. 13.5** Women from the culinary team carrying bamboo tubes from the kitchen in traditional woven bags (Saloi)

### *Participatory Rural Appraisal (PRA)*

Participatory Rural Appraisal (PRA) techniques follow a bottom-up approach to inquiry where participants are actively involved in the data collection, discussion and evaluation of the data. PRA activities therefore enable information to be shared and owned by local people and gathered in a more relaxed research context (Chambers, 1994a). The inherent shift in power dynamics makes PRA techniques more conducive to forming mutual respect between outsider and local populations, which improves accuracy and reliability of the information shared (Chambers, 1994a). PRA is particularly useful in understanding the status of different capitals that can be brought to community development, including community-based tourism, while allowing for objective measurement. Through the eyes of the community, the conversion of capitals goes through social and cultural screens that can pre-empt what is possible for community investment in a CBT initiative, also filtered by community aspirations and expectations. This involves being relaxed and not rushing, hearing rather than informing, clarifying and exploring instead of moving on to another topic of theoretical interest, and seeking whole-of-community perspectives by engaging with marginalised and disempowered groups that may include women, youth, the less able and the aged (Chambers, 1994b).

The PRA approach to identifying and quantifying social and cultural phenomena within a community was essential for identifying the success indicators of a CBT project, as well as the capitals applied in developing and operating the initiative. The methods also facilitate reflection of process and outcomes and what investments might be adjusted to ensure ongoing success. PRA allows both the facilitator and local people to explore areas of concern and potential growth, and to empower decision-making through the process of data collection. It is within this framework that CBT can both be appropriately implemented and sustained to achieve social, cultural and economic growth (Nair & Hamzah, 2015).

PRA methods used to identify the variables that have influenced the success at CAGS included: transect walks, participatory mapping and timelines, focus groups and one-on-one interviews (Chambers, 1994a, 1994b). Each participatory activity included a small group of men or women from the community to identify the gendered dimensions of the CAGS project. The data generated were then used to identify emergent consensus themes regarding success factors and impacts.

Responses from PRA workshops, focus group discussions and interviews were compiled and subjected to thematic analysis to identify patterns or themes emerging from the qualitative data (see Clarke & Braun, 2017). While the tourism literature on CBT success factors and benefits provided a template for the probing questions, for analysis we chose not to categorise responses to the template, but rather let themes emerge from the data to identify the personal and social meaning attributed to the CBT initiative (see Clarke & Braun, 2017). Within the emergent themes, key variables were identified that were considered by community members to be influential in achieving positive outcomes for the community and contributing to the success of CAGS. The main themes identified were: a reinforcement of cultural identity, increased education and awareness through engaging visitors, increased economic activity, a shift in gender identities and roles and the identification of key CAGS champions.

## Results

Since the inception of the CAGS project in 2017–2018, there has been a positive shift in the community's appreciation of their own cultural identity. This was evidenced by participants reporting how important it was for the community to be promoting their traditional Indonesian cuisine, music, dance and spice forests.

> The people of Ternate are passionate about their unique history and culture. We believe Cengkeh Afo works in harmony with Ternate's character by exposing the art and culture that exists in the community…The goal is to teach the indigenous culture that has existed since the Sultans' era so that it may be resurrected. To have our culture hold society together and to ensure that our culture will be renewed by our elders. Culture plays a significant role in society. (Cengkeh Afo Village Leader, Jauhar A. Mahmud)

> In addition to visiting the spice forest, we encourage tourists to hear our traditional music, visit the former Dutch plantations…learn how to peel cinnamon tree skin, harvest nutmeg, propagate vanilla, visit the villager's gardens, see over 14 species of bamboo, meet the local community, as well as visit the CAGS restaurant at Cengkeh Afo. (CAGS member)

Participants also emphasised a sense of pride and a willingness to conserve traditional skills and practices that they believed were slowly disappearing. All identified the importance of independence and ownership of the project as a motivation for the community. Even with limited government support and intervention, they believed they had the capacity to create something both feasible and sustainable.

> We want to educate people through eco-tours and our restaurant, to teach people how to evolve and sustain life, the North Maluku way. When tourists come to the Cengkeh Afo & Gamalama Spice Project they can see many things, like culture, people and the unity of people. Cooperation is our foundation. When we don't have money, we have cooperation. (Cengkeh Afo Village Leader, Jauhar A. Mahmud)

The community also had the view that government involvement was only necessary in supporting roles, such as providing public infrastructure (e.g. roads to their village) and skills training:

> It is important to have skills for a project like this. I have financial skills, I know how to manage money, how to divide salary and I have administration skills. I would to like to learn how to modify the current menu to make the food even more interesting. I don't know what training is needed. I know the government could provide this but unsure on the name of the department. (Norma Badu, Head Chef)

The respondents emphasised how communication, community spirit, cooperation and happiness had increased and strengthened community relationships. For example, respondents explained how families had shifted their focus from day-to-day tasks just for the family, to shared communal activities to support the CAGS initiative. This built a sense of shared responsibility for both family and community tasks, resulting in families working together to support and progress the project.

> Both men and women have been important. Men fix things and women talk with each other. The women involved with CAGS want to contribute to the project. Other women in the village that are not involved with CAGS are happy with gardening and taking care of their house and children. CAGS attracts older women because they can have a work-life balance. They do not have the energy to farm anymore, so they are happy

to cook for CAGS in the day and go home to complete housework and rest in the afternoons. (Norma Badu, Head Chef)

This community spirit impacted on the community's positive mental well-being and social cohesion, transforming the community into one large family.

It is my goal that people will be able to learn about the CAGS's community. Participation of all members of the community is our priority and in every single development, the community works together. (Cengkeh Afo Village Leader, Jauhar A. Mahmud)

For example, the project engages local farmers who provide vegetables to the restaurant. The older women and youth collect spices from the forest, and children and youth visit the local library for a weekly reading group, with books supplied by visitors to the village (Fig. 13.6).

Men from the local community generate an income through their authentic craftsmanship of bamboo furniture and traditional huts, which

**Fig. 13.6** The CAGS project built a library where weekly reading groups are offered to children in the community

the customers sit in when they visit the CAGS restaurant (Fig. 13.7). The CAGS team also participated in community service activities, such as rebuilding infrastructure after flooding, sweeping leaves off the streets, and collecting plastic bottles for recycling:

> Cans and plastic bottles are recycled, and the money raised goes to the community fund, which helps the local community during times of necessity. (Nilam Sari, CAGS member)

In this way, the sharing of responsibilities motivated the community to work together towards progressing sustainable community development. This was summarised as a metaphor by one respondent who described the community as a broomstick: "if there is only one stick you cannot sweep the ground, but with many sticks you can".

## Increased Education/Awareness Through Engaging Visitors

The theme of co-learning and enhanced community pride through engagement with international and domestic visitors was particularly evident during the participatory activities. It was clear from discussions that the community recognised the benefits that flow when visitors from a range of cultures visit the project site. Respondents described how visitors often spend time engaging with the community; transferring knowledge that benefits both the visitor and community. The respondents described how they had adapted their practices as a result of ideas generated from the exchanges with outsiders. For example, the practice of keeping the project site waste and plastic free was adopted when the Balinese NGO 'Bye Bye Plastic' came to discuss the benefits. As a result, the project uses renewable resources and plant-based products for much of the restaurant's daily activities. As the head chef of CAGS explained:

> When the 'Bye Bye Plastic Bag' ambassador came to visit; I was able to understand more about plastic waste. Since meeting with the NGO, the project has pledged to become a zero plastic initiative. (Norma Badu, CAGS Head Chef)

Respondents also explained how visitors gave the community a sense of pride because it made them realise that they have something very special that is respected by others. They also described how they were learning different languages, customs and behaviours from interaction

**Fig. 13.7** CAGS men with Professor Bill Carter (centre) contributing their building skills at the initiation of the project

with outsiders, and how adapting to unfamiliar cultures contributes to project success.

> I teach English to the community's women and children. They are always eager to learn and are excited about the possibility of being able to communicate with tourists in English (Halida Nuria Ma'rus, English Lecturer from Khairun University).

The respondents expressed pride in catering for and entertaining large gatherings, including Indonesian government officers, executives from Indonesian banks, foreign diplomats from Jakarta and Indonesian celebrities including Miss Indonesia. They were also very proud that they had been invited to Jakarta and Bali to demonstrate their spice cuisine, including to the Bali Food Festival in 2019. In 2020, they were invited to represent Indonesia at the *Village International de la Gastronomie,* a food festival in Paris, although the trip had to be cancelled due to the COVID-19 pandemic.

## Increased Economic Activity

All CAGS members indicated an increase in cash flow for families involved in the project and how this was immensely beneficial. Subsistence farmers, who usually sold surplus at the local markets, explained how this was now a supplementary income. For some, the income from the project more than met the basic needs of the family.

> The project has impacted our families immensely. We can have more pocket money for school supplies and snacks, our parents are now employed and have additional income to support us. As we see it, there have been no negative effects, only good things. (Young person, CAGS)

> I have noticed the increase in guests coming to Cengkeh Afo. I have also noticed that my income has increased. (Ibu Nilum, CAGS member)

> When I save my income from CAGS, I am able to afford to look after my family sufficiently. (Fatima Ibrahim, CAGS member)

> I harvest Cassava in September, November, December and sell it to the restaurant. As a result, I can earn a living to support my family. (Rakina Mahmud, CAGS member)

The financial benefits were often proudly mentioned in the context of no longer needing local government assistance (e.g. rice rations), and how community ownership and independence meant that income stayed within the community.

> We are not a commercialised tourism enterprise; we believe in sharing our earnings with the CAGS community and with local student tour guides. (Ratna Robo, CAGS member)

Some respondents revealed entrepreneurial aspirations of opening stores in town to sell produce associated with the project, such as bamboo furniture and cutlery.

> If the bamboo is still fresh, it must be dried out before using it for building purposes. The beautiful batik bamboo design is visible after it dries. Green bamboo in particular is used for cooking and straws and ripe bamboo is used to build furniture, furnishings, hold up structures and for souvenirs. (Asman Ali (bamboo craftsman), CAGS member)

The respondents observed that as the number of visitors grew, children of the families who worked on-site were able to gain employment. The younger children managed parking, collecting a small fee, while youth who had completed senior high school provided a motorbike taxi service (*ojek*), collecting visitors from town and bringing them to CAGS. The young respondents identified the benefits of gaining practical skills, including learning English and Japanese from tourists, but also the capacity to save money for the option of going to college.

> I would like to increase our income so we have enough money to improve infrastructure and don't have to wait for donations. I want to increase my income through nutmeg and clove. Cengkeh Afo provides enough for me to pay for daily needs. I have 6 kids with 5 kids in school. I have been able to afford schooling with CAGS income alone. (Norma Badu, Head Chef)

> I would like the kids to contribute to CAGS though learning different languages to communicate with tourists. I want them to choose what they do in the future so they are not obliged to take over roles at CAGS. I hope my kids can add to the world through the knowledge they have learnt through school. (Norma Badu, Head Chef)

## Women's Empowerment

Many respondents emphasised how the traditional gender roles of women involved in the project had transformed as a result of their work with CAGS. This had a positive effect on the mental well-being of the women, and the success of the CBT project. Many female respondents described how, prior to the project, they relied heavily on their husbands for support, whereas now they had greater independence, and the capacity to work alongside their husbands in complementary positions within the CBT project.

One female respondent expressed the importance of mothers—as well as fathers—being able to meet their children's physical needs, which they felt had more of a positive impact on the mental health and well-being of each family. Many female respondents also stated that they were more active, and that this positively affected their mental well-being.

The women explained how they were able to plan for a different future and to discuss ideas, as a family, rather than simply planning for their family's day-to-day needs. All female respondents expressed a greater sense of happiness, due to their roles within CAGS. Female respondents described how when they went into town people recognised them and often called out in recognition, due to the popularity and success of the CBT project. They expressed how this enhanced their sense of pride in what they were doing. In this respect, participants acknowledged the changing roles for women in the community, which motivated the younger female generation to engage with traditional ecological and cultural knowledge, and to develop skills to work in the CBT project. A small number of young women helped to serve food and drinks to visitors, and learnt traditional recipes and cooking processes, to ensure the protection of traditional cultural knowledge and heritage:

> Our parents have taught us so much about Cengkeh Afo and its history, and Ternate too. One thing is for certain: our community will continue to protect Cengkeh Afo in the future. This is because it is built into the consciousness of our people to conserve Cengkeh Afo and its history. Our culture is deep-rooted in the natural world, and we are very conscious of that. (Young female participant, CAGS member)

It has been documented in other women-empowering community development initiatives how men have sometimes felt marginalised or unhappy with the stronger position their wives have assumed in a community, as a result of a successful CBT project. It was clear in the case of the CAGS project, however, that there was minimal gender conflict as a result of the CBT initiative, and that the positive effects on the women also benefited the men. To clarify, traditional male dominance in village decision-making was still evident, but the project embraced women's real and possible contribution, which in the past had only been related to their harvesting and preparation of cloves and nutmeg for sale. The perceived value for this activity had been relatively low and was only seen as a contribution to family and village return. However, the creation of the CAGS restaurant with its clear economic benefits, social networks and national recognition was seen as a significant contribution to the community. The project provided a reliable income during traditionally difficult times, before the clove and nutmeg and harvesting season. The male participants openly acknowledge the economic benefits of the CAGS project for family support, as well as cultural enrichment and reinforcement. It appeared that the men were proud of what the women have achieved beyond their traditional roles of family raising and household management. They also felt proud that the women were networking outside of the community, and that the project had given them new social skills, and skills in hospitality, which they said had made them more "productive and healthy" (Pak Apin, CAGS construction worker).

The male respondents also expressed how they felt included in the initiation of the CAGS project and in the ongoing decision-making. The project appeared to have evoked male pride in what the women had achieved. They conveyed their joy in seeing their wives and family happier with the change in their responsibilities and work, which created positive outcomes in the household and the community. The male participants also expressed how pleased they were that while the women were making food in the kitchen, and serving at table, they could provide entertainment (playing traditional music) for the guests, which they enjoyed doing. The men and male youth also had the opportunity to participate in the CAGS project as tour guides, hiking in the forest, and providing tourists with the history of Cengkeh Afo. There appeared to be a genuine sense of gendered partnership (Fig. 13.8).

In summary the CAGS project revealed an equal distribution of work and benefits between the men and the women in the community, which created a relaxed environment for the tourists to visit. Women's empowerment was demonstrated by the women's request that they were the recognised leaders in the CBT initiative, but they also felt that they could call on the men for support, which appeared to be willingly given. It is also important to point out that the CAGs initiative does not replace traditional livelihood activity, but instead complements and supplements ongoing traditional activities. Male and female cooperation in the CBT

**Fig. 13.8** CAGS members with CAGS Founder Kris Syamsudin (white shirt) and Dr Roswita Aboe (first author, bottom right)

project was seen as being immensely beneficial to progression and success for the community and each household, not only economically, but also socially, culturally and from a mental health perspective.

## The Importance of Mentors and Facilitators

> The CAGS project has the potential to preserve ancient traditions and traditional food systems; and to provide household income for community members, particularly for women, the youth, the elderly and the disabled. The project is increasing the awareness of the benefits of indigenous knowledge for sustainable development. Such an awareness can improve the everyday lives of households in rural areas across the nation and the well-being of people across the globe. (Kris Syamsudin, CAGS founder)

All respondents acknowledged the influence of key women leaders in the project, but at the forefront was identification of an external champion from Ternate, Pak Kris Syamsudin, who encouraged and facilitated the establishment of CAGS. All respondents praised Pak Kris as the unofficial leader because he encouraged the community to explore how the significance of the site and its history could be presented to visitors. Pak Kris spoke with the village leaders and the community through participatory workshops, to discuss their ideas for a CBT project that would benefit them, while maintaining a balance between environmental conservation and economic productivity (Fig. 13.9). He suggested possible approaches, but the choice of action remained with the community and the few that provided leadership. In the search of authenticity to make the new CBT project stand out, a vision was created to build on the historical, cultural and environmental significance of the site. This vision was linked to the historic colonial trade in spices, particularly clove (*Cengkeh Afo*) and nutmeg. The community understood that they had the natural, human and social capital to convert the clove plantation, and other local spices growing in the forest, into a viable commercial tourism attraction that demonstrated their cultural heritage.

Most respondents reported that they did not know what creating a CBT would involve when it was first suggested by Pak Kris and some were hesitant, while others from neighbouring villages became jealous. However, Pak Kris consulted with those from nearby villages and explained how a CBT project site in their vicinity would also benefit them, as more local materials, crops and resources would be purchased to satisfy an increase in local, national and international tourist numbers. Due to this dialogue between the community and neighbouring villages, any differences or disagreements were shared and addressed before the

**Fig. 13.9** Kris Syamsudin (second left) meets with village leaders to discuss the potential CBT project

project's inception, leading to more harmony and cooperation within and between different communities.

The respondents subsequently identified greater cohesion between their community and neighbouring villages, which resulted from group discussions where other ideas and proposals were raised, with Pak Kris as a mediator. They said that the success of the CBT was their way of thanking Pak Kris for changing their lives.

> CAGS is successful because of the support we received from Pak Kris. He taught us how to make CAGS a tourism place. (Ibu Imba, CAGS member)

> CAGS is successful because we had great ideas from Pak Kris. That's why we have been successful. (Ibu Ani, CAGS member)

## Discussion

The success of CAGS was initiated with the guidance of a key local champion, and the support of the village head. The result was a shift towards a more positive cultural identity, improved status and enhanced roles for women in the community, an increase in education and cultural awareness through engaging visitors, and a significant expansion in economic activity and livelihood diversification.

Community participation proved to be an effective means for the local population to be involved in sustainable CBT development. Community participation in the CAGS project has not only been about efficient and equitable redistribution of resources, but also about the transfer and sharing of knowledge between community members. It is the transformation in the processes of knowledge building and learning that has facilitated the community's self-development. Community engagement with visitors has also led awareness and appreciation of other cultures, including for children and youth.

The CAGS project has also resulted in women's empowerment, enabling the community to equitably redistribute benefits and costs, with ongoing women's engagement significantly contributing to the success of the project. Specifically, there has been a shift in gender roles in the village, resulting in the social, cultural, economic, educational and psychological empowerment of the women who manage the CAGS operation (c.f. Dolezal & Novelli, 2021; Phan, 2016). Women are often overlooked in the decision-making process in CBT projects due to their supposed submissive roles and status within the community (see Flacke-Neudorfer, 2007). However, in the case of CAGS, the roles and responsibilities of women have eclipsed those of the men, enhancing their status in the community. The women have enjoyed increased social status and recognition in the community, through regular interaction with international and domestic guests, which extends their knowledge and cultural awareness. These experiences have both motivated and empowered the women to maintain the growth of CAGS, and increase their desire for maintaining and expanding the enterprise.

Although there is evidence of negative outcomes from CBT projects in case studies around the world (Bulilan, 2014; Zielinski et al., 2020), they do not appear to apply to CAGS. The local people have embraced community participation, have no reliance on outside organisations, or interventions by local government, and argue that this approach has strengthened their belief in their community and their motivation to succeed. While participants in this study admitted that they have faced barriers surrounding education, business inexperience and insufficient investment capital, they managed to overcome these barriers and build strength and resolve through facing adversity. While it is true that prior to commencement of the CAGS project the community were not aware of the concept of CBT, and many did not understand what was being proposed, the addition of a key champion was one of the most important variables in enabling the CBT project to flourish.

In addition, the cultural authenticity of the CAGS CBT product was critical to attracting tourists (Dolezal, 2011; Ernawati et al., 2017; Khanom et al., 2019). Cultural authenticity is often sought by discerning tourists as the product offered within some CBT projects is seen as inauthentic, and not a true reflection of the cultural identity of the host community (Somnuek, 2010). The issue of authenticity reinforces the importance of community autonomy, participation and decision-making, so that the inherent social and cultural traits of a community are at the forefront of project design, to ensure its ongoing success (Khanom et al., 2019; Kim et al., 2019). The CAGS project is a product of community participation and decision-making, and a celebration and show case of the local spices and cuisine which are unique to Ternate's cultural heritage. The traditional cuisine and entertainment, set in a unique landscape, are a reflection of genuine authenticity.

A key aspect for a successful CBT, therefore, is generating autonomy within a community to encourage community-led participation. Nevertheless, active engagement with relevant government bodies is also a key factor when developing a sustainable community-led tourism product (Aquino et al., 2018). This applies to the development of any tourism product beyond the extremely small-scale, usually family-owned, enterprise. For example, Saufi et al.'s (2014) study in Lombok, Indonesia, identified perceived inhibitors to active community participation in a CBT. These included top-down approaches that constrained community decision-making processes, precluding innovation and the benefits derived from tourism. This suggests that the CAGS approach to limiting government engagement and dependency has proven to be successful in retaining its decision-making capabilities while maintaining net benefits for the community. However, members of CAGS acknowledge that the provincial government does have a role to play in supporting the CBT initiative, including ensuring an environment attractive to both domestic and international tourists (Kontogeorgopoulos et al., 2014). The CAGS community therefore seeks practical support where government has a mandated role (e.g. public infrastructure), but not involvement in business development and operations.

Finally, the importance of multiple leaders, with different roles, is often ignored in the literature as a key component of a successful CBT. It is clear from this study that CAGS had a local champion who actively participated throughout the processes of planning and development, initiating community workshops, encouraging holistic communal input and facilitating effective, relevant decision-making processes (see Blackman et al., 2004; Kontogeorgopoulos et al., 2014). The leadership and commitment of key women, supported by men from the village, turned a concept and vision into a reality.

## Conclusion

The *Cengkeh Afo & Gamalama Spices* community-based tourism initiative experienced significant success from its inception in 2017–2018 until 2020 when COVID-19 stalled progress. Although COVID-19 has decimated business since 2020, commitment to the re-establishment of the ecotourism restaurant is strong, both within the village and from public and private supporters, who all acknowledge what has been achieved. Importantly, the women leaders steadfastly refuse to relinquish their ability to determine the future of their ecotourism enterprise. The positive feedback from within the community, and from domestic and international visitors, together with national and international recognition, has reinforced community commitment to the project.

Overall, CAGS is characterised by self-motivation and entrepreneurship, minimal assistance from government, with mentoring by a key champion who has supported community decisions from the outset. This approach has resulted in increased individual incomes, cultural awareness and positive flows to the community associated with diversity in economic activity. It has also led to the social, cultural, economic, educational and psychological empowerment of women and young people involved in the project. In this respect, power dynamics within the community have transformed due to a transition away from traditional gendered roles. Increased family incomes have improved capacity to meet immediate family needs for food, children's education and health related costs. The success of CAGS has also resulted in increased community and cultural pride, and a change in aspirations for the future at the individual, family and community levels. Resilience to future economic shocks has been built on enhanced human, social and economic capitals, and a greater sense of security and well-being. There is also an emerging attitude of entrepreneurship and confidence to take risks when investing capitals.

These outcomes fulfil the rationale promulgated in the literature for a successful CBT approach. The principles, processes and mechanisms adopted by CAGS from initiation, through to establishment and operation are equally advocated in the literature. What distinguishes this CBT initiative is how the community overcame what might be perceived as impediments to success and expected need for external input. Other unique factors which contribute to CAGS success include the significant internal investments made at the individual and community levels, the shared leadership and adjustment of power relationships when focused on the project, and the reinforcement of commitment with each small achievement, fostered by a mentor but celebrated by the community. Limitations of human and economic capital were addressed through drawing on the collective of community skills, especially the traditional, and community investments. What emerged was community ownership and pride that fostered a bounded attitude of 'we can do this by ourselves' and each achievement reinforced commitment.

When reviewing a successful CBT case study in northern Thailand, Kontogeorgopoulos et al. (2014) identify fortunate geographical conditions, external support, transformational leadership and luck as key ingredients for success. However, the CAGS experience is somewhat divergent. Fortunate geographical conditions possibly align with a latent heritage context and tourist attractor. External support can be substituted with external recognition, guidance and facilitation, but mainly internal commitment and willingness to invest. Transformational leadership seems essential because of the change in livelihood based on inexperience with the tourism sector, but for CAGS leadership was shared and tasks distributed and not constrained by traditional power structures. The CAGS community made their own luck, but perhaps the substitute was taking advantage of serendipity or actively exploiting opportunities as they arose.

While the variables for success have led to immense positive outcomes for the CAGS community, it may not be necessarily realistic to suggest that CBT projects elsewhere, which pursue the same approach, will also experience success. Although "copy-cat" tourism is common in Indonesia and could benefit other CBT projects in the region, communities are not static in their aspirations, goals, aims and ambitions and what works for one community, might not for another. While the information generated by this small case study of a community-based tourism operation is useful for understanding some of the key factors that led to its success, more in-depth qualitative research surrounding the success of CBT projects is needed to reinforce arguments for, and variables that lead to, success for the host community. Furthermore, although in the CAGs case the community found power and independence in a lack of government involvement, this does not suggest that local and national governments do not have a role to play in sustainable and community-based tourism.

The CAGS project has the potential to preserve a cultural heritage inherited over multiple generations, in the form of traditional food systems, and to provide household income for community members, particularly for women, the youth, the elderly and the disabled. The project is also increasing local and international awareness of the benefits of indigenous knowledge for sustainable development. Such an awareness can improve the everyday lives of households in rural areas across the nation and the well-being of people across the globe.

**Acknowledgements** This project was made possible by the Australian Government's Australia Awards Short Course programme and the Endeavour Leadership Program through short-term mobility grants for students enrolled in the Master of International Development at the University of the Sunshine Coast (UniSC). We would like to thank the Governor of North Maluku, Abdul Ghani Kasuba, and the Mayor of Ternate Burhan Abdurrahman for their support of our visit to Ternate. We are also very grateful to the Rector of Khairun University Professor Husen Alting for their support of UniSC staff and students, especially for organising Khairun University student facilitators, and for providing transport, advice and support while in Ternate. A special thanks is extended to the Khairun University student facilitators and translators during the workshops and interviews: Riecko Salam, Surya Jumsar, Indra Ohorella and Ririn Fahmi. Last but not least we offer special thanks to the CAGS community for allowing UniSC staff and students to visit their unique part of the world, and for participating in the activities.

## References

Acharya, B. P., & Halpenny, E. A. (2013). Homestays as an alternative tourism product for sustainable community development: A case study of women-managed tourism product in rural Nepal. *Tourism Planning & Development, 10*(4), 367–387. https://doi.org/10.1080/21568316.2013.779313

Andriansyah, A., Taufiqurokhman, T., & Wekke, I. S. (2019). Impact of environmental policy factors on tourism industry: A study from Indonesia over last three decades. *International Journal of Energy Economics and Policy, 9*(3), 360–365. https://doi.org/10.32479/ijeep.7754

Ansari, S., Munir, K., & Gregg, T. (2012). Impact at the 'bottom of the pyramid': The role of social capital in capability development and community empowerment. *Journal of Management Studies, 49*(4), 813–842. https://doi.org/10.1111/j.1467-6486.2012.01042.x

Arintoko, A., Ahmad, A., Gunawan, D., & Supadi, S. (2020). Community-based tourism village development strategies: A case of Borobudur Tourism Village Area, Indonesia. *Geojournal of Tourism and Geosites, 29*(2), 398–413. https://doi.org/10.30892/gtg.29202-477

Aquino, R. S., Lueck, M., & Schänzel, H. A. (2018). A conceptual framework of tourism social entrepreneurship for sustainable community development. *Journal of Hospitality and Tourism Management, 37*, 23–32. https://doi.org/10.1016/j.jhtm.2018.09.001

ASEAN. (2016). *ASEAN Community Based Tourism Standard, Cambodian Ministry of Tourism, Phnom Penh, Cambodia.* Accessed 15 May 2020. https://www.asean.org/wp-content/uploads/2012/05/ASEAN-Community-Based-Tourism-Standard.pdf

Ashley, C., & Garland, E. (1994). *Promoting community-based tourism development: Why, what and how.* In: Directorate of Environmental Affairs, Ministry of Environment and Tourism Research Discussion Paper, No. 4, Directorate of Environmental Affairs, Ministry of Environment and Tourism., Windhoek, Namibia 3 https://ashleyinsight.co.uk/wp2020/wp-content/uploads/2020/04/community-based-tourism-Nam.pdf

Blackman, A., Foster, F., Hyvonen, T., Jewell, B., Kuilboer, A., & Moscardo, G. (2004) Factors contributing to successful tourism development in peripheral regions. *Journal of Tourism Studies, 15*(1), 59–70. https://doi.org/10.3316/ielapa.200501369

Bulilan, C. (2014). Exhausted incentives and weakening commitment: The case of community-based tourism in Pamilacan, Bohol, Philippines. *Philippine Quarterly of Culture and Society,* 42(1/2), 16–40. Retrieved August 4, 2021, from http://www.jstor.org/stable/44512010.

Chambers, R. (1994a). Participatory rural appraisal (PRA): Analysis of experience. *World Development, 22*(9), 1253–1268. https://doi.org/10.1016/0305-750X(94)90003-5

Chambers, R. (1994b). The origins and practice of participatory rural appraisal. *World Development, 22*(7), 953–969. https://doi.org/10.1016/0305-750X(94)90141-4

Chambers, R. (2006) *Revolutions in development inquiry* (232 pp.) Earthscan. ISBN: 978-1-84407-625-3 (paperback).

Clarke, V., & Braun, V. (2017). Thematic analysis. *The Journal of Positive Psychology, 12*(3), 297–298. https://doi.org/10.1080/17439760.2016.1262613

Dolezal, C. (2011). Community-based tourism in Thailand: (Dis-)illusions of authenticity and the necessity for dynamic concepts of culture and power. *Austrian Journal of South-East Asian Studies (ASEAS), 4*(1), 129–138. https://doi.org/10.4232/10.ASEAS-4.1-7

Dolezal, C., & Novelli, M. (2021). Power in community-based tourism: Empowerment and partnership In Bali. *Journal of Sustainable Tourism*. https://doi.org/10.1080/09669582.2020.1838527

Ernawati, N. M., Sanders, D., & Dowling, R. (2017). Host-guest orientations of community-based tourism products: A case study in Bali, Indonesia. *International Journal of Tourism Research, 19*(3), 367–382. https://doi.org/10.1002/jtr.2119

Flacke-Neudorfer, C. (2007). Tourism, gender and development in the third world: A case study from Northern Laos. *Tourism and Hospitality Planning & Development, 4*(2), 135–147. https://doi.org/10.1080/14790530701554173

Giampiccoli, A., & Kalis, J. H. (2012). Tourism, food, and culture: Community-based tourism, local food, and community development in Mpondoland. *Culture Agriculture, Food, and Environment: the Journal of Culture and Agriculture, 34*(2), 101–123. https://doi.org/10.1111/j.2153-9561.2012.01071.x

Hiwasaki, L. (2006). Community-based tourism: A pathway to sustainability for Japan's protected areas. *Society & Natural Resources, 18*(8), 675–692. https://doi.org/10.1080/08941920600801090

Karacaoğlu, S., & Birdir, K. (2017). Success factors of Community Based Tourism (CBT) perceived by local peoples: The case of % 100 Misia project. *International Rural Tourism and Development Journal (IRTDJ), 1*(2), 53–61. http://www.turizmvekalkinma.org/index.php/irtadjournal/article/view/72/64

Khanom, S., Scott, N., Kennelly, M., & Moyle, B. (2019). Conceptual model for mutual (host-guest) authentication of intangible cultural heritage. *E-Journal of Tourism*, 160–177. https://doi.org/10.24922/eot.v6i2.53471

Kim, S., Whitford, M., & Arcodia, C. (2019). Development of intangible cultural heritage as a sustainable tourism resource: The intangible cultural heritage practitioners' perspectives. *Journal of Heritage Tourism, 14*(5–6), 422–435. https://doi.org/10.1080/1743873X.2018.1561703

Kline, C., McGehee, N., & Delconte, J. (2019). Built capital as a catalyst for community-based tourism. *Journal of Travel Research, 58*(6), 899–915. https://doi.org/10.1177/0047287518787935

Kodir, A. (2018). Tourism and development: land acquisition, achievement of investment and cultural change (case study tourism industry development in Batu City, Indonesia. *GeoJournal of Tourism and Geosites, 21*(1), 253–265. https://doi.org/10.30892/gtg.21120-285

Kontogeorgopoulos, N., Churyen, A., & Duangsaeng, V. (2014). Success factors in community-based tourism in Thailand: The role of luck, external support, and local leadership. *Tourism Planning & Development, 11*(1), 106–124. https://doi.org/10.1080/21568316.2013.852991

Krohn, G. M. (1993). *The spice islands to circa 1600: Society, politics and trade* (University of New England Masters Thesis). https://hdl.handle.net/1959.11/18675

McCall, C. E., & Mearns, K. F. (2021). Empowering women through community-based tourism in Western Cape, South Africa. *Tourism Review International, 25*, 157–171. https://doi.org/10.3727/154427221X16098837279967

Musavengane, R., & Kloppers, R. (2020). Social capital: An investment towards community resilience in the collaborative natural resources management of community-based tourism schemes. *Tourism Management Perspectives, 34*, 100654. https://doi.org/10.1016/j.tmp.2020.100654

Nara, V., & Irawan, N. (2020). Managing women empowerment through participation in sustainable tourism development in Kampong Phluk, Siem Reap, Cambodia. *International Journal of Economics, Business and Accounting Research, 4*(2), 262–269.

Nair, V., & Hamzah, A. (2015). Successful community-based tourism approaches for rural destinations: The Asia Pacific experience. *Worldwide Hospitality and Tourism Themes, 7*(5), 429–439. https://doi.org/10.1108/WHATT-06-2015-0023

OECD. (2020). OECD tourism trends and policies 2020: Indonesia. *OECD iLibrary*. Retrieved from https://www.oecd-ilibrary.org/sites/6c4fc1cd-en/index.html?itemId=/content/component/6c4fc1cd-en

Phan, L. D. (2016). *Women's empowerment and fertility preferences in Southeast Asia* (Unpublished PhD Thesis). University of Sydney.

Pearce, D. (1992). Alternative tourism: Concepts, classifications and questions. In V. L. Smith & W. R. Eadington (Eds.), *Tourism alternatives: Potentials and problems in the development of tourism* (pp. 18–30). Wiley.

Prasad, M., & Syamsudin, K. (2022). *The heart of the spice forest, PT Peopleprint-works Indografika, Jakarta*. ISBN: 9786027167032.

Radcliffe, S. A., & Laurie, N. (2006). Indigenous groups, culturally appropriate development, and the socio-spatial fix of Andean development. In S. Radcliffe (Ed.), *Culture and development in a globalizing world: Geographies, actors and paradigms* (pp. 83–107). Routledge. https://doi.org/10.4324/9780203641019

Reed, M. G. (1997). Power relations and community-based tourism planning. *Annals of Tourism Research, 24*(3), 566–591. https://doi.org/10.1016/S0160-7383(97)00023-6

Saufi, A., O'Brien, D., & Wilkins, H. (2014). Inhibitors to host community participation in sustainable tourism development in developing countries. *Journal of Sustainable Tourism, 22*(5), 801–820. https://doi.org/10.1080/09669582.2013.861468

Sen, A. (1999). *Development as Freedom*. Alfred Knopf.

Sohnrey, F., & Hegemann, K. (2021). Core community capitals and impact on successful community development. In N. Walzer (Ed.), *Community owned businesses: International entrepreneurship, finance and economic development* (pp. 82–102). Routledge.

Somnuek, P. (2010). The development of community-based tourism business In Phu Khratae Island, Nonklang Sub-district, Phiboonmangsaharn District, Ubonratchathani Province, Thailand. In C. A. Brebbia & F. D. Pineda (Eds.), *Sustainable tourism IV, WIT transactions on ecology and the environment* (Vol. 139, pp. 525–535).

UNESCO. (2003). *Text of the convention for the safeguarding of the intangible cultural heritage*. https://ich.unesco.org/en/convention

United Nations World Tourism Organization. (2020). *Global report on women in tourism* (2nd ed.). https://www.unwto.org/publication/global-report-women-tourism-2-edition

Westoby, P., & Shevellar, L. (2018). Wicked problem and community development. In L. Shevellar & P. Westoby (Eds.), *Routledge handbook of community development research* (ebook). Taylor and Francis.

Windarti, Y. (2016). Local government attitudes toward sustainable tourism development (Case of Bandung City, Indonesia). *International Journal of Social Science and Humanity, 6*(7), 551–556. http://www.ijssh.org/vol6/709-SH025.pdf

Zielinski, S., Kim, S.-I., Botero, C., & Yanes, A. (2020). Factors that facilitate and inhibit community-based tourism initiatives in developing countries. *Current Issues in Tourism, 23*(6), 723–739. https://doi.org/10.1080/13683500.2018.1543254

CHAPTER 14

# A Review of Coastal Resources Co-management in Southeast Asia

*Nga Ho*

## Introduction

The last three decades experienced a rapid decline of coastal resources in Southeast Asia. The degradation has been widely documented for different types of marine resources. The deterioration of the resources is evidenced in fisheries (Abdullah & Kuperan, 1997; Beddington et al., 2007; Cinner, 2011; Pomeroy, 2012; Pomeroy et al., 2007; Rosenberg, 2003); in coral reefs (L. M. Burke et al., 2002a, 2002b; Dunning, 2018; Heery et al., 2018; Shiva, 1991; Burke et al., 2002a, 2002b; Heber Dunning, 2015; Wilkinson et al., 1996) and in mangrove forests (DasGupta & Shaw, 2017; Polidoro et al., 2010; Richards & Friess, 2016; Spalding et al., 2010; Thomas et al., 2017). The decline of the coastal resources in Southeast Asia has been affected by many factors

N. Ho (✉)
School of Agriculture and Food Sciences, University of Queensland, Brisbane, QLD, Australia
e-mail: n.ho2@uq.edu.au



C. Baldwin and S. van Bommel (eds.), *Rural Development for Sustainable Social-ecological Systems*,
Palgrave Studies in Environmental Sustainability,
https://doi.org/10.1007/978-3-031-34225-7_14

including rapid growth of population putting more pressure on food demand and leading to overexploitation of resources; the competition in land uses in coastal areas; coastal environmental pollution; the uncertainty of the ecosystems under the impact of climate change; and the change in resource uses and resource exploitation technology (Dunning, 2018; Muawanah et al., 2012; Pomeroy et al., 2007; Williams, 1996). However, the inefficiency of the management systems has been widely blamed for the collapse of the coastal resources in Southeast Asia (Friess et al., 2016; Nielsen et al., 2004; Pomeroy, 1995). The top-down management systems that were dominant in the region could not effectively handle the rapid decline of the resources and the complexity of the systems in the context of environmental changes (Kuperan & Abdullah, 1994; Pomeroy, 1995; Satria & Matsuda, 2004; Siry, 2006). To deal with these problems, co-management systems (mostly initiated in the 1990s in Southeast Asia) have recognized the importance of active participation of resource users in management. Co-management has been widely studied and practised in Southeast Asia, more than any other region in the world (Wamukota et al., 2012), with more than 250 co-management projects being carried out in the Philippines alone between 1990 and the 2000s (Ratner et al., 2012). Co-management is most used for fisheries management due to the open access status of these types of resources that make them vulnerable to depletion (Wilson et al., 2013).

Co-management is about power and responsibility sharing between different parties such as government and local people or resource users (Berkes, 2009, 2010). It is "a logical approach to solving resource management problems by partnership" (Carlsson & Berkes, 2005, p. 71). A co-management system can be determined by (i) level of sharing power and decision-making between a government and a resource user group; (ii) the participation level of resource users in decision-making processes; and (iii) the support from government to enable resource users to carry out the shared responsibility and power. Co-management is employed when there are conflicts in resource access and use, when there is an increase in the complexity and uncertainty of resource management, when the interest in good governance principles and processes becomes visible, and when decentralization is happening (Borrini, 2007).

Though there is a wide range of literature on coastal resource co-management in Southeast Asia, most of the published documents focused on case studies in one or two countries. Some other studies reviewed the co-management practices in the region, but the focus of these reviews was on one or two aspects of the co-management system only. This chapter reviews the co-management systems and practices of coastal resources in Southeast Asia, with a focus on fisheries co-management, by looking at different elements of co-management systems across the Southeast Asian countries. This scoping review aims to analyse the evolution, arrangements of and factors affecting co-management in Southeast Asia to provide a broad picture of co-management in Southeast Asia. This review places more focus on identifying and mapping the characteristic of co-management processes/arrangements and factors affecting coastal resources co-management in Southeast Asia, rather than analysing the feasibility, appropriateness, meaningfulness or effectiveness of the co-management systems.

The review was based on peer-reviewed journal articles. A search of published journal articles was done by using the University of Queensland library web-base. There were two sets of key words for the search. The first set of key words were Co-management, fisheries co-management, marine resource (with "s" and without "s") co-management, coastal resource co-management and one of the following words Asia, Southeast Asia, Vietnam, Indonesia, Malaysia, Philippines, Thailand, Cambodia in the title. The second set of key words were natural resources management decentralization, coastal management decentralization (with "s" or "z" for decentralization), fisheries management decentralization (with "s" or "z" for decentralization), marine management decentralization and one of the following words Asia, Southeast Asia, Vietnam, Indonesia, Malaysia, Philippines, Thailand, Cambodia in the title. There were 379 publications from the search using the first set of key words, and 230 publications from the search using the second set of key words. The review also used the references found in the list of references of the searched papers, the report of co-management projects in Southeast Asia and conference proceedings on coastal and fisheries co-management in Southeast Asia.

Though most of the reviewed articles/documents were published more than 10 years ago (from the early 1990s to the late 2010s), the review, based on empirical studies on co-management in Southeast Asia, provides insights for stakeholders of coastal resource management systems including policymakers, resource users, NGOs and donors to better plan and manage the systems. It is expected to contribute to developing an approach to management of the coastal resources systems to cope and adapt with the rapid change, uncertainty and complexity of the coastal resources systems in future.

## Evolution of Coastal Co-management in Southeast Asia

Co-management arrangements between governments and resource users in Southeast Asia were initiated in response to the fact that neither the government nor the community alone can resolve resource depletion problems. Therefore, there has been a shift in coastal resources management in the region (Fig. 14.1). The transition of coastal management in Southeast Asia happened in association with administrative decentralization in natural resources management and the devolution of management responsibility and rights to communities (Berkes, 2010; Ratner et al., 2012; White, 2008).

Historically, coastal resources in Southeast Asian countries are state property and managed under centralized policies (Pomeroy, et al., 2009; Satria & Matsuda, 2004; Siry, 2006). The governments have exclusive control over coastal resource governance. However, the governments were not capable of dealing with the rapid degradation of resources due to understaffing that resulted in poor monitoring of resource use and conflict between resource users (Pomeroy et al., 2009; Satria & Matsuda, 2004). Top-down management did not support the strong connection between government and community to effectively resolve the conflicts between resource users, and between government and resource users (Pomeroy et al., 2007). Consequently, management solutions were based

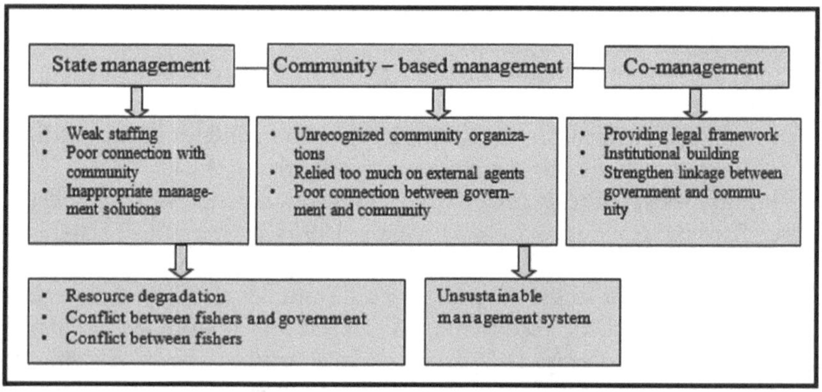


**Fig. 14.1** The evolution of co-management in Southeast Asia

on government administrative power rather than on the understanding of the status of the ecological and social systems (Satria & Matsuda, 2004; Siry, 2006). These problems led to advocacy for community-based resource management in many Southeast Asian countries. Community-based management featured cooperation between NGOs, foreign aid programmes and the private sector with local organizations (Ratner et al., 2012). However, the community-based management systems could not succeed in isolation because they lacked legal support (Pomeroy & Viswanathan, 2003). This resulted in unsustainable management systems. To fix the problems of the central management and community-based management systems, over time, coastal management in Southeast Asia moved to a new approach of co-management that emphasized the partnership between government and the local community of resource users (Pomeroy, 1995; Pomeroy & Ahmed, 2006).

The evolution of co-management in Southeast Asia has proceeded through two "generations" (Ratner et al., 2012). In the first generation, the co-management practices and research placed more focus on the community level through building capacity for stakeholders; development of resource user groups; design of co-management processes and tools; facilitating the collaboration among stakeholders; and advocating legal changes to support community rights. The second generation of co-management had a broader look at the systems by shifting the focus on inter-sectoral resource access and use competition such as marine fisheries, aquaculture, agriculture, mangrove forest, networking institutions across the scale of the co-management systems. The second-generation perspective of co-management emphasized the importance of politics, power and the role of the state in building and sustaining co-management.

The shift of coastal resource management to co-management in Southeast Asia is linked to major reforms of legislative systems that recognized the rights of resource users' groups in resource management and the partnership between these groups and government (Pomeroy, 1995; Satria & Matsuda, 2004; Siry, 2006). Table 14.1 shows some examples of the legal frameworks for co-management in some Southeast Asian countries.

The reforms have been made in different types of legal documents including laws, policies, degrees, and directives (Ho et al., 2015; Marschke, 2012a; Pomeroy, 1995; Pomeroy et al., 2009; Siry, 2006; Sok et al., 2012). The co-management systems have also been integrated and promoted in the national and local socio-development plans (Marschke, 2012a; Pomeroy, 1995). These reforms have cleared the way

**Table 14.1** Legal framework and integration of co-management into socio-economic development plans

| *Countries* | *Legal framework* | *Integration of coastal resource co-management into socio-economic development plans* |
|---|---|---|
| Vietnam | Fishery Law 2003; Directive 03/2006/CT-BTS of The Ministry of Fisheries; Decree 123/2006/NĐ-CP and 33/2010/NĐ-CP | 20 years fisheries management plan 2001–2010 |
| Malaysia | National Policy on Coastal Resources Management 1992 | |
| Indonesia | Local Autonomy Law (Law 22/1999), law 25/1999 and Law 32/2004 | |
| Philippines | Local government code 1991 Fisheries Code 1998 | 1993–1998 Medium term Philippine Development Plan |
| Thailand | Marine and Coastal Resources Management (2015) Fisheries Act 2015 | The seventh five year national Economic and Social Development Plan (1991–1996) |
| Cambodia | Fisheries Policy Reforms (2000–2001); National Fisheries Policy Statement 2005 Fisheries Law 2006 Sub-decree on Community Fisheries Management 2005; 2007 Guidelines for Community Fisheries 2007 | The Strategic Planning Framework for Fisheries: 2010–2019 |

for the decentralization of management rights to a local authority and for the local authority to devolve management functions to community. As a result, the legitimacy of resource users' groups has been recognized in the management system. Box 14.1 provides an example of the decentralization and policy reform in fisheries co-management in Vietnam.

**Box 14.1 Decentralization of fisheries management and evolution of fisheries co-management in Vietnam**

The decentralization of fisheries management in Vietnam was undertaken in the mid 1990s when the evidence of over-exploitation of coastal fisheries and coastal resource degradation was widely recognised by the government. This was when the concept of co-management was first mentioned, but it was not officially recognised until 2003 when fisheries law was reformed to facilitate the decentralization of fisheries management rights to lower levels. The new law supports co-management by stating that "The provincial People's Committees are responsible for ....... organizing local people to exercise the rights to monitor, detect and prosecute violations committed to fisheries activities in fishing grounds" (Clause number 3, article number 15, Fisheries Law 2003). During the next few years since the new law was passed by the National Assembly, some decrees (e.g. Decree 123/2006/NĐ-CP and 33/2010/NĐ-CP) have been promulgated to provide guidelines for the decentralization and devolution of management rights to local level and community. Accordingly, the decentralization of coastal resource management has occurred at three levels: (1) from Central Government to Provincial Government; (2) from Provincial Government to District/Commune Government; and (3) from District/Commune Government to community. The importance of the fisheries community in fisheries management was emphasised in the national 20-year fisheries management plan between 2001 and 2020. The role of community as one of the key partners working with government management of fisheries resources was recognised by government in the Prime Minister's decision 131/2004/QĐ TTg in 2004. In 2006, for the first time ever, the government supported the allocation of management rights to community through fishers' organizations in the Directive 03/2006/CT-BTS of The Ministry of Fisheries. Then in 2009, the term "Fisheries co-management" was first mentioned in an official administrative document (Document 1700 /BNN-KTBVNL) issued by the General Department of Fisheries Resources Protection and Exploitation, Ministry of Agriculture and Rural Development. In the same year (2009) the first ever property rights, territorial use rights for fisheries (TURFs) were granted to fishers' organizations by a government authority (Ho et al., 2015; Pomeroy et al., 2009)

The strategic planning framework for fisheries (2010–2019) in Cambodia is a good example of the integration of marine resource co-management into socio-economic development plans. This framework aimed to support the achievement of Cambodia's Millennium Development Goals and implementation of the national strategic development plan as "the fisheries sector has for many years contributed significantly to the employment and livelihoods of the poor, to food security, and to GDP and foreign exchange balance" (Cambodia Ministry of Agriculture, Forestry and Fisheries, 2010, p. 2). In this framework, one of the indicators of successful management of fisheries resources that would contribute to the success of the implementation of the national strategic development plan was "The strengthening of Community Fisheries to conduct effective Natural Resource Management". The strengthening consisted of "demarcation, establishment and maintenance of fish sanctuaries, patrolling and networking. Should other natural resource co-management processes and/or systems then Community Fisheries be created, new targets for them will be set" (Cambodia Ministry of Agriculture, Forestry and Fisheries, 2010, p. 28).

## Co-management Arrangements in Southeast Asia

Table 14.2 summarizes key elements of coastal resource co-management arrangements in Southeast Asia. These concepts are expanded below.

## Incentives for Co-management

"Co-management does not come about automatically but requires some impetus. Most commonly, it is the recognition of a resource management problem that triggers co-management" (Pomeroy & Berkes, 1997, p. 476). In Southeast Asia the strong motivation for fishers to work with government and other partners has been the depletion of resources (Kuperan & Abdullah, 1994; Muawanah et al., 2012; Pomeroy et al., 2007; Sudtongkong & Webb, 2008; Van Tuyen et al., 2010). While most fishers' livelihoods relied heavily on natural resources, by joining

**Table 14.2** Key elements of coastal resource co-management arrangements in Southeast Asia

| *Co-management elements* | *Description* |
|---|---|
| Motivation for/Incentive of co-management | Resource degradation, conflict resolution, economic incentive, social capital |
| Co-management initiators | International organizations/agencies through research and development projects, resource users and government |
| Resource users' organization for co-management | Fishing/Fishery Associations, Village Management Committee, Resource Management Committee, Fishing Cooperatives |
| Co-management forms | – Formal and informal<br>– Complete and incomplete |
| Property rights | Territorial Use Rights for Fisheries |

co-management, fishers expected an improvement in resource health that would contribute to better social-economic conditions (Katon et al., 1998) (Katon, 1998; Kuperan & Abdullah, 1994; Masae et al., 1998; Pomeroy et al., 2001; van Mulekom, 1998). Another motivation for people to participate in co-management was to enforce the rules and improve legitimacy for better use of resources and conflict resolution (Katon, 1998; Marschke, 2012b; Pomeroy et al., 2007; Sunderlin & Gorospe, 1997; van Mulekom, 1998). In these cases, the elimination of illegal and destructive fishing activities and equal access to and use of resources have been the incentives for fishers to manage jointly with government. In some cases, the direct benefits from participating in co-management activities have motivated fishers to co-operate with government. For example, by co-managing, local people could improve their livelihoods through ecotourism activities (Alcala & Russ, 2006) or make extra income by joining resource protection activities (Pham, 1999). Another motivation for resources users to participate in co-management (e.g. through joining resource users' organizations) is social capital (Gurney et al., 2016; Ho et al., 2015). For example, one of the reasons that many fishers in Tam Giang Lagoon, Vietnam gave for joining the Fishery Association (FA)—a community partner in fisheries co-management—was social capital. Being a member of an FA, fishers expected to improve their social relations within an FA and among FAs, learning from each other. A feeling of being empowered, being involved, being an "owner" and being heard have been the motivations for resource users to work with government (Armitage et al., 2011; Ho et al., 2015).

Governments in Southeast Asia come to co-management of fisheries resources with local people mainly to seek sustainable coastal resource management that they were not able to handle by themselves (Katon, 1998; van Mulekom, 1998). The governments have chosen co-management as a means of responding to a management crisis (Pomeroy & Berkes, 1997; Pomeroy et al., 2007). In some cases, co-management was set up as a government action for promoting democracy at a grassroots level (Pomeroy et al., 2009).

## Initiating Co-management

Co-management initiators refer to those who start the idea of co-management and/or introduce and promote co-management concepts into practice. The idea of building co-management was, in some cases, initiated by government and resource users and in other cases by external agents. The promotion of co-management can be classified as external beginnings and internal beginnings (Pomeroy & Rivera-Guieb, 2006, pp. 46–51). A co-management arrangement initiated externally to the community is one in which the ideas for co-management originated with an external agent. Co-management arrangements could be initiated internally, i.e. within the community, beginning from the resource users who are experiencing problems of resource degradation and conflict in resource use.

Co-management in Southeast Asia can be classified as having an external beginning because there was an intensive involvement of external agencies in the early stages of co-management development in the region. Concepts of co-management that were introduced through funded projects, were identified as a useful intervention approach by a large donor-assisted or a component of a large development or research programme (Pomeroy, 2006). Most of these projects were carried out between the 1990s and 2010s and funded by Western countries or organizations. Different types of donors funded co-management research and practices in Southeast Asia. They included international development agencies (DANIDA, SIDA, CIDA / IDRC, USAID), UN organizations (FAO, GEF),[1] World Bank, Private Funds (The Ford Foundation, The MacArthur Foundation, The Rockefeller Foundation) and environmental conservation organizations (WWF, Conservation International, The Nature Conservancy) (Agrawal & Gibson, 1999). In some cases, the same donor funded in different countries and sites. For example, IDRC funded the same kind of co-management projects in several Southeast Asian countries including Vietnam, The Philippines and Cambodia almost at the same time (Baticados & Agbayani, 2000; Marschke & Nong, 2003; Van Tuyen et al., 2010). In other cases, there were several donors involved in development of co-management at the same site in the same period. Taking Vietnam for an example, in 2008, there were 18 co-management pilot projects operating in Vietnam and these projects were funded by 14 donors (Lai, 2008). The development of fisheries co-management in Tam Giang Lagoon involved 6 donor-funded projects working at this site at the same time (Ho et al., 2016). One of the largest and earliest co-management projects in Southeast Asia was an ICLARM/IFM project. This was a 10-year project staring in 1994 and focusing on fisheries co-management Asia and Africa including 5 Southeast Asian countries. This project has conducted co-management research and practice in 40 sites across the two continents. DANIDA was the principal donor of this project with additional funding from SIDA, IDRC and USAID (Pomeroy, 2006).

Another situation was where government defined a specific protected area and set up co-management to protect that area (Pomeroy, 2006; Sudtongkong & Webb, 2008). In addition, co-management arrangements in some cases in Southeast Asian countries were initiated internally, i.e. within the community, from the resource users experiencing problems of resource degradation and conflict in resource use (Kongkeaw et al., 2019). For this, resource users need advice and support from government officials and/or external agents to solve problems (Katon et al. 1999).

The interventions of the donor-funded projects in building and promoting co-management were approached in different ways. Some

[1] DANIDA: Danish International Development Agency; SIDA: Swedish International Development Cooperation Agency; CIDA/IDRC: Canadian International Development Agency/International Development Research Centre; USAID: US Agency for International Development; FAO: Food and Agriculture Organization, GEF: Global Environment Faculty.

projects funded pilot co-management arrangements and based on the outcomes of the pilot projects, the co-management concepts and approach were legitimized by governments through law and policy reform. This approach of intervention can be found in the case studies in Vietnam (Armitage et al., 2011), Cambodia (Marschke, 2012b) and the Philippines (Pomeroy, 2006). In other cases, the donor-funded projects helped government to implement, or promote a co-management approach that was already in place and/or based on traditional management systems such as in Indonesia (Armitage, 2003), Malaysia (Siry, 2006), and Thailand (Kongkeaw et al., 2019; Pomeroy, 1995).

## Co-management Form

Co-management between government and community can be made through formal or informal agreements. In other words, there are customary or non-notarized agreements and formal legal agreements, project-based agreements between donors and recipient communities and relevant authorities. Informal agreements, such as norms (customary law and practice) and procedures (conflict management and dispute settlement) to govern natural resources, are often unique having evolved through generations of a traditional management system (Borrini, 2007; Satria & Matsuda, 2004). Informal agreements can be unspoken agreements and/or customary agreements. Unspoken agreements embedded in local culture, history, social systems and cultural practices are based on people's belief and religion where people perceive "nature" as a living part in the earth (Pomeroy & Pido, 1995). Customary agreements can be unwritten rules, customs, conventions and norms that are developed and agreed upon by resource users in a community or between communities (Borrini, 2007). Informal agreements often have low legitimacy with government, and are usually enforced through social sanctions according to customary law, with decision-making in the hands of local institutions. The recognition of such agreements by governmental agencies can foster very effective co-management systems (Hauck & Sowman, 2001; Pomeroy et al., 2001). In contrast to informal agreements, formal agreements have high legitimacy with government. Sometimes a formal agreement can be reached outside legal definition in which different stakeholders are willing to sit together for an agreement on allocation, use and management of resources (Jordan, 1989).

In Southeast Asia, both formal and informal co-management exist. However, many co-management cases can be classified as informal. In many cases the resource users' organizations were established and collaborated with government to carry out co-management activities without any formal agreements especially financial commitments (Marschke & Nong, 2003; Marschke, 2012b; Sunderlin & Gorospe, 1997). Even in the case where TURF was granted, in some cases community could not perform its co-management responsibility and rights due to lack of legal commitments at higher levels (Ho et al., 2015). As a result, many communities shared the management responsibility with government including patrolling, rule enforcement and resource monitoring but their operation relied on external support instead of government (Marschke, 2012a; Sunderlin & Gorospe, 1997). This situation made co-management in Southeast more like a community-based co-management system (Pomeroy, 2006) than co-management in Western definitions.

The co-management between government and community can be on a spectrum from "incomplete" to "complete" degrees depending on the extent to which the management functions are performed jointly (Pinkerton, 1989). A co-management system is considered as complete if all management functions are performed jointly between government and community. If the sharing of management functions between government and community is not carried out fully, the co-management is seen as incomplete. In southeast Asia, the case of co-management in Tam Giang, Vietnam (in theory) can be considered as "complete", where Fishery Associations collaborated with government to: do data collection to analyse resource status; prepare fishery plans; make fishery allocation decisions; protect resources; regulate and practice enforcement; and contribute to long-term planning and broader policy decision-making (Ho et al., 2015; Marschke et al., 2012). The co-management systems in Malaysia, on the other hand, are considered incomplete because the responsibility for monitoring, control and surveillance is shared between government and fishers' organizations but the government controls the law and enforcement (Siry, 2006).

## Co-management Organization: Community Representation in Co-management

The early definitions of co-management during the 1980s and early 1990s only emphasized the involvement of two primary stakeholders—government and local community/local resource users—in co-management (Berkes et al., 1991; Duncan et al., 1989). By the late 1990s and 2000s, the definition expanded to include other stakeholders (Borrini-Feyerabend et al., 2000; Borrini, 2007; Jentoft, 2003). Under any definition, community is always a key partner in co-management as it is crucial to have an organization at a local level to collaborate with government. In Southeast Asia, such community organizations in co-management vary in formation and organization in different countries. In some cases, new organizations representing resource users were created to partner with the government. The formation of these resources users' organizations and building capacity for those organizations were done through donor-funded projects (Armitage et al., 2011; Baticados & Agbayani, 2000; Marschke & Nong, 2003). In other cases, the community organization was based on traditional resource management systems (Satria et al., 2006). Likewise, the institutional organization of the resource users' organizations vary between the cases. In some situations, these organizations were considered as a government administrative body that worked under the local council (Sunderlin & Gorospe, 1997). In other cases, these organizations are independent social-professional organizations, e.g. Fishery Associations (Van Tuyen et al., 2010).

## Property Rights: Management Units and Management Rights Arrangement

Management unit refers to the "scale" of co-management arrangements. "Scale" is defined as the spatial, quantitative or jurisdictional dimensions used to study how and to what level co-management is arranged. The scale of co-management varies from place to place but is defined by the physical size of the area to be managed and numbers of members involved in management of an area (Pomeroy, 2006). The co-management case studies in Southeast Asia did not provide a clear description of the management unit or scale for co-management. In most cases, the case studies focused on the decentralization and devolution of management rights and responsivity to community, and institutional building for co-management, without showing clearly how the co-management units are defined and identified. So, it is not clear how a large area of coastal resources such as fisheries that extend across several communities can be managed by community, especially where a community claimed the rights over a part of the common pool area. In a few cases, the management units were clearly defined by setting official geographical boundaries of the resource area that is under co-management. These cases were reported in Malalison Island (Philippines) and Tam Giang Lagoon (Vietnam). In Malalison Island, the boundaries of marine areas belonged to fishers' associations on the island allowing the fishers to control and manage fishing activities in the area (Baticados & Agbayani, 2000). In the case of Tam Giang Lagoon, the lagoon was divided into different areas that were allocated to Fishery Associations (FAs) which are responsible for managing the area granted to them by the local authority (Van Tuyen et al., 2010).

Similar to management units, property rights for open access coastal resources were not discussed clearly in the literature of co-management in Southeast Asia. Property rights determine how a resource is used and owned. Right to property such as a fishery area can be held by an individual or shared by a group of people (Mascia & Claus, 2009). If a property right is shared by a group of people, for example members of a Fishery Association, it is called a collective property right. One form of property right in coastal resource management is territorial use rights for fisheries (TURF). TURF is held by a group of fishers for fisheries management, mostly in small-scale fisheries (Wilen et al., 2012). Though TURF has a long history in fisheries management in the world, it was not widely applied in Asia until recent decades (Nguyen et al., 2017). In Southeast Asia, TURFs have been documented in the long history of traditional resource management systems such as *Sasi* in Indonesia (Harkes & Novaczek, 2002). This concept was still applied in the new era of devolution and decentralization of fisheries management in Indonesia that shifted the systems towards a new stage of community-based or co-management (Metzner, 2008). Granting TURF was also documented in the case of Malalison Island in Central Philippines, where the Fisher's Association of the Island held the right to 1km$^2$ of marine resources (Baticados & Agbayani, 2000). In Southeast Asia, the case of Tam Giang Lagoon, Vietnam showed a clear example of how collective property rights work in fisheries co-management. In 2009, the first time ever in Vietnam, TURFs were granted to Fishery Associations (FAs) by a local authority. FAs hold the right to the area of lagoon allocated to them under TURF. With TURF, FAs have bundles of rights including access (the right to enter the allocated lagoon area), extraction (the right to fish in the allocated area), management (the right to regulate the use of the allocated resources) and exclusion (the right to determine who have an access to the resources (Marschke et al., 2012).

## Factors Affecting the Sustainability of Co-management

Based on the experiences gained from more than 20 co-management studies, Pomeroy et al. (2001) reviewed and analysed the conditions and factors affecting co-management in Southeast Asia. They classified the factors into three groups: supra-community level, community level and individual and household level. Under supra-community level, two factors were identified: enabling policy and legislation, and external agents. Within the community level, 15 conditions for successful co-management were presented. These conditions focused on the issues of scale and boundaries, government support, institutional organization, property rights and capacity building. The factors at individual and household level mainly focused on analysing individual incentives for co-management. They also reviewed the interaction between the factors. Though it was a comprehensive review, there was a lack of in-depth analysis of each factor.

In Southeast Asia, some studies showed that the reasons behind the failure or ineffective operation of co-management systems are lack of legal framework, lack of financial support and donor-funded dependency. Pomeroy and Berkes (1997) claimed that

> Legislation and policy for co-management are embedded in a broader network of laws, policies and administrative procedures, at both national and local government levels.... Government administrative and institutional structures and fisheries laws and policies will, in most cases require restructuring to support these initiatives [for co-management and power sharing]. (Pomeroy & Berkes, 1997, p. 470)

This claim shows the importance of legal frameworks and legislative reforms to support co-management. However, the absence of a legal framework has affected the possibility and sustainability of co-management in Southeast Asia. In some cases, reforms of legislative systems were not carried out at the same time or in tandem with co-management. In that situation, the law and sub-degree were pending to be approved by government while the co-management activities were carried out informally and sometime illegally (Marschke & Nong, 2003). There was also another case where the legal framework was in place to support co-management, but the co-management functions were not performed legally by community due to low legitimacy of the shared power and the mismatch in giving directions, rule and regulation between different government administration levels (Ho et al., 2015).

Lack of financial support has been one of the core factors affecting the continuous operation of many co-management schemes in Southeast Asia. In many cases (Marschke & Nong, 2003; Siry, 2006; Sunderlin & Gorospe, 1997), co-management was arranged with the rights and responsibility decentralized to community but without any financial support from the government for long-term operation of the system. As a result, the operation of co-management depended on the voluntary contribution of community members, the support from external agencies such as NGOs and donor-funded projects. Lack of a regular budget from the government affected the frequency and effectiveness of resource management activities such as patrolling, regulation enforcement or violation sanction. This made co-management, in some cases, more project-oriented and problem-driven than a long-term operation (Siry, 2006).

## Conclusion

It can be said that co-management in most cases in Southeast Asia is a donor-driven system due to the intensive involvement of donor-funded projects in the development and implementation of co-management systems. The co-management projects funded by international organizations have promoted the evolution of co-management, decentralization processes or turned the decentralization and devolution processes into practice. The donor-funded projects played an important role in establishing the co-management arrangements, institutional development and building capacity for co-management stakeholders (Pomeroy et al., 2001). However, to some extent the reliance on external support has made these systems unsustainable. Community relied on external support to conduct the resource management activities under co-management and when the external funds finished those activities discontinued (Ho et al., 2016). The approach that donor-funded projects used to build co-management also had impact on the sustainability of the systems. Some projects had their own project teams who designed and implemented all activities, with low participation of community (Ho et al., 2016). In these cases, the "ownership" of a project by the community was low and the outcomes of those projects were not maintained by community and/or stakeholders (Lai, 2008). Most of the co-management projects were funded by Western donors and the co-management concepts introduced by the consultants/experts from those countries were not suitable for the Asian context because the concepts of power sharing in Western co-management systems applied in some Asian countries were not supported by laws at the time when co-management was established (Ho, 2016; Lai, 2008). Those co-management systems based on traditional management systems and existing institutions at community level using a participatory approach (e.g. Participatory action research) were more likely to be sustainable (Lai, 2008).

## References


Abdullah, N. M. R., & Kuperan, K. (1997). Fisheries management in Asia: The way forward. *Marine Resource Economics, 12*(4), 345–353. https://doi.org/10.1086/mre.12.4.42629208

Agrawal, A., & Gibson, C. C. (1999). Enchantment and disenchantment: The role of community in natural resource conservation. *World Development, 27*(4), 629–649. https://doi.org/10.1016/S0305-750X(98)00161-2

Alcala, A. C., & Russ, G. R. (2006). No-take marine reserves and reef fisheries management in the Philippines: A new people power revolution. *Ambio, 35*(5), 245–254. https://doi.org/10.1579/05-A-054R1.1

Armitage, D., Marschke, M., & van Tuyen, T. (2011). Early-stage transformation of coastal marine governance in Vietnam? *Marine Policy, 35*(5), 703–711. https://doi.org/10.1016/j.marpol.2011.02.011

Armitage, D. R. (2003). Traditional agroecological knowledge, adaptive management and the socio-politics of conservation in Central Sulawesi, Indonesia. *Environmental Conservation, 30*(1), 79–90. https://doi.org/10.1017/S0376892903000079

Baticados, D. B., & Agbayani, R. F. (2000). Co-management in marine fisheries in Malalison Island, central Philippines. *International Journal of Sustainable*

*Development and World Ecology, 7*(4), 343–355. https://doi.org/10.1080/13504500009470053

Beddington, J. R., Agnew, D. J., & Clark, C. W. (2007). Current problems in the management of marine fisheries. *Science, 316*(5832), 1713–1716. https://doi.org/10.1126/science.1137362

Berkes, F. (2009). Evolution of co-management: Role of knowledge generation, bridging organizations and social learning. *Journal of Environmental Management, 90*(5), 1692–1702. https://doi.org/10.1016/j.jenvman.2008.12.001

Berkes, F. (2010). Devolution of environment and resources governance: Trends and future. *Environment Conservation, 37*(4), 489–500. https://doi.org/10.1017/S037689291000072X

Berkes, F., George, P. J., & Preston, R. J. (1991). Co-management: The evolution in theory and practice of the joint administration of living resources. *Alternatives, 18*(2), 12–18.

Borrini-Feyerabend, G., Farvar, M. T., Nguinguiri, J. C., & Ndangang, V. A. (2000). *Co-management of natural resources: Organising, negotiating and learning-by-doing*. GTZ and IUCN, Kasparek Verlag.

Borrini, G. (2007). *Sharing power: Learning-by-doing in co-management of natural resources throughout the world*. Earthscan.

Burke, L., Selig, L., & Spalding, M. (2002a). Management of coastal resources. *Reefs at risk in Southeast Asia*. World Resources Institute.

Burke, L. M., Spalding, M. D., & Selig, E. (2002b). *Reefs at risk in southeast Asia*. World Resources Institute Kogan; Page.

Cambodian Ministry of Agriculture, Forestry, and Fisheries (2010). The Strategic Planning Framework for Fisheries: 2010–2019. Kingdom of Cambodia. https://www.fao.org/faolex/results/details/en/c/LEX-FAOC143042/#:~:text=This%20national%20Strategic%20Planning%20Framework,compliance%20with%20the%20Law%20on

Carlsson, L., & Berkes, F. (2005). Co-management: Concepts and methodological implications. *Journal of Environmental Management, 75*(1), 65–76. https://doi.org/10.1016/j.jenvman.2004.11.008

Cinner, J. E. (2011). Social-ecological traps in reef fisheries. *Global Environmental Change, 21*(3), 835–839. https://doi.org/10.1016/j.gloenvcha.2011.04.012

DasGupta, R., & Shaw, R. (2017). Mangroves in Asia-Pacific: A review of threats and responses. *Participatory Mangrove Management in a Changing Climate*, 1–16.

Duncan, C. A. M., McCay, B. M., & Acheson, J. M. (1989). The question of the commons: The culture and ecology of communal resources. *Labour/Le Travail, 24*(24), 340–342. https://doi.org/10.2307/25143301

Dunning, K. H. (2018). *Managing coral reefs: An ecological and institutional analysis of ecosystem services in Southeast Asia*. Anthem Press.

Friess, D. A., Thompson, B. S., Brown, B., Amir, A. A., Cameron, C., Koldewey, H. J., Sasmito, S. D., & Sidik, F. (2016). Policy challenges and approaches for the conservation of mangrove forests in Southeast Asia. *Conservation Biology, 30*(5), 933–949. https://doi.org/10.1111/cobi.12784

Gurney, G. G., Cinner, J. E., Sartin, J., Pressey, R. L., Ban, N. C., Marshall, N. A., & Prabuning, D. (2016). Participation in devolved commons management: Multiscale socioeconomic factors related to individuals' participation in community-based management of marine protected areas in Indonesia. *Environmental Science and Policy, 61*, 212–220. https://doi.org/10.1016/j.envsci.2016.04.015

Harkes, I., & Novaczek, I. (2002). Presence, performance, and institutional resilience of sasi, a traditional management institution in Central Maluku, Indonesia. *Ocean and Coastal Management, 45*(4), 237–260. https://doi.org/10.1016/S0964-5691(02)00057-1

Hauck, M., & Sowman, M. (2001). Coastal and fisheries co-management in South Africa: An overview and analysis. *Marine Policy, 25*(3), 173–185. https://doi.org/10.1016/S0308-597X(01)00007-0

Heber Dunning, K. (2015). Ecosystem services and community based coral reef management institutions in post blast-fishing Indonesia. *Ecosystem Services, 16*, 319–332. https://doi.org/10.1016/j.ecoser.2014.11.010

Heery, E. C., Hoeksema, B. W., Browne, N. K., Reimer, J. D., Ang, P. O., Huang, D., Friess, D. A., Chou, L. M., Loke, L. H. L., Saksena-Taylor, P., Alsagoff, N., Yeemin, T., Sutthacheep, M., Vo, S. T., Bos, A. R., Gumanao, G. S., Syed Hussein, M. A., Waheed, Z., Lane, D. J. W., … & Todd, P. A. (2018). Urban coral reefs: Degradation and resilience of hard coral assemblages in coastal cities of East and Southeast Asia. *Mar Pollut Bull*, *135*, 654–681. https://doi.org/10.1016/j.marpolbul.2018.07.041

Ho, N. (2016). Fisheries co-management in Vietnam: From theory to practice. Case study in Tam Giang Lagoon. In *The University of Queensland, School of Agriculture and Food Sciences*.

Ho, N. T. T., Ross, H., & Coutts, J. (2015). Power sharing in fisheries co-management in Tam Giang Lagoon, Vietnam. *Marine Policy, 53*, 171–179. https://doi.org/10.1016/j.marpol.2014.12.006

Ho, N. T. T., Ross, H., & Coutts, J. (2016). Can't three tango? The role of donor-funded projects in developing fisheries co-management in the Tam Giang Lagoon system, Vietnam. *Ocean and Coastal Management*, *121*, 97–106. https://doi.org/10.1016/j.ocecoaman.2015.12.011

Jentoft, S. (2003). Co-management—The way forward. In J. N. P. D. D Wilson (Ed.), *The fisheries co-management experience: Accomplishments, challenges and prospects*. Kluwer Academic Publishers.

Jordan, D. (1989). 'Negotiating salmon management on the Klamath river. In E. Pinkerton (Ed.), *Co-operative management of local fisheries: New directions for improved management and community development*. University of British Columbia Press.

Katon, B., Pomeroy, R, Ring, M & Garces, L. (1998). *Mangrove rehabilitation and coastal resource management of Mabini-Candijay: a case study of fisheries co-management arrangements in Cogtong Bay, Philippines* (Working Paper No. 33 of the Fisheries Co-management Research Project).

Kongkeaw, C., Kittitornkool, J., Vandergeest, P., & Kittiwatanawong, K. (2019). Explaining success in community based mangrove management: Four coastal communities along the Andaman Sea, Thailand. *Ocean and Coastal Management, 178*, 104822. https://doi.org/10.1016/j.ocecoaman.2019.104822

Kuperan, K., & Abdullah, N. M. R. (1994). Small-scale coastal fisheries and co-management. *Marine Policy, 18*(4), 306–313. https://doi.org/10.1016/0308-597X(94)90045-0

Lai, T. P. (2008). *Fisheries co-management in Vietnam: Issues and approach.*

Marschke, M. (2012a). Desiring local resource governance. In *Life, fish and mangroves* (pp. 13–38). University of Ottawa Press. http://www.jstor.org.ezproxy.library.uq.edu.au/stable/j.ctt5vkc9f.7

Marschke, M. (2012b). Villagers pursuing local resource governance, 1998–2010. In *Life, fish and mangroves* (pp. 79–106). University of Ottawa Press. http://www.jstor.org.ezproxy.library.uq.edu.au/stable/j.ctt5vkc9f.10

Marschke, M., Armitage, D., Van An, L., Van Tuyen, T., & Mallee, H. . (2012). Do collective property rights make sense? Insights from central Vietnam. *International Journal of the Commons, 6*(1), 1–27. https://doi.org/10.18352/ijc.334

Marschke, M., & Nong, K. (2003). Adaptive co-management: Lessons from Coastal Cambodia. *Canadian Journal of Development Studies/Revue Canadienne D'études Du Développement, 24*(3), 369–383. https://doi.org/10.1080/02255189.2003.9668927

Masae, A., Nissapa, A., & Boromtanarat, S. (1998). *An analysis of fisheries co-management arrangements: The case of Ban Laem Makham, Sikao District, Trang Province, South Thailand* (Working Paper No. 37 of the Fisheries Co-management Research Project).

Mascia, M. B., & Claus, C. A. (2009). A property rights approach to understanding human displacement from protected areas: The case of marine protected areas. *Conservation Biology, 23*(1), 16–23. https://doi.org/10.1111/j.1523-1739.2008.01050.x

Metzner, R. (2008). Property rights and institutional arrangements in southeast Asian fisheries. In *Poverty reduction through sustainable fisheries: Emerging policy and governance issues in southeast Asia* (pp. 233–260). ISEAS–Yusof Ishak Institute.

Muawanah, U., Pomeroy, R. S., & Marlessy, C. (2012). Revisiting fish wars: Conflict and collaboration over fisheries in Indonesia. *Coastal Management, 40*(3), 279–288. https://doi.org/10.1080/08920753.2012.677633

Nguyen Thi Quynh, C., Schilizzi, S., Hailu, A., & Iftekhar, S. (2017). Territorial Use Rights for Fisheries (TURFs): State of the art and the road ahead. *Marine Policy, 75*, 41–52. https://doi.org/10.1016/j.marpol.2016.10.004

Nielsen, J. R., Degnbol, P., Viswanathan, K. K., Ahmed, M., Hara, M., & Abdullah, N. M. R. (2004). Fisheries co-management—An institutional innovation? Lessons from South East Asia and Southern Africa. *Marine Policy, 28*(2), 151–160. https://doi.org/10.1016/S0308-597X(03)00083-6

Pham, V. P., H. G. (1999). *Case study of community-based coastal resource management in Vietnam* (Working Paper, Fisheries Co-management Research Project, Issue).

Pinkerton, E. (1989). Introduction: Attaining better fisheries management through co-management—Propects, problems and propositions. In E. Pinkerton (Ed.), *Co-operative Management of local fisheries: New directions for improved management and community development*. University of British Columbia Press.

Polidoro, B. A., Carpenter, K. E., Collins, L., Duke, N. C., Ellison, A. M., Ellison, J. C., Farnsworth, E. J., Fernando, E. S., Kathiresan, K., Koedam, N. E., & Livingstone, S. R. (2010). The loss of species: Mangrove extinction riskand geographic areas of global concern. *PLOS ONE, 5*. https://doi.org/10.1371/journal.pone.0010095

Pomeroy, B., & Ahmed, M. (2006). *Fisheries and coastal resources co-management in Asia: Selected results from a regional research project*.

Pomeroy, R., Parks, J., Pollnac, R., Campson, T., Genio, E., Marlessy, C., Holle, E., Pido, M., Nissapa, A., Boromthanarat, S., & Thu Hue, N. (2007). Fish wars: Conflict and collaboration in fisheries management in Southeast Asia. *Marine Policy, 31*(6), 645–656. https://doi.org/10.1016/j.marpol.2007.03.012

Pomeroy, R., Nguyen, K. A. T., & Thong, H. X. (2009). Small-scale marine fisheries policy in Vietnam. *Marine Policy, 33*(2), 419–428. https://doi.org/10.1016/j.marpol.2008.10.001

Pomeroy, R. A., M. (2006). *Fisheries and Coastal Resources Co-management in Asia: Selected Results from a Regional Research Project* (WorldFish Center Studies and Reviews 30, Issue).

Pomeroy, R. S. (1995). Community-based and co-management institutions for sustainable coastal fisheries management in Southeast Asia. *Ocean & Coastal Management, 27*(3), 143–162. https://doi.org/10.1016/0964-5691(95)00042-9

Pomeroy, R. S. (2012). Managing overcapacity in small-scale fisheries in Southeast Asia. *Marine Policy, 36*(2), 520–527. https://doi.org/10.1016/j.marpol.2011.10.002

Pomeroy, R. S., & Berkes, F. (1997). Two to tango: The role of government in fisheries co-management. *Marine Policy, 21*(5), 465–480. https://doi.org/10.1016/S0308-597X(97)00017-1

Pomeroy, R. S., Katon, B. M., & Harkes, I. (2001). Conditions affecting the success of fisheries co-management: Lessons from Asia. *Marine Policy, 25*(3), 197–208. https://doi.org/10.1016/S0308-597X(01)00010-0

Pomeroy, R. S., & Pido, M. D. (1995). Initiatives towards fisheries co-management in the Philippines: The case of San Miguel Bay. *Marine Policy, 19*(3), 213–226. https://doi.org/10.1016/0308-597X(94)00008-G

Pomeroy, R. S., & Rivera-Guieb, R. (2006). *Fishery co-management: A practical handbook Fishery co-management a practical handbook*. CABI Pub.

Pomeroy, R. S., & Viswanathan, K. K. (2003). Experiences with fisheries co-management in southeast Asia and Bangladesh. In D. C. Wilson, J. R. Nielsen, & P. Degnbol (Eds.), *The fisheries co-management experience: Accomplishments, challenges and prospects* (pp. 99–117). Springer Netherlands. https://doi.org/10.1007/978-94-017-3323-6_7

Ratner, B. D., Oh, E. J. V., & Pomeroy, R. S. (2012). Navigating change: Second-generation challenges of small-scale fisheries co-management in the Philippines and Vietnam. *Journal of Environmental Management, 107*, 131–139. https://doi.org/10.1016/j.jenvman.2012.04.014

Richards, D. R., & Friess, D. A. (2016). Rates and drivers of mangrove deforestation in Southeast Asia, 2000–2012. *Proceedings of the National Academy of Sciences of the United States of America, 113*(2), 344–349. https://doi.org/10.1073/pnas.1510272113

Rosenberg, A. A. (2003). Managing to the margins: The overexploitation of fisheries. *Frontiers in Ecology and the Environment, 1*(2), 102–106. https://doi.org/10.1890/1540-9295(2003)001[0102:MTTMTO]2.0.CO

Satria, A., & Matsuda, Y. (2004). Decentralization of fisheries management in Indonesia. *Marine Policy, 28*(5), 437–450. https://doi.org/10.1016/j.marpol.2003.11.001

Satria, A., Matsuda, Y., & Sano, M. (2006). Questioning community based coral reef management systems: Case study of Awig-Awig in Gili Indah, Indonesia. *Environment, Development and Sustainability, 8*(1), 99–118. https://doi.org/10.1007/s10668-005-0909-9

Shiva, V. (1991). Conserving the world's biological diversity. *Futures: The Journal of Policy, Planning and Futures Studies, 23*(6), 663–667. https://doi.org/10.1016/0016-3287(91)90089-K

Siry, H. Y. (2006). Decentralized coastal zone management in Malaysia and Indonesia: A comparative perspective 1. *Coastal Management, 34*(3), 267–285. https://doi.org/10.1080/08920750600686679

Sok, S., Yu, X., & Wong, K. K. (2012). Impediments to community fisheries management: Some findings in Kompong Pou commune, Krakor District in Cambodia's Tonle Sap: Impediments to community fisheries management. *Singapore Journal of Tropical Geography, 33*(3), 398–413. https://doi.org/10.1111/sjtg.12007

Spalding, M., Kainuma, M., & Collins, L. (2010). *World Atlas of Mangroves*. Taylor & Francis Group. http://ebookcentral.proquest.com/lib/uql/detail.action?docID=585447

Sudtongkong, C., & E. L. Webb. 2008. Outcomes of state- vs. community-based mangrove management in southern Thailand. *Ecology and Society, 13*(2), 27. [online] http://www.ecologyandsociety.org/vol13/iss2/art27/

Sunderlin, W. D., & Gorospe, M. L. G. (1997). Fishers' organizations and modes of co-management: The case of San Miguel Bay, Philippines. *Human Organization, 56*(3), 333–343. https://doi.org/10.17730/humo.56.3.4576w8444 51k342t

Thomas, N., Lucas, R., Bunting, P., Hardy, A., Rosenqvist, A., & Simard, M. (2017). Distribution and drivers of global mangrove forest change, 1996–2010. *PLoS ONE, 12*(6), e0179302.

van Mulekom, L. T., E 1998. (1998). *Community-based coastal resource management in Orion, Bataan, Philippines: a case study on the development of a municipal-wide community-based fisheries co-management system* (Working Paper No. 38 of the Fisheries Co-management Research Project, Issue).

Van Tuyen, T., Armitage, D., & Marschke, M. (2010). Livelihoods and co-management in the Tam Giang lagoon, Vietnam. *Ocean and Coastal Management, 53*(7), 327–335. https://doi.org/10.1016/j.ocecoaman.2010.04.001

Wamukota, A. W., Cinner, J. E., & McClanahan, T. R. (2012). Co-management of coral reef fisheries: A critical evaluation of the literature. *Marine Policy, 36*(2), 481–488. https://doi.org/10.1016/j.marpol.2011.09.001

White, A. T. (2008). Status of coastal and marine resources: Implications for fisheries management and poverty in Southeast Asia. In *Poverty reduction through sustainable fisheries: Emerging policy and governance issues in Southeast Asia* (pp. 199–232). ISEAS–Yusof Ishak Institute.

Wilen, J. E., Cancino, J., & Uchida, H. (2012). The economics of territorial use rights fisheries, or TURFs. *Review of Environmental Economics and Policy, 6*(2), 237–257. https://doi.org/10.1093/reep/res012

Wilkinson, C. R., Chou, L. M., Gomez, E., Ridzwan, A. R., Soekarno, S. & Sudara, S. (1996). Status of coral reefs in Southeast Asia: Threats and responses in Proceedings of the colloquium on global aspects of coral reefs,

Miami, 1993, ed R. N. Ginsburg & F. G. W. Smith, (University of Miami, RSMAS), 1994, pp. 311–317. *Biological Conservation*, *76*(2), 217. https://doi.org/10.1016/0006-3207(96)83270-8

Williams, M. (1996). *The transition in the contribution of living aquatic resources to food security* (Vol. 13). The International Food Policy Research Institute (IFPRI).

Wilson, D. C., Degnbol, P., & Nielsen, J. R. (2013). *Fisheries Co-management Experience: Accomplishments, Challenges and Prospects* (Vol. 26). Springer.

CHAPTER 15

# Communities on Indonesian Shorelines: Ocean Plastic Through the Eyes of Local People

*Anya Phelan and Novie A. Setianto*

## Introduction

Globally, plastic waste continues to escape after-use collection systems and flow into the natural environment leading to environmental, economic, and social impacts (Beaumont et al., 2019; Geyer et al., 2017; Jambeck et al., 2015). Approximately eight billion kilograms per year of plastic waste comes from coastal regions (Jambeck et al., 2015). This is an overwhelming issue, particularly for countries like Indonesia which has

A. Phelan (✉)
Department of Business Strategy and Innovation, Business School, Griffith University, Brisbane, QLD, Australia
e-mail: a.phelan@griffith.edu.au

N. A. Setianto
Faculty of Animal Science, Jenderal Soedirman University, Purwokerto, Indonesia



C. Baldwin and S. van Bommel (eds.), *Rural Development for Sustainable Social-ecological Systems*,
Palgrave Studies in Environmental Sustainability,
https://doi.org/10.1007/978-3-031-34225-7_15

81,000 km of coastline (Harris, 2019), and more than half of its population, approximately 190 million people, living near the coast (Dahuri, 2007; Harris, 2019). By some estimates, Indonesia is also rated the second largest contributor to ocean plastic behind China (Garcia et al., 2019; Jambeck et al., 2015).

Numerous studies have confirmed that marine plastic poses a significant threat to ocean life, livelihoods, and marine ecosystem services (Beaumont et al., 2019; Lamb et al., 2018; Worm et al., 2017; Wright et al., 2013). Well-known impacts include ingestion (Karbalaei et al., 2019; Liboiron et al., 2019; Rochman et al., 2013, 2015), entanglement (Walker et al., 1997; Wright et al., 2013; Zhao et al., 2016), chemical contamination (Karami et al., 2016), dispersal of invasive species (Kirstein et al., 2016), and release of persistent, bio-accumulating and toxic substances (PBTs) (Beaumont et al., 2019; Demirdjian & Mokatsian, 2017).

Remote, coastal, and island communities are particularly vulnerable to the issue of plastic pollution since they depend directly on healthy marine ecosystems for food, livelihoods, income, cultural, recreational, and spiritual needs (Duraiappah et al., 2005; Liquete et al., 2013; Worm et al., 2006). Although coastal communities contribute to plastic pollution to some extent, they are also recipients of large volumes of marine plastic debris brought in on ocean currents. Furthermore, exports of recyclable materials from developed to developing countries have resulted in significant transfer of waste pollution (Liu et al., 2018; Walker, 2018).

Many middle- and low-income countries lack both effective waste collection and waste management systems, resulting in these countries being the main sources of global plastic pollution due to the mismanagement of plastic waste (Ritchie & Roser, 2018). While the majority of global plastics waste historically has been generated in the Global North, a large portion of manufacturing and consumption of single-use plastic packaging has shifted to the Global South, particularly Asia (Demirdjian & Mokatsian, 2017). Infrastructure for waste management and disposal, however, has not kept up (Glaeser & Glaser, 2010; Shuker & Cadman, 2018). The situation has been further amplified by the proliferation of single-use 'sachet' size products, often directed towards low socio-economic communities and low-income families who buy most of their food and other household items in small daily portions (Ang & Sy-Changco, 2007; Shah, 2012; Sheth et al., 2016; Singh et al., 2009). Waste from small single-use packaging is notoriously difficult to recycle

(Linder, 2017) and is particularly prevalent in remote and rural communities which have less sophisticated waste management infrastructure (Williams et al., 2019).

At a systems level, social and economic costs are often borne by those affected rather than those responsible (Mouat et al., 2010; Ten Brink et al., 2009). Strategies to reduce plastic waste pollution are typically targeted at consumers (Leal Filho et al., 2019), with a focus on behaviour change (Fowler & Close, 2012; Minton et al., 2012). Meanwhile the rate of marketing and distributing Fast Moving Consumer Goods (FMCG) to low-income developing countries continues to grow (Anderson & Billou, 2007). Improving waste management in densely populated urban centres has been a key focus of many waste management studies (Agamuthu & Masaru, 2014; Damanhuri et al., 2014; Meidiana & Gamse, 2010; Zurbrugg, 2002). Less is known, however, about the factors contributing to ocean plastic pollution in remote coastal communities, especially from a systems perspective and how the local communities are perceiving and responding to ocean plastic. Additional information is needed at local and regional scales to develop appropriate capability and mitigation strategies (Farhan & Lim, 2011).

This chapter describes a study conducted with Professor Helen Ross, and which examined knowledge, understanding, and mental models of community members in remote Indonesian communities on the issue of ocean plastic (Phelan et al., 2020). Using a systems thinking approach, we explored how community mental models link factors contributing to ocean plastic pollution with their key livelihoods, such as fisheries, aquaculture, and tourism. In this chapter, we describe the methods used, and present findings from coastal communities in two archipelagos of southern Sulawesi—Selayar and Wakatobi. Both areas are in highly biodiverse marine sensitive areas and are part of Indonesia's marine national park network. We discuss factors contributing to the use and disposal of single-use plastics in coastal and island communities in Eastern Indonesia and offer new insights into unmanaged plastic waste and the support that is required to solve it. We conclude with a discussion on the need for more responsible supply chains, and the integrated role of producers and manufacturers. We close with a reflection on Professor Ross's influence on our life and career.

## Locations and Background Information

We sought locations that represent typical remote coastal communities in Indonesia. The majority of coastal communities in Indonesia are impacted by both ocean plastic debris transported from other locations, and locally sourced waste from land-based leakage and disposal in the sea. The criteria for choosing the study sites included: communities that were remote from commercial centres or urban population; close to sensitive marine areas, (e.g. areas rich in coral reefs and biodiversity); subject to seasonal storms and thus plastic marine debris; and communities with minimal tourism. Note that the majority of the remote, coastal communities in Indonesia do not have extensive tourism. Areas with tourism tend to have more organised beach clean-ups, as proprietors initiate waste mitigation efforts with or without involving the local communities.

The study was conducted in two coastal zones in Sulawesi (Fig. 15.1); first, Selayar, a coastal regency, under the administration of South Sulawesi province, and second, Wakatobi, a coastal regency under the administration of Southeast Sulawesi province (Fig. 15.2). Both regencies are situated within the tropical marine waters of the Coral Triangle, a 6 million $km^2$ marine area located in the western Pacific Ocean which encompasses waters of Indonesia, Malaysia, the Philippines, Papua New Guinea, Timor Leste, and Solomon Islands (Fig. 15.1). The Coral Triangle is considered to be the world's epicentre for marine biodiversity, supporting more than 600 of the world's coral reef species (Veron et al., 2009), including the highest diversity of reef fish, seagrass, and mangroves (Green et al., 2008). The rich marine life in the Coral Triangle also supports the livelihoods of more than 120 million people and provides resources for millions more. Unfortunately, many marine ecosystems in the Coral Triangle, and in the waters surrounding South Sulawesi, are under threat from anthropogenic impacts including coastal development, pollution, illegal fishing, over-exploitation, and climate change (Hoegh-Guldberg, 2015).

Both Selayar and Wakatobi are relatively remote from Indonesia's main island, Java, and distant from their respective regional centres with reliance on slow boat and air travel: Selayar is 173 km from Makassar, and Wakatobi is 273 km from Kendari and 520 km from Makassar. The nearest recycling facility is in Makassar. The majority of the population at both locations lives on the coast and depends on coastal ecosystems and small-scale fisheries for food and income. Most fishers are artisanal (traditional), with small boats without engines, or with small outboard

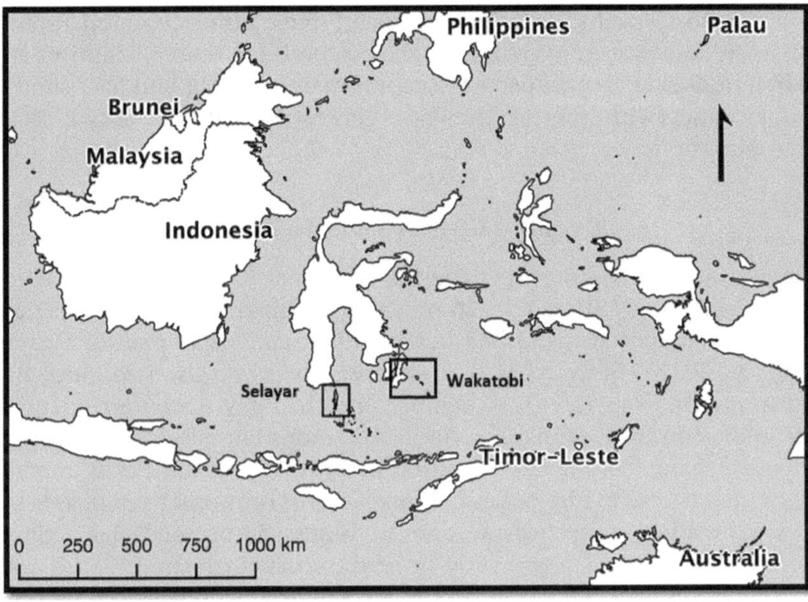


**Fig. 15.1** The research was conducted at two coastal zones of Sulawesi

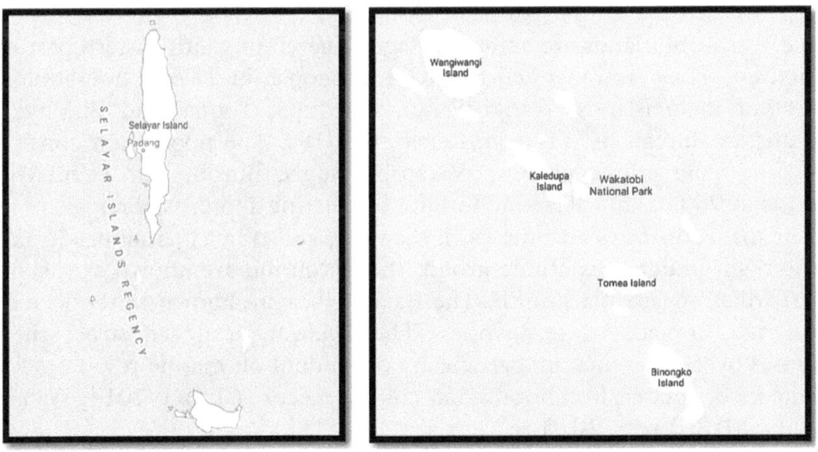


**Fig. 15.2** The Selayar Island and the Wakatobi island zoomed in

motors. Most small-scale fishers use hand-lines; other common fishing gear includes spear-guns, raft lifts, gillnets, and fish traps (Adhuri et al., 2016). Both sites are prospective ecotourism destinations and have similar characteristics with respect to waste generation patterns and lack of disposal options.

### *Selayar—Western South Sulawesi*

The Selayar islands, situated in south-western part of South Sulawesi, are an archipelago of 130 islands, 26 of which are inhabited. The population is 134,280 people with 33,713 households (Statistics Bureau of Selayar Regency, 2018). Selayar has two seasons. The dry season includes the east monsoon from July to November, and the rainy season includes the west monsoon from January to April. The monsoon periods are particularly significant. During the monsoon periods large quantities of marine plastic debris, carried by ocean currents, arrive from other regions. Selayar's population comprises the Selayar, Bugis, Buton, and Bajo ethnic groups, all of whom live primarily along the coast (Liebner, 1998). There are 88 villages in Selayar (Statistics Bureau of Selayar Regency, 2018).

### *Wakatobi—Southeast Sulawesi*

The Wakatobi islands are an archipelago situated in south-eastern part of Sulawesi. They are inhabited by 103,450 people, in 27,631 households, on four main islands: Wangi-Wangi, Kaledupa, Tomia, and Binongko (Statistics Bureau of Wakatobi Regency, 2018). The population consists of four main ethnic groups; Wakatobi, Bugis, Buton, and Bajo, who reside in 90 coastal villages and utilise the marine and coastal resources as their main source of income (fish, seaweed, and other marine products). The main indigenous ethnic group, the Wakatobi, are known as tukang besi which means blacksmith. The Bajau tribes are known as the former seafaring nomads or sea–gypsies. The Bajau typically construct their houses over water and are particularly dependent on marine resources for their food, shelter, livelihoods, and cultural needs (Clifton, 2014; Wahab et al., 2018; Yakin, 2013).

Wakatobi also has two seasons—the dry season, which lasts from April to August and includes the east monsoon (June–September); and the rainy season, which lasts from September to April and includes the west monsoon (December–March).

## Methods

The study used mixed quantitative and qualitative methods (Bryman, 2008; Creswell et al., 2003), in which a survey offered statistically valid representation of the extent of knowledge about plastics and particular behaviours, and ability to relate these to demographic variables; semi-structured interviews provided more detailed information on household waste generation and disposal to expand upon that collected in the survey; and focus group discussions (facilitated with use of SESAMME) allowed detailed elicitation and exploration of villagers' mental models (Jones et al., 2011) about the place of plastics in the villagers' lifestyles and livelihoods. In this study, we focused on the conceptual component of the systems modelling framework (Sterman, 2002), through the lens of community members' mental models of the causes and effects of ocean plastic. To describe and analyse the perception and behaviour of the coastal communities regarding plastic waste and pollution we adopted the first two steps in Stermans' framework; structuring the problem and discovering the causal structure (Sterman, 2002).

### *Village Selection*

For logistical reasons, we sampled sets of villages that could be reached from a single accommodation base in each study region. All selected sites are representative of typical coastal villages in the region, and have minimal tourism. A total of 10 villages were selected as sample: three villages in Selayar and seven villages in Wakatobi.

### *Household Survey*

The survey assessed demographic variables, householders' levels of knowledge and understanding about plastic waste and ocean plastic, as well as community and personal waste disposal behaviour. Owing to lack of a reliable address lists to use as sampling frame, we identified houses on location and through Google© maps. We used a systematic random sampling approach with the aim of sampling one-third of the village households. The final sample for the survey was 473, consisting of 210 men (44%) and 263 women (56%). Ages of the participants range from 18 to 78+, with varied levels of education, 57% completing middle school or below. The majority of the sample were long-term residents of the

**Table 15.1** Sample characteristics

| | n | *% of sample* |
|---|---|---|
| Total Sample | 473 | – |
| Female | 263 | 56 |
| Male | 210 | 44 |
| Education–Primary and Middle School | 269 | 57 |
| Education—High School | 134 | 28 |
| Education—University | 70 | 15 |
| Long-term residents of the region (+15 years) | 421 | 89 |
| Fishers and/or farmers | 194 | 41 |
| Housewife | 132 | 28 |
| Average household weekly income: <US$28 | | |
| Average number of people per household: 4.5 | | |
| Average age: 41.2 | | |

region (15+ years) and were primarily fishers and/or farmers. The sample characteristics are shown in Table 15.1.

The questionnaire comprised 38 questions that assessed beliefs, knowledge, and behaviour in relation to plastic waste and waste management more generally as well as demographic variables. Fourteen of the knowledge questions formed a plastic knowledge scale, see Table 15.2 in Results section. The knowledge questions were coded 0 for an incorrect answer and 1 for a correct answer. There were three questions where scores for the questions could be greater than 1 because more than one of the options was correct. The questionnaire was pilot tested during pre-project scoping visits to three remote parts of Eastern Indonesia, including Wakatobi (in different villages to the full survey).

### *Semi-Structured Household Interviews*

In order to gauge the amount of household waste produced daily by households of a variety of sizes and to discuss waste disposal practices in greater detail than was possible in the survey, short semi-structured interviews of 20–30 minutes were conducted at 11 households in Selayar and 14 households in Wakatobi. These used convenience sampling, approaching people at home, aiming for a range of household sizes and prosperity. None of these households were in the survey sample. The household sizes ranged from one to 10 people. The participants were

**Table 15.2** Individual knowledge scale questions with percent of correct responses

| *Single response questions* | *Correct answer(s)* | *N* | *% correct* |
|---|---|---|---|
| Organic waste (e.g. food waste, plant litter) thrown on the ground will quickly break down and disappear (become part of the soil) | Agree | 473 | 335 (**70.8%**) |
| Snack food wrappers and other plastic packaging thrown on the ground will quickly break down or disappear | Disagree | 472 | 300 (**63.6%**) |
| The closest recycling facility (plastic processing factory) is located in? | Makassar | 473 | 21 (**4.4%**) |
| Do fish and other marine animals eat plastic waste? | Yes | 471 | 55 (**11.7%**) |
| What effect does plastic waste have on the environment? | Negative effect | 471 | 273 (**58.0%**) |
| Does burning rubbish, including plastic, affect human health? | Yes | 471 | 350 (**74.3%**) |
| Rubbish left on the ground will eventually make its way into the ocean | Agree | 470 | 223 (**47.4%**) |
| In the ocean, how long does the plastic that makes up a plastic bag last? | Up to 20 years | 469 | 143 (**30.5%**) |
| In the ocean, how long does the plastic that makes up a plastic bottle last? | UP to 400 years | 469 | 66 (**14.1%**) |
| In the ocean, how long does discarded fishing line does last? | Up to 600 years | 467 | 59 (**12.6%**) |
| Have you heard about 'microplastics'—tiny pieces of plastic floating in the ocean? | Yes | 470 | 51 (**10.9%**) |
| *Multi-response questions* | | *N* | *1 correct* |
| Which of the following items can be re-processed in a factory? | Plastic water bottle/sealed cup, coca-cola can, glass bottle | 473 | 223 (47.1%) |
| Do any of the following types of rubbish affect marine life? | Plastic litter, cigarette butts, discarded fishing line | 473 | 315 (**66.6%**) |

asked about where and how often they shop and what they buy, what they spend, how much waste they generate each day (usually showing us a container), where they dispose of it, and how often. During the interviews, we were shown waste piles and household burning sites, and were sometimes taken indoors to see waste receptacles. We were also able to observe household behaviour, such as children requesting pocket money or returning from the kiosk with sweet drinks and packaged snacks. Interview notes and photographs were used to identify common themes and practices. Scales were used to weigh household waste, when possible.

### *Focus Group Discussions (FGD) Using System Dynamics Application SESAMME—Socio-Ecological App for Mental Model Elicitation*

The FGD use customised software SESAMME (Smith & Richards, 2018) to record the perceived causes and effects of ocean plastics. Eleven FGDs were conducted, five in Selayar and six in Wakatobi. The groups were separated by gender, where possible, to allow for the possibility of men and women having different mental models, and to compensate for the possibility of women being less likely to speak out in a mixed gendered group. The participants were asked to identify activities, resources, pressures, as well as their interaction that influence the problem of plastic waste in the ocean, including plastic waste that arrives on their shorelines by ocean currents. Afterward, participants were asked to identify actions which could be taken to address the problem. Output of the FGD is a combined Causal Loop Diagram (CLD).

The statistics menu of the SESAMME apps helps to identify common interactions and structure of the mental models across the groups. From this a combined CLD, incorporating those already produced on-screen during the focus groups, can be generated using Vensim® software (developed by Ventana Systems Inc.). A CLD is a map of a system showing the causal relationships between connected variables (Maani & Cavana, 2007; Sterman, 2002). It represents causal relationships between variables that lead to feedback loops that determine the behaviour of the system. A variable is a 'condition, situation, action or decision' that can influence, and/or be influenced by another variable (Maani & Cavana, 2007). Variables are connected by directed arcs or arrows, which indicate the cause-to-effect relationship. Polarities (plus or minus) are then used to characterise the nature of the relationship between any two connected variables. The polarity is defined as positive (+) if the connected variables move in the same direction; i.e. as the cause variable increases (or decreases), the effect variable also increases (or decreases). The polarity is defined as negative (−) if the connected variables move in opposite directions; as the cause variable increases (or decreases), the effect variable also decreases (or increases).

CLDs, by definition, are focused on representing feedback loops within the system structure, that reveal the dynamic behaviour of a system (Maani & Cavana, 2007). A feedback is comprised of a chain of cause-effect connections that connects back to the initial 'cause' variable in the chain. In systems modelling, there are two types of feedback loops. Reinforcing loops are positive feedback systems that 'reinforce' or amplify change in a system over time. Balancing loops counteract change in a system over time; they seek stability or return to control (Maani & Cavana, 2007). Male and female groups were analysed separately, but since there were minimal differences, the results were merged. To simplify output where variables refer to similar activity, resources or pressures were combined as one common variable. The analysis of the CLD was then followed by identification of the system archetypes which characterise the system behaviour (Maani & Cavana, 2007; Sterman, 2000), and help to determine the leverage points, the places where an intervention should have the most influence on systems behaviour (Senge et al., 2007).

## Results

### *'Plastic Literacy'—Understanding of Environmental Impacts and Recycling*

The mean of the knowledge items was used to measure plastic knowledge and the scale had adequate reliability ($\alpha = 0.63$). The mean score was $M = 7.12$ ($SD = 3.01$) with scores ranging from 0 to 16, suggesting a low level of plastic knowledge overall. Inspection of responses to the individual questions comprising the knowledge scale (see Table 15.2) shows the lowest levels of knowledge on the following questions: only 4.4% of respondents knew the location of the closest recycling facility, 88.3% did not think that fish and other marine animals eat marine plastic, 89.1%, were also not familiar with the term 'microplastics', and a majority (ranging from 69.5 to 87.4%) did not know how long the plastic in plastic bags, plastic bottles and fishing line last in the ocean. Moreover, just over half of respondents (52.6%) did not believe that rubbish left on the ground would eventually make its way into the ocean. On the other hand, a majority (63.6%) of respondents did understand that plastic packaging does not break down when thrown on the ground, and recognised that plastic has a negative effect on the environment (58%), and that burning plastic rubbish can affect human health (74.3%). Responses to the open-ended questions in the survey suggested that respondents perceive plastic waste to be inert or not threatening and that those who perceive plastic to have a negative effect on the environment stipulated that plastic waste has a negative effect on the environment primarily for aesthetic reasons ('because it makes the village look dirty').

The univariate demographic comparisons showed that householders from Wakatobi had higher knowledge (M = 7.32, SD = 2.24) than those in Selayar (M = 6.47, SD = 3.29), $t(471) = -3.22$, $p = 0.001$, and males (M = 7.47, SD = 3.07) had higher knowledge than females (M = 6.81, SD = 2.92), $t(471) = 2.38$, $p = 0.018$. Knowledge levels were also significantly different between each level of education, $F(2, 470) = 65.21$, $p < 0.001$. Posthoc comparisons showed that those with university education had significantly higher knowledge (M = 9.40, SD = 2.50) than those with middle or high school education (M = 7.66, SD = 2.79), and they were significantly higher than those with primary education (M = 5.41, SD = 2.55). We also explored whether villages that differed in their culture varied in knowledge levels. For example, it may be that Bajo villages—those populated by seafaring nomadic cultures where the researchers observed more plastic and rubbish overall—may exhibit lower levels of knowledge. A comparison of Bajo (i.e. Appatanah, Somabano, and Mantigola) versus non-Bajo villages, however, did not reveal significant differences in knowledge levels across culture, t(471) = 1.26, p = 0.209 (Bajo villages: M = 6.81, SD = 3.02; non-Bajo: M = 7.21, SD = 3.00). Knowledge did not differ depending on the income level of the households, t(427) = 0.063, p = 0.960, (households earning less than 300,000 Rupee M = 6.99, SD = 2.98; households earning more than 300,000 Rupee: M = 6.97, SD = 3.08). Finally, age was significantly negatively correlated with knowledge, $r(472) = -0.319$, $p < 0.001$. The linear regression showed that only gender, age, and education emerged as significant predictors of plastics knowledge when controlling for all other demographic variables.

### *Purchasing and Disposal*

Household interviews and observations showed that typical householders purchase staples (such as coffee, rice, oil, sugar, power milk, etc.), snacks, and other household goods at local kiosks and small village shops daily. Most shops do not have refrigeration. The majority of products come in small 'sachet' size single-use plastic packaging and are typically processed or instant food items. Fish is either caught or purchased at the local market. Fishers prefer to sell good quality fish to fish collectors and keep smaller lesser quality fish for their own consumption. Very few vegetables are grown on the islands due to poor soil quality. Vegetables are brought in by boat and sold at the local markets. Chickens and goats are common in villages and help to dispose of kitchen waste. Many households reported that packaged foods from stores were cheaper and more convenient than fresh foods, and so were preferred. The most popular items are processed snacks which are primarily purchased by children with pocket money that their parents readily provide, despite low incomes. Some of the respondents with boats would travel long distances, and frequently, to shop at cheaper stores, finding the savings outweighed fuel costs and inconvenience. Most, however, preferred daily local shopping.

None of the villages that we focused on in our research have regular waste collection services or a 'garbage bank'. A garbage bank is a term used in Indonesia for a facility where empty plastic containers can be sold and are sorted, shredded, and moved down the value chain. Interviews and observations revealed that households dispose of their rubbish, including plastic, either by burning in small frequent bonfires in the yard, throwing it directly into the ocean (often in a plastic bag), or piling it behind the house. Some is piled up and then burnt. Mangrove areas and other wetlands are also used for rubbish dumping and disposal. Rubbish is also used to fill in mangrove areas as part of a process for converting wetlands to dry land. In other communities observed in our scoping visits, rubbish is also dumped in drains and streams, later to be carried to the ocean.

Survey results showed that convenience is important, with 92% of respondents confirming that they are only prepared to walk very short distances from their house to dispose of their daily rubbish. Observations and interviews showed that plastic waste from snack products is typically dropped on the ground, in the street, or yard. If dropped in the yard, it is swept up every one to two days and then disposed of as above.

Most households burn their household waste, including plastic and leaf litter, once or twice a week. Some households burn rubbish every day. Some householders report taking care about timing, avoiding winds and breezes.

Disposable diapers or nappies are widely used, a relatively recent phenomenon both in Selayar and Wakatobi. Kiosks and local stores sell nappies packaged individually. Parents interviewed reported using 2–3 nappies per day. Nappies are disposed of directly in the ocean, occasionally they are dumped with other rubbish. In some parts of Sulawesi, it is considered taboo to burn human waste, as a result most used nappies make their way into the ocean.

In the survey, only 30% of respondents said that plastic waste (bottles, wrappers, and other packaging) should be put in a bin. Instead, 59% said it should be burned, 3.8% said it should be buried, and 13.5% said it should be put in the ocean (or closest waterway). Although the majority of survey respondents (74.3%) said that burning plastic affects human health, especially as it 'causes coughing and makes it difficult to breathe', interviews and focus group discussions confirmed that most community members will continue to burn plastic as there are no other options for disposal. Household interviews revealed that plastic waste is not prioritised in the community or household clean-ups, because, unlike organic waste, it does not have a smell and so is considered benign. The interviews revealed that the smell from decaying organic rubbish is often associated with ill-health. At the time of the study, bins for street collection had just been introduced in two Kaledupa (Wakatobi) villages, but collections were not well established and households were not accustomed to using them yet. There was no managed landfill for disposal of the wastes once collected. Waste that is collected is taken to dumps that are typically close by and are uncontained.

During the twenty-five semi-structured household interviews, household waste volumes were estimated by weighing household rubbish, through observation, and through discussion. Average daily household waste was estimated at 2 kg/day or 14 kg/week. Typical daily household waste included: kitchen waste, garden leaf litter, cardboard packaging, plastic bottles and packaging, cigarette packaging, glass bottles, paper, and miscellaneous items. Waste from recreational activities (such as picnics and sporting events) and community gatherings (such as weddings and parties) were estimated at 0.25/kg of plastic waste per person per event (based on observation, discussions with informants, and by weighing resulting plastic bottles and packaging from observed community events). This number was multiplied by the average household size, which the survey indicated was 4.5, resulting in an estimated total of 1.125 kg/household/week. Fishing and boat travel was estimated in the same way based on the assumption a typical household will do at least one inter-island round trip per week (resulting in 0.25/kg of plastic waste per person or 1.125 kg/household/week). Respondents reported that all waste generated at sea is typically thrown overboard. Estimated household waste flows are presented in Table 15.3.

Multiplying the estimated total household waste (17.375 kg/household/week), and the mean number of households in the villages studied, 230, we estimate that at the household level a typical village generates approximately 4,000 kg of rubbish per week including plastic. In our survey, nearly half (48%) of respondents stated that in their community most of the rubbish is burned, and 25% stated that most of the rubbish is thrown in the ocean. Applying the assumption that the rubbish that is not burned makes its way into the ocean, we estimate that approximately half of the rubbish produced by a typical coastal village, or 2,000 kg/week, leaks into the ocean. This result is relevant because there are thousands of typical coastal villages in the Indonesian archipelago.

**Table 15.3** Estimated household-level waste flows per week

| | *kg per week* |
|---|---|
| Average waste from a typical household (4–5 people) based on 2 kg/day per household | 14 |
| Waste from recreational activities produced by members of household | 1.125 |
| Waste from community gatherings, weddings, and other special events; Aggregated per household | 1.125 |
| Fishing and boat travel—estimated per household | 1.125 |
| **Total household waste** | **17.375 kg/household/week** |

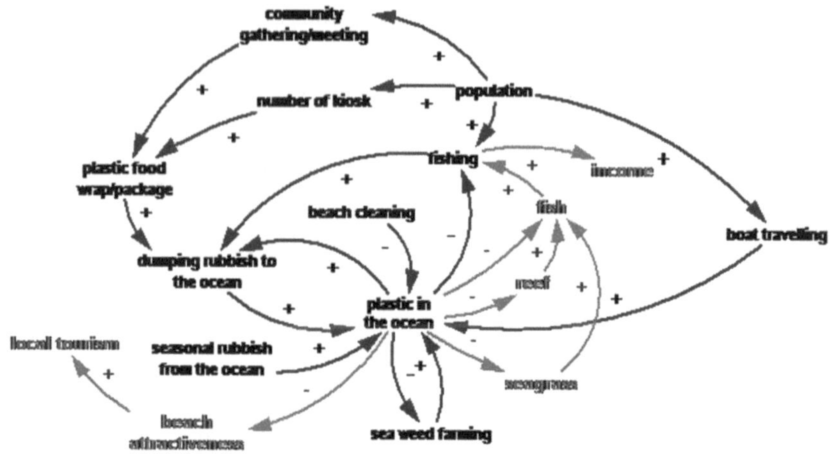


**Fig. 15.3** Mental model of all the activities and resources contributing to and affected by plastic in the ocean

### *Mental Models of the Plastics 'System'*

Focus group discussions using SESAMME allowed community members to build collective mental models of the causes and consequences of ocean plastics.

All the activities and resources contributing to and affected by plastic in the ocean are presented in Fig. 15.3.

### *System Archetypes*

Archetypes are generic systems structures which describe the common dynamic processes of the systems (Maani & Cavana, 2007; Sterman, 2002). Finding archetypes helps to identify the leverage points for intervention in the system (Senge & Sterman, 1992). Analysis of the model identifies three types of archetypes; 'limits to growth', 'escalation', and 'fixes that fail'.

*Limits to Growth*

This archetype describes a process in which a period of accelerating growth is followed by a period of deceleration (Senge & Sterman, 1992).

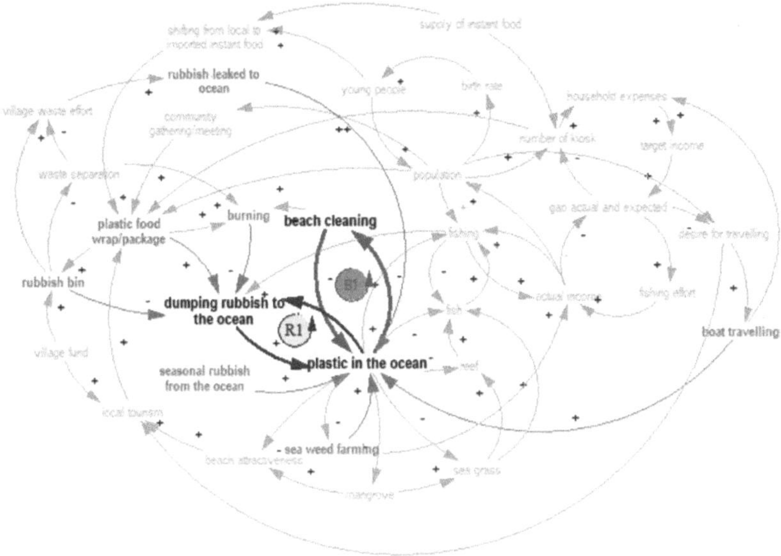


**Fig. 15.4** The limits to growth archetype

The limits to growth as related to plastic are shown in Fig. 15.4, with full model shown in grey.

The reinforcing loop (R1) explains that increasing plastic in ocean means people will throw even more rubbish into the ocean. R1 is followed by a balancing loop (B1), which suggests that beach cleaning efforts will reduce plastic in the ocean.

*Escalation Archetype*

The model in Fig. 15.5 shows that plastic in the ocean will keep increasing as a result of an escalation archetype (B3 and B5). More income will enable households to buy more groceries and do more travelling, which leads to more plastic waste. The archetype indicates that people will try to increase or at least maintain their level of income which results in producing more plastic waste.

*Fixes-That-Fail Archetype*

Figure 15.6 shows the fixes that fail in the model.

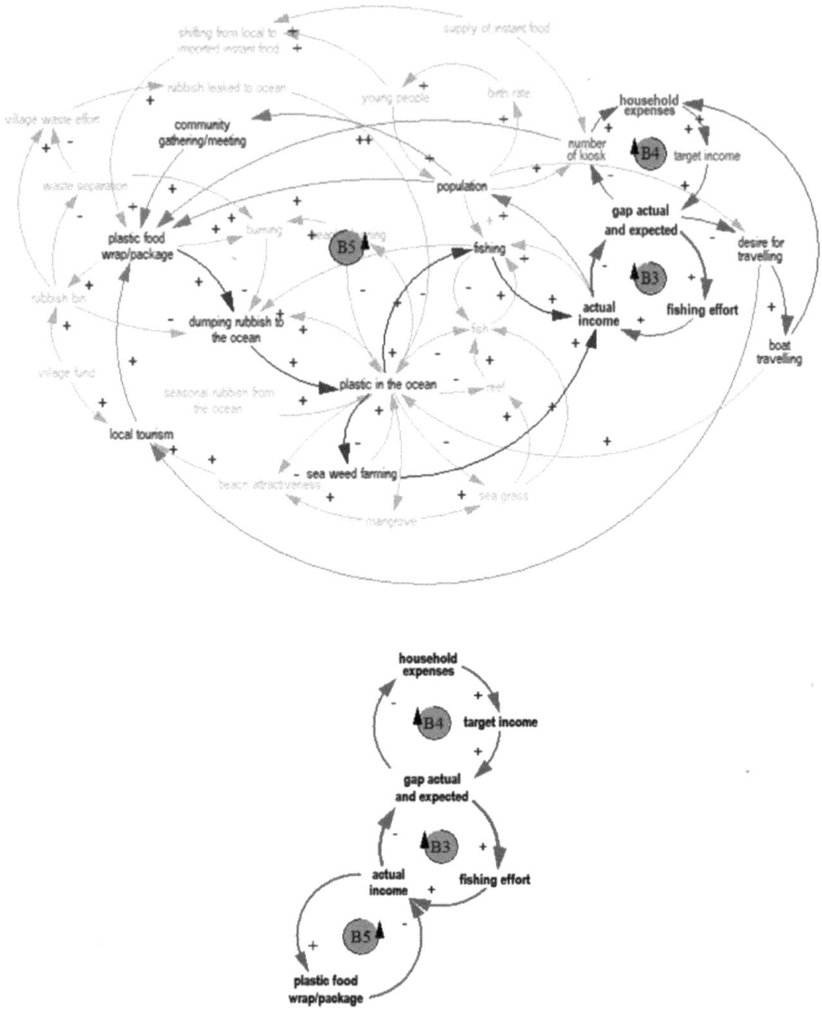


**Fig. 15.5** The escalation archetype

There are two activities currently practised by these communities to manage their waste: (1) burning the 'dry' waste (such as leaves, plastics, paper); and (2) providing rubbish bins at the front of the house (newly introduced, in some Wakatobi villages only). Burning will reduce the

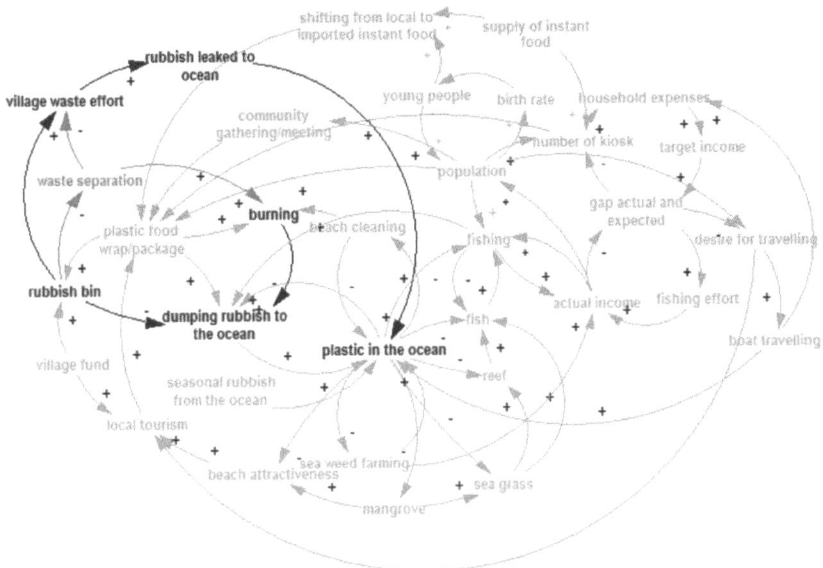


**Fig. 15.6** The Fixes-that-fail archetype

amount of rubbish dumped to the ocean. However, it has negative effects such as air pollution, irritating eyes, noses, and lungs. Providing rubbish bins, with a collection service, should reduce the dumping of rubbish in the ocean, but as there is no waste treatment or landfill on any of the islands studied, all of the waste collected will be dumped at an unmanaged dump site located at some distance from the settlement. Observation showed that some of these disposal areas are close to mangroves and houses. More rubbish sent to poorly controlled dump sites could result in a higher proportion of the rubbish flowing to the ocean. Thus, providing rubbish bins, which seems a good solution initially, in the long term could create more problems.

From the shared mental model, we are able to identify that although communities suffer from plastic waste, some of them keep dumping their rubbish in the ocean because they feel that their contribution is insignificant compared to the amount of rubbish already there, visible in the quantities washed up on their beaches annually. Also, the model shows that an obvious intervention such as distributing rubbish bins to the

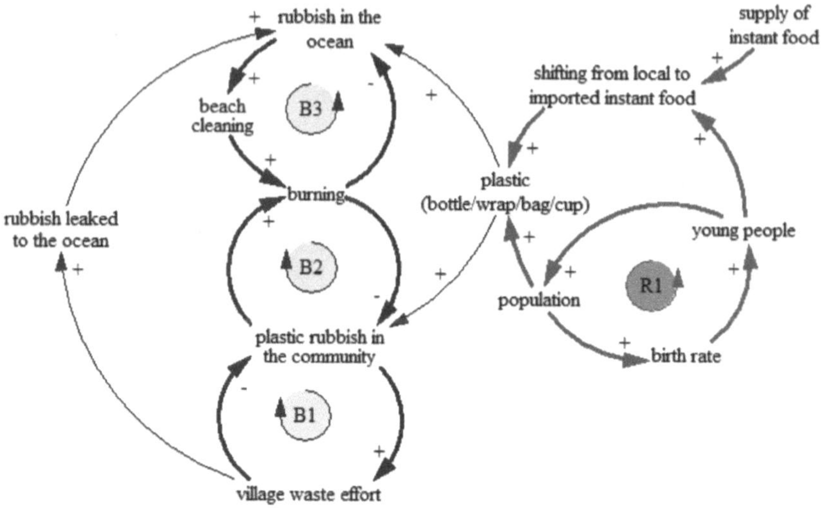


**Fig. 15.7** Plastic waste dynamic for both sites

community is not always the best solution (and even sometimes creates more problems) unless attention is paid to more fundamental aspects of waste management. Figure 15.7 demonstrates the plastic waste dynamic for both sites.

## Discussion

The study aimed to understand the factors that contribute to the use and management of single-use plastics in coastal and island communities in Eastern Indonesia, and to offer new insights into unmanaged plastic waste and the support that is required to reduce the problem. This is a critical issue when considering that small remote villages account for a large proportion of the coastline in Indonesia, and such communities have far less capacity to address the issues than their urban counterparts. Results from the survey indicate that knowledge about plastic and how to manage it is relatively low. This finding was mirrored in the household interviews and observations, although these latter methods also highlighted the reasons for the increase in plastics in villages and the barriers that exist to appropriate management of plastic waste. Villagers reported that foods that come in plastic packaging are often cheaper and more convenient than fresh foods, and in all the villages that were part of our study, there were no regular waste collection services. This lack of infrastructure meant that villagers must find alternatives: burning or dumping their waste, sometimes in piles that can be washed away in heavy rains, sometimes directly in the ocean. Our estimates from household interviews suggest that almost half of the rubbish produced by the typical village may 'leak' into the ocean. Importantly, the findings from the focus groups allowed us to tap into the mental models of villagers so that we can better understand the causes and consequences of ocean plastics. This method provided deeper understanding of the role of plastics in people's lifestyles, livelihoods, and environment, and also of their awareness, than the other methods did. Villagers identified activities that contribute to ocean plastic including boat travel, weather (seasonal monsoons), fishing, seaweed farming, kiosks, dumping, and community gatherings. On the other hand, beach cleaning can help to reduce the problem. Villagers also identified that the livelihood activities of fishing and seaweed farming are affected negatively by ocean plastic.

### *What Mental Models Reveal*

The conceptual models represent shared mental models, built from collective discussion among local men 506 and women in the participating communities. A mental model is a 'model in the mind' (Jones et al., 2011).

The interactions presented through the closed-loop diagrams have the advantage of showing levels of local nuance, insider revelations about how plastics in the ocean relate to the everyday lives of the relatively poor. In doing so they complement the growing literature on physical aspects of the problem (the extent and locations of plastics in the ocean, the need to improve waste management) (Beaumont et al., 2019; Geyer et al., 2017).

While the shared mental model is based on local people's understandings of causes and effects of plastic pollution, this matches well to the team's observations in the field, household interviews, and survey results. It shows a complex system, in which there are multiple causes of the extent of plastics in the ocean, inter-related with factors such as increasing income and population growth. Interestingly, but perhaps not surprisingly (as noted above), community awareness is not mentioned specifically in the mental models, though it is somewhat apparent in the noting of behaviours such as throwing rubbish overboard during boat trips. Shared models also showed that the vast quantity of seasonal rubbish brought ashore in the storm seasons discourages the community from cleaning the beaches, and encourages more disposal of rubbish to the ocean as individuals perceive their additions as insignificant compared to the piles of rubbish brought in by the currents.

Lack of choice in relation to disposal options underlies parts of the model. Food shopping options are biased towards packaged and processed foods, and there is a severe lack of safe and convenient waste disposal options for households and communities. Our observation suggests the latter is not easily resolved. None of the islands appears to have land suitable to turn into landfills, recycling is not easily viable so far from the main Indonesian recycling stations (and does not deal with the many other types of waste), and lack of affordable small-scale technologies so far precludes safe incineration in villages. Even if effective waste management were available, collection services at a household level would also need to be established—survey results and the household interviews suggest that convenience plays an important role in waste disposal behaviour.

We elucidate further on the system archetypes revealed through focus group discussions to help further explain the resulting dynamics. The limits to growth archetype, shown in Table 15.7, suggests that the plastic waste in the ocean will keep increasing. Rubbish brought ashore by the ocean currents, broken floats (typically Styrofoam or PET bottles) used for seaweed farming, single used plastics dumped from boats, and rubbish leaked from the unmanaged village disposal, significantly outweighs the small-scale and infrequent community beach cleaning efforts. Additionally, the ever-increasing use of single-use plastic in daily life creates more pressures to help accelerate the reinforcing loop, R1. Furthermore, more plastic in the ocean discourages villagers from collecting and managing plastic waste, and leads to further dumping in the ocean.

The escalation archetype, shown in Table 15.8, focuses on the household need for greater income (B4). As most of the community members are fishers, we use the example of fishers needing to increase their fishing effort to gain more income (B3). More income leads to more plastics waste (particularly in the context where there are no waste management systems) (B5). In instances when marine plastic debris interferes with fishing activities through propeller entanglements and polluted landing sites, fishing effort also increases.

The fixes-that-fail archetype captures the quick fix solution with potential unintended consequences. The mental models reveal that village beach clean-ups, although beneficial in ameliorating ocean plastic, have led to increased burning of plastic waste (due to lack of other disposal options). The resulting localised air pollution is creating respiratory concerns, such as coughing, itchy, and burning eyes (shown in the survey results), with potential long-term health implications. Similarly, the very recent introduction of bins by some villages has affected traditional disposal practices—separation of dry waste (paper, leaf litter, plastic) and organic kitchen waste (typically left out for chickens and other animals), and has led to overflowing unmanaged village dump sites, which also present a health hazard.

### *Driving Factors for Plastic Waste Outpacing Mitigation Efforts*

The findings from the focus group discussions, household interviews, and observation indicate that daily use of single-use plastics is outpacing standard disposal methods, such as burning and dumping (either directly in the ocean or behind the house). This supports existing literature which states that availability of fast-moving consumer goods and the rising standard of living is driving plastic waste accumulation rates in emerging economies (Geyer et al., 2017; Mahnke et al., 2016; Mangla et al., 2018; Ritchie & Roser, 2018). Our findings also confirm that seasonal storms and high transportation costs make waste management a difficult issue for remote Indonesian communities. The accumulation rates of plastic on the coastlines, in some regions, are reaching catastrophic proportions (Eriksen et al., 2014; Tullo, 2018). When plastic is improperly discarded, it will find its way into waterways and the ocean (Hardesty et al., 2017; Jambeck et al., 2015; Lebreton et al., 2017). In most coastal communities waste never makes it to a landfill, or anywhere near a recycling facility. Villages studied for this research did not have effective infrastructure to collect and recycle plastic waste.

In remote coastal communities in Indonesia waste management entirely depends on village funds and provincial government support (and our observation is that this is not a vote-winning issue for political candidates). Waste management infrastructure necessary to keep up with waste generation rates is unattainable for small low-resource communities. Large-scale centralised waste management is very challenging due to complex coastal geography, distances between islands and high transportation costs. Furthermore, relationship with the ocean and dependence on the ocean to meet daily needs, including waste disposal, is linked to long-standing cultural traditions and beliefs (Yakin, 2013). The plastic literacy survey conducted as part of this study demonstrates that understanding of the environmental impacts of plastic is low. Better understanding of environmental impacts on the part of the community would not necessarily help, however, as effective waste disposal options are limited. Nevertheless, the low plastic literacy suggests that even if adequate waste management was introduced, there would be a need to accompany it with education and awareness raising.

As findings from the focus groups suggest, seasonal monsoons play an important role in the amount of plastic villagers encounter. During the east and west monsoons, the situation reaches extreme proportions. A monsoon season lasts for several months. The high winds and storms bring high volumes of marine plastic debris on ocean currents inundating the coastlines of island communities. Large volumes of marine plastic debris interfere with fishing and other livelihoods. Most communities are not equipped to manage large-scale clean up. As noted in the focus groups, weekly beach clean-ups in a few of the communities, mostly done by hand, are unable to deal with high volumes. This, coupled with lack of disposal options with burning being the only alternative, results in most of the marine plastic debris left on the coastlines. The villages studied left the majority of marine debris on beaches. A handful of tourism and dive operators encountered in the adjacent national parks confided that they struggle to keep up, spending considerable amounts of time, money, and effort to stem the plastic tide.

## Conclusion

This chapter summarises the results of a study in low-resource coastal communities in Selayar and Wakatobi regions of south Sulawesi, Indonesia. This study was developed and conducted with Professor Helen Ross, and demonstrates that remote coastal communities are forced to shoulder the impacts of the ocean plastic crisis. There are tens of thousands of similar coastal communities in Indonesia, all struggling to cope with their own plastic waste, plus overwhelming quantities of waste brought in by ocean currents.

Findings from the study demonstrate that ocean plastic affects the daily lives of people living on remote coastal communities in numerous ways.

The shared mental models, surveys, and interviews show that members of the community perceive and have experience of marine plastics having a negative impact on fishing, seaweed farming, and emerging regional industries including tourism. The system results show that communities are caught in a perpetual reinforcing loop. Although our survey results show that plastic literacy is low, even if it were higher there is little the coastal communities could do to manage plastic waste effectively unless presented with better choice architecture, both on the supply side and in disposal options.

We recognise that plastic pollution has to be solved on multiple fronts, our data suggests that producer responsibility is imperative and a circular plastic economy is greatly needed. For coastal communities in emerging economies, the ocean plastic crisis cannot be abated without responsible supply chains and non-plastic alternatives. Producers and manufacturers can no longer focus only on low-cost packaged products, without taking responsibility for the outcomes. Without access to degradable, environmentally friendly products, and a circular plastic system, coastal communities and surrounding marine ecosystems will continue to drown in plastic waste.

## Closing Remarks

From Dr. Anya Phelan

I first met Professor Helen Ross as a Postdoctoral Research Fellow for the CCRES project in Indonesia. Although part of a different working group I was drawn to Helen's sincere and impeccable approach to working with people and communities. She was patient and kind, and generously shared her knowledge, experience, and deep understanding of culture and customs. I had the privilege of working on several projects with Helen and learned a great deal from her; she is undoubtedly the best mentor I have ever had.

When conducting fieldwork, Helen always urged us to practice deep listening, never to forget that good research takes time. She always extended genuine consideration and respect to everyone, and readily used humour and kindness as the source of creative thinking. Above all, she dedicated herself to see beyond what is readily apparent, to take conventional wisdom and turn it on its side, and to take nothing for granted. I am forever grateful for the opportunity to work with Helen and to learn from her. I deeply admire her adventurous spirit, she inspires me to be a better person, and a better researcher.

From Dr. Novie A. Setianto

I admire Prof Ross' passion and commitment in carrying out this research. I learned a lot about how to build an effective team and how to train our thinking. Prof. Ross gives many examples of the importance of systems thinking and the need to elaborate on multiple perspectives. I learned that it is necessary to have 'drone-view' perspectives. Sometimes we need to fly up to understand the big picture and fly down to observe more details. In this work, I learned a social perspective that plastic waste not only pollutes the environment but also damages the communities' minds that throwing rubbish into the ocean is ok.

## References


Adhuri, D. S., Rachmawati, L., Sofyanto, H., & Hamilton-Hart, N. (2016). Green market for small people: Markets and opportunities for upgrading in small-scale fisheries in Indonesia. *Marine Policy, 63*, 198–205.

Agamuthu, P., & Masaru, T. (2014). *Municipal solid waste management in Asia and the Pacific Islands: Challenges and strategic solutions*. Springer.

Anderson, J., & Billou, N. (2007). Serving the world's poor: Innovation at the base of the economic pyramid. *Journal of Business Strategy, 28*(2), 14–21.

Ang, R. P., & Sy-Changco, J. (2007). The phenomenon of sachet marketing: Lessons to be learned from the Philippines. *Enhancing Knowledge Development in Marketing*, 5.

Beaumont, N. J., Aanesen, M., Austen, M. C., Börger, T., Clark, J. R., Cole, M., Hooper, T., Lindeque, P. K., Pascoe, C., & Wyles, K. J. (2019). Global ecological, social and economic impacts of marine plastic. *Marine Pollution Bulletin, 142*, 189–195.

Bryman, A. (2008). Why do researchers integrate/combine/mesh/blend/mix/merge/fuse quantitative and qualitative research. *Advances in mixed methods research*, 87–100.

Clifton, J. (2014). Maritime ecocultures: Bajau communities of eastern Indonesia. In *Ecocultures* (pp. 27–43). Routledge.

Creswell, J. W., Clark, V. P., & Garrett, A. (2003). Advanced mixed methods research. In C. A. Thousand Oaks (Ed.), *Handbook of mixed methods in social and behavioural research* (pp. 209–240). Sage.

Dahuri, R. (2007). Pre-and post-tsunami coastal planning and land-use policies and issues in Indonesia. *area planning and management in Asian tsunami-affected countries*, 111.

Damanhuri, E., Handoko, W., & Padmi, T. (2014). Municipal solid waste management in Indonesia. In *Municipal Solid Waste Management in Asia and the Pacific Islands* (pp. 139–155). Springer.

Demirdjian, Z., & Mokatsian, Z. (2017). The Death of the Oceans: Implications for Business and Socieity. *ASBBS Proceedings, 24*(1), 167.

Duraiappah, A. K., Naeem, S., Agardy, T., Ash, N. J., Cooper, H. D., Diaz, S., Faith, D. P., Mace, G., McNeely, J. A., & Mooney, H. A. (2005). Ecosystems and human well-being: biodiversity synthesis; a report of the Millennium Ecosystem Assessment.

Eriksen, M., Lebreton, L. C., Carson, H. S., Thiel, M., Moore, C. J., Borerro, J. C., Galgani, F., Ryan, P. G., & Reisser, J. (2014). Plastic pollution in the world's oceans: More than 5 trillion plastic pieces weighing over 250,000 tons afloat at sea. *PLoS ONE, 9*(12), e111913.

Farhan, A., & Lim, S. (2011). Resilience assessment on coastline changes and urban settlements: A case study in Seribu Islands, Indonesia. *Ocean & Coastal Management, 54*(5), 391–400.

Fowler, A. R., III., & Close, A. G. (2012). It ain't easy being green: Macro, meso, and micro green advertising agendas. *Journal of Advertising, 41*(4), 119. https://doi.org/10.1080/00913367.2012.10672461

Garcia, B., Fang, M. M., & Lin, J. (2019). Marine plastic pollution in Asia: All hands on deck! *Chinese Journal of Environmental Law, 3*(1), 11–46.

Geyer, R., Jambeck, J., & Law, K. L. (2017). Production, use, and fate of all plastics ever made. *Science Advances, 3*(7), e1700782.

Glaeser, B., & Glaser, M. (2010). Global change and coastal threats: The Indonesian case. An attempt in multi-level social-ecological research. *Human Ecology Review*, 135–147.

Green, A., Petersen, N., Cross, A., & MacLoed, E. (2008). Coral triangle facts, figures, and calculations: Part ii: Patterns of biodiversity and endemism. *The Nature Conservancy December*, *16*, 2008.

Hardesty, B. D., Lawson, T., van der Velde, T., Lansdell, M., & Wilcox, C. (2017). Estimating quantities and sources of marine debris at a continental scale. *Frontiers in Ecology and the Environment, 15*(1), 18–25.

Harris, P. G. (2019). *Climate change and ocean governance: Politics and policy for threatened seas*. Cambridge University Press.

Hoegh-Guldberg, O. (2015). *Reviving the Ocean Economy: the case for action.*

Jambeck, Geyer, R., Wilcox, C., Siegler, T. R., Perryman, M., Andrady, A., Narayan, R., & Law, K. L. (2015). Plastic waste inputs from land into the ocean. *Science, 347*(6223), 768–771.

Jones, N., Ross, H., Lynam, T., Perez, P., & Leitch, A. (2011). Mental models: An interdisciplinary synthesis of theory and methods. *Ecology and Society, 16*. https://doi.org/10.5751/ES-03802-160146

Karami, A., Romano, N., Galloway, T., & Hamzah, H. (2016). Virgin microplastics cause toxicity and modulate the impacts of phenanthrene on biomarker responses in African catfish (Clarias gariepinus). *Environmental Research, 151*, 58–70.

Karbalaei, S., Golieskardi, A., Hamzah, H. B., Abdulwahid, S., Hanachi, P., Walker, T. R., & Karami, A. (2019). Abundance and characteristics of microplastics in commercial marine fish from Malaysia. *Marine Pollution Bulletin, 148*, 5–15.

Kirstein, I. V., Kirmizi, S., Wichels, A., Garin-Fernandez, A., Erler, R., Martin, L., & Gerdts, G. (2016). Dangerous hitchhikers? Evidence for potentially pathogenic Vibrio spp. on microplastic particles. *Marine Environmental Research, 120*, 1–8.

Lamb, J. B., Willis, B. L., Fiorenza, E. A., Couch, C. S., Howard, R., Rader, D. N., True, J. D., Kelly, L. A., Ahmad, A., & Jompa, J. (2018). Plastic waste associated with disease on coral reefs. *Science*, *359*(6374), 460–462.

Leal Filho, W., Saari, U., Fedoruk, M., Iital, A., Moora, H., Klöga, M., & Voronova, V. (2019). An overview of the problems posed by plastic products and the role of extended producer responsibility in Europe. *Journal of Cleaner Production, 214*, 550–558.

Lebreton, L. C., Van Der Zwet, J., Damsteeg, J.-W., Slat, B., Andrady, A., & Reisser, J. (2017). River plastic emissions to the world's oceans. *Nature Communications, 8*, 15611.

Liboiron, M., Melvin, J., Richárd, N., Saturno, J., Ammendolia, J., Liboiron, F., Charron, L., & Mather, C. (2019). Low incidence of plastic ingestion among three fish species significant for human consumption on the island of Newfoundland, Canada. *Marine Pollution Bulletin, 141*, 244–248.

Liebner, H. (1998). Four oral versions of a story about the origin of the Bajo people of southern Selayar. *Living through histories: Culture, history and social life in South Sulawesi* (pp. 107–133).

Linder, M. (2017). Ripe for disruption: Reimagining the role of green chemistry in a circular economy. *Green Chemistry Letters and Reviews, 10*(4), 428–435.

Liquete, C., Piroddi, C., Drakou, E. G., Gurney, L., Katsanevakis, S., Charef, A., & Egoh, B. (2013). Current status and future prospects for the assessment of marine and coastal ecosystem services: a systematic review. *PLoS One*, *8*(7), e67737.

Liu, Z., Adams, M., & Walker, T. R. (2018). Are exports of recyclables from developed to developing countries waste pollution transfer or part of the global circular economy? *Resources, Conservation and Recycling, 136*, 22–23.

Maani, K., & Cavana, R. Y. (2007). *Systems thinking, system dynamics: Managing change and complexity*. Prentice Hall.

Mahnke, H., Ballent, M., Baumann, S., Imperiale, F., von Bergen, M., Lanusse, C., Lifschitz, A. L., Honscha, W., & Halwachs, S. (2016). The ABCG2

Efflux transporter in the mammary gland mediates veterinary drug secretion across the blood-milk barrier into milk of dairy cows. *Drug Metabolism and Disposition, 44*(5), 700–708. https://doi.org/10.1124/dmd.115.068940
Mangla, S. K., Luthra, S., Mishra, N., Singh, A., Rana, N. P., Dora, M., & Dwivedi, Y. (2018). Barriers to effective circular supply chain management in a developing country context. *Production Planning & Control, 29*(6), 551–569.
Meidiana, C., & Gamse, T. (2010). Development of waste management practices in Indonesia. *European Journal of Scientific Research, 40*(2), 199–210.
Minton, E., Lee, C., Orth, U., Kim, C.-H., & Kahle, L. (2012). Sustainable marketing and social media: A cross-country analysis of motives for sustainable behaviors. *Journal of Advertising, 41*(4), 69–84. https://doi.org/10.1080/00913367.2012.10672458
Mouat, J., Lozano, R. L., & Bateson, H. (2010). *Economic impacts of marine litter*. Kommunenes Internasjonale Miljøorganisasjon.
Phelan, A., Ross, H., Setianto, N. A., Fielding, K., & Pradipta, L. (2020). Ocean plastic crisis—Mental models of plastic pollution from remote Indonesian coastal communities. *PLoS ONE, 15*(7), e0236149.
Ritchie, H., & Roser, M. (2018). Plastic pollution. *Our world in data*.
Rochman, C. M., Hoh, E., Kurobe, T., & Teh, S. J. (2013). Ingested plastic transfers hazardous chemicals to fish and induces hepatic stress. *Scientific Reports, 3*, 3263.
Rochman, C. M., Tahir, A., Williams, S. L., Baxa, D. V., Lam, R., Miller, J. T., Teh, F.-C., Werorilangi, S., & Teh, S. J. (2015). Anthropogenic debris in seafood: Plastic debris and fibers from textiles in fish and bivalves sold for human consumption. *Scientific Reports, 5*, 14340.
Senge, P. M., Lichtenstein, B. B., Kaeufer, K., Bradbury, H., & Carroll, J. S. (2007). Collaborating for systemic change. *MIT Sloan Management Review, 48*(2), 44.
Senge, P. M., & Sterman, J. D. (1992). Systems thinking and organizational learning: Acting locally and thinking globally in the organization of the future. *European Journal of Operational Research, 59*(1), 137–150.
Shah, A. M. (2012). Business strategies in the emerging markets. *Journal of Asia-Pacific Business, 13*(1), 4–15.
Sheth, J. N., Sinha, M., & Shah, R. (2016). *Breakout strategies for emerging markets: Business and marketing tactics for achieving growth*. FT Press.
Shuker, I. G., & Cadman, C. A. (2018). *Indonesia-Marine debris hotspot rapid assessment: Synthesis report (English)*. Marine Debris Hotspot Rapid Assessment (Synthesis Report). World Bank Group. Washington: World Bank Group.
Singh, R., Ang, R. P., & Sy-Changco, J. A. (2009). Buying less, more often: An evaluation of sachet marketing strategy in an emerging market. *The Marketing Review, 9*(1), 3–17.

Smith, C., & Richards, R. (2018). *Using the SESAMME app: Scripts for guiding focus group discussions, Capturing Coral Reef and Related Ecosystem Services (CCRES) Project*.

Statistics Bureau of Selayar Regency. (2018). Kepulauan selayar regency in figures. Statistics Bureau of Selayar Regency.

Statistics Bureau of Wakatobi Regency. (2018). Kepulauan Wakatobi regency in figures. Wakatobi.

Sterman, J. (2000). *Business dynamics*. McGraw-Hill, Inc.

Sterman, J. (2002). *System dynamics: Systems thinking and modeling for a complex world*. MIT Sloan School of Management.

Ten Brink, P., Lutchman, I., Bassi, S., Speck, S., Sheavly, S., Register, K., & Woolaway, C. (2009). *Guidelines on the use of market-based instruments to address the problem of marine litter*. Institute for European Environmental Policy (IEEP), Sheavly Consultants.

Tullo, A. H. (2018). Fighting ocean plastics at the source. *Chemical & Engineering News, 96*(16), 29–34.

Veron, J., Devantier, L. M., Turak, E., Green, A. L., Kininmonth, S., Stafford-Smith, M., & Peterson, N. (2009). Delineating the coral triangle. *Galaxea, Journal of Coral Reef Studies, 11*(2), 91–100.

Wahab, M. R. A., Zakaria, R. M. A., Hadrawi, M., & Ramli, Z. (2018). *Selected topics on archaeology, history and culture in the Malay world*. Springer.

Walker, T., Reid, K., Arnould, J. P., & Croxall, J. (1997). Marine debris surveys at Bird Island, South Georgia 1990–1995. *Marine Pollution Bulletin, 34*(1), 61–65.

Walker, T. R. (2018). China's ban on imported plastic waste could be a game changer. *Nature, 553*(7689), 405–405.

Williams, M., Gower, R., Green, J., Whitebread, E., Lenkiewicz, Z., & Schröder, P. (2019). *No time to waste: Tackling the plastic pollution crisis before it's too late*.

Worm, B., Barbier, E. B., Beaumont, N., Duffy, J. E., Folke, C., Halpern, B. S., Jackson, J. B., Lotze, H. K., Micheli, F., & Palumbi, S. R. (2006). Impacts of biodiversity loss on ocean ecosystem services. *Science, 314*(5800), 787–790.

Worm, B., Lotze, H. K., Jubinville, I., Wilcox, C., & Jambeck, J. (2017). Plastic as a persistent marine pollutant. *Annual Review of Environment and Resources, 42*, 1–26.

Wright, S. L., Thompson, R. C., & Galloway, T. S. (2013). The physical impacts of microplastics on marine organisms: A review. *Environmental Pollution, 178*, 483–492.

Yakin, H. S. M. (2013). Cosmology and world-view among the Bajau: The supernatural beliefs and cultural evolution. *Mediterranean Journal of Social Sciences, 4*(9), 184.

Zhao, S., Zhu, L., & Li, D. (2016). Microscopic anthropogenic litter in terrestrial birds from Shanghai, China: Not only plastics but also natural fibers. *Science of the Total Environment, 550*, 1110–1115.
Zurbrugg, C. (2002). Urban solid waste management in low-income countries of Asia: How to cope with the garbage crisis. *Presented for: Scientific Committee on Problems of the Environment (SCOPE) Urban Solid Waste Management Review Session, Durban, South Africa* (pp. 1–13).

# PART III

# Social Aspects of Resilience

CHAPTER 16

# Introduction to Section 3—Social Aspects of Resilience

*Katrina Brown*

The last two decades have seen a rise in interdisciplinary and social science interrogations of resilience. Resilience provides a dynamic way to conceptualise change and responses to change. But in this field, resilience was initially derived from an ecological model of adaptation, which is not immediately transferable to understanding complex ecological-social interactions and processes of change. As outlined in the introduction, in Chapter 1, the analysis of social aspects of resilience initially focused on critiques of how social ecological systems approaches understood resilience. This was seen as apolitical and undertheorized, overlooking power and social difference, and making assumptions about behaviours and using simplistic economistic explanations of how people were linked to resources through utilitarian and extractive relationships. Analyses were based on limited—and often socially shallow—empirical analysis.

K. Brown (✉)
University of Exeter, Exeter, UK
e-mail: katrina.brown@exeter.ac.uk



C. Baldwin and S. van Bommel (eds.), *Rural Development for Sustainable Social-ecological Systems*,
Palgrave Studies in Environmental Sustainability,
https://doi.org/10.1007/978-3-031-34225-7_16

People—society—were seen as 'add-ons' to ecosystems, and communities were assumed to be contiguous with ecosystems and often as homogenous and physically bounded. The methods used by SES scholars to understand social resilience were often descriptive and varied in their focus—individuals, households, and communities. During the 2000s, these views were scrutinised and often critiqued, but few constructive and conceptually rich and methodologically rigorous investigations were presented.

Prof. Helen Ross was amongst a small group of social science scholars who entered this space and produced conceptually sophisticated analysis which took the field forward in innovative, accessible, and constructive ways. Helen's work on social aspects of resilience has been significant in three dimensions. Firstly, it has been innovative in integrating cross-disciplinary perspectives in new ways, adding conceptual breadth and depth that particularly exploits Helen's background in psychology to produce interdisciplinary insights informed by social as well as natural sciences. Secondly, it provides methodological advances which build on this integrated approach and that are often interdisciplinary and participatory, deeply rooted in principles of active involvement and engagement with research subjects as partners. Thirdly, it is grounded in empirical analysis and in the wealth of experience that Helen and her co-authors and collaborators bring from field-based work across the world. This ensures that these conceptual and methodology insights can be applied in the real world—that they have relevance and applicability.

The chapters in this section demonstrate these attributes, and build on, illustrate, and advance many of the themes developed in Helen Ross' work. In tracing the impacts of Berkes and Ross' 2013 paper on community resilience, in Chapter 17, Katherine Witt and co-authors show the tremendous reach of this work, across diverse fields and disciplines, and applied in a range of contexts, including climate change, ecosystem services, health, agriculture and food systems, and disaster management. The framework proposed by Berkes and Ross integrated core social science concepts such as agency, structure, power, and culture, giving prominence to place attachment, social networks, and cross-scale interactions. The application of this work is further elaborated in Chapter 18, by Maclean and Witt examining six attributes of social resilience, synthesised from a suite of studies in northern Queensland, thus effectively demonstrating the application of these concepts. Chapter 19 presents an analysis of community resilience in the mining sector, with a case study of three communities in Bolivia's altiplano, showing how this framework travels. One key aspect is that it is a strength-based approach to resilience, so can inform 'resilience building' programmes. The study reveals some of the potential tensions and trade-offs between individual or family resilience, and collective or community resilience. Thus, households are required to make trade-offs when deciding whether to invest resources and effort into community-based efforts. It also tackles the thorny issue of how resilience is distinct from sustainability, and points to how resilience analysis such as this framework can inform sustainability, and particularly how sustainability is not just about environment.

These are followed by three further chapters which look at what aspects of individual and broader organisational features might support resilience and provide examples of the innovative ways in which we learn about this. Chapter 20, authored by Jones and Baldwin, explores some of the cognitive constructs that underpin our relationship to the world. Mental models represent how we make sense of complexity and how we understand cause and effect in social ecological systems—and how we recognise and understand change. In turn, these are underpinned by values that in part determine how we navigate conflicts and trade-offs, for example between social, economic, and environmental outcomes. Understanding how values influence peoples' actions and decisions is core to policy and natural resource management, but until recently was under-represented in study of social ecological systems and resilience. Chapter 21 by Li Liu builds on these insights and examines the role of transformative collaboration and joint knowledge creation necessary for systems change. This kind of collaboration involves interdependence, continuous contact, and generous information exchange. Both of these chapters reflect Helen Ross' commitment to working in partnership with communities, and in seeing social and ecological resilience as intrinsically linked to innovation, knowledge co-creation, and empowerment. Chapter 22 by Valery and colleagues expands the scale of analysis and provides insights into the governance of resilience. Governance itself is presented here as a dynamic and evolving system, and boundary organisations are seen as important to support sustainable governance of social ecological systems. This is especially the case in protected areas—in this instance in the Galapagos Islands—where complexity is manifest as conflicts, tensions, and trade-offs between different social groups, between different agencies and organisations with differing aims and objectives. Boundary organisations can provide a legitimate space for collaboration, particularly between governments and communities.

These chapters provide rich insights to elaborate and apply core concepts and approaches that Prof. Helen Ross has pioneered in her research. They demonstrate the interdisciplinary nature of the work, the application of core social science theory, employed in diverse settings. A deep and enduring commitment to engagement, collaboration, transformation, empowerment evident in Prof. Helen Ross' work, is represented here. But throughout, a respect for people and places is paramount, and a sense of why we do this work—to nurture and shape change to benefit people and the planet.

CHAPTER 17

# Towards an Integrated Approach to Community Resilience: Ten Years On

*Katherine Witt, Tracy Cheung, and Jonathan Simpson*

## Introduction

When Professor Helen Ross first began in the School of Natural and Rural Systems Management (NRSM) at The University of Queensland's Gatton Campus over 20 years ago, I was an enthusiastic undergraduate student passionate about sustainable use of natural resources. I can remember the excitement of having a new 'lady professor' (not common in regional agriculture-focused campuses at the time). Helen, a 'proper social scientist' was to bring an entirely new skill set to the School as it shifted away from its traditional plant, animal and soils science approach towards teaching a broader interdisciplinary approach to natural resources

K. Witt (✉) · J. Simpson
University of Queensland, Brisbane, QLD, Australia
e-mail: k.witt@uq.edu.au

T. Cheung
Centre for Natural Gas, The University of Queensland, Brisbane, QLD, Australia



C. Baldwin and S. van Bommel (eds.), *Rural Development for Sustainable Social-ecological Systems*,
Palgrave Studies in Environmental Sustainability,
https://doi.org/10.1007/978-3-031-34225-7_17

management. Helen brought expertise in social systems, sustainable rural and regional development and something called resilience—we were not really sure what that meant.

At around the same time, my new favourite book had arrived at the UQ Gatton campus library. It was Fikret Berkes, Carl Folke and Johan Colding's *Linking Social and Ecological Systems: Management Practices and Social Mechanisms for Building Resilience* (1998). The detailed case studies whet my appetite for future research and I found the subtlety and complexity of connections between human and natural systems endlessly fascinating. I was immediately drawn to Helen and her work and have had the privilege of working with her and the benefit of her guidance and advice since that time.

Today, complex socio-ecological systems (SES) approaches and concepts of community resilience underpin most of my work including understanding multiple values in resource use, cumulative impacts of extractive resources and energy development, impacts of energy transition and facilitating community-led change/development. There is one paper that has had significant influence for me and this is the 2013 paper by Fikret Berkes and Helen Ross titled *Community Resilience: Toward an Integrated Approach*. In this chapter, we aim to highlight the breadth and nature of the impact the Berkes and Ross 2013 article has had for others over time and across the world.

## Methods

The purpose of this paper is to highlight the nature and breadth of the impact of the Berkes and Ross 2013 article. To achieve this, a systematic database search was performed using the Scopus database to identify the publications that cited Berkes and Ross 2013. This search, conducted in February 2021, resulted in 578 publications. Based on this result, we employed a review approach involving three stages.

First, we wanted to identify the geographic spread of impact, or the extent to which Berkes and Ross 2013 has influenced scholars around the world. To do so, we extracted the corresponding author addresses provided within the articles and created a count for each country. There was no corresponding author information provided for 71 articles. The data was then used to create a hotspot map of where papers citing Berkes and Ross 2013 originated (Fig. 17.1) using Microsoft Excel's inbuilt map chart tool.

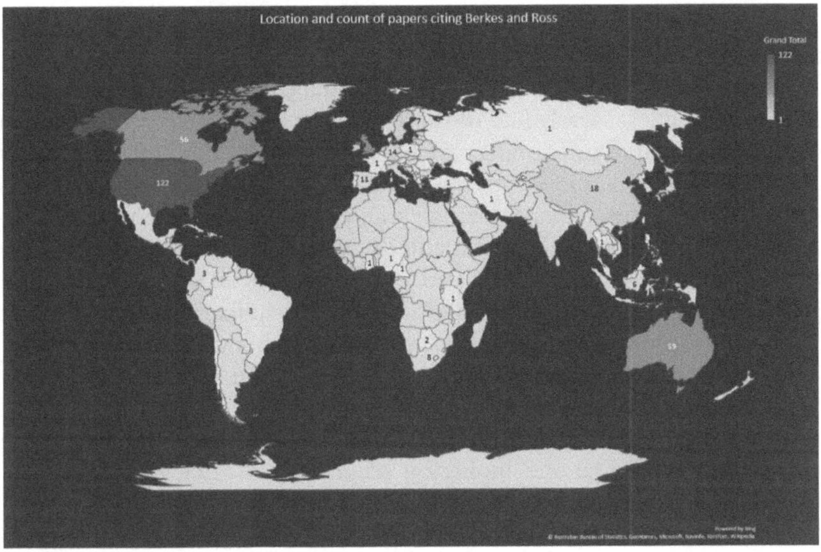


**Fig. 17.1** Hotspot map of the publications citing Berkes and Ross 2013

Nearly ten years on, Berkes and Ross 2013 had been cited by researchers from 51 countries across the world. Not surprisingly, the most articles originate from English-speaking countries in the global North, e.g. the United States (n = 122), the United Kingdom (n = 60), Australia (n = 59) and Canada (n = 56). The second-largest group of articles are written by researchers from Europe, including the Netherlands (n = 16), Sweden (n = 14), Germany (n = 14) and Spain (n = 11). While the articles reviewed might reflect Anglo-American and Eurocentric interests, we can also observe a growing body of literature emerging from research institutes in Asia, such as China (n = 18), Japan (n = 12), South Korea (n = 7) and more recently Indonesia (n = 6).

Second, we wanted to explore where Berkes and Ross (2013) has had an impact in terms of discipline and subject area. From the articles reviewed, we identified 251 different journal titles. We then mapped the journal titles into subject areas using the categories information from Scimago Journal & Country Rank. If the journal covers more than one subject area, only the subject area(s) with the highest value in 2013–2021 is mapped. When there is more than one relevant subject area, the journal is considered as multidisciplinary.

Berkes and Ross (2013) has been engaging more scholars and more diverse disciplines. Nearly half of the articles (n = 220) were published in journals with a multidisciplinary focus. For example, the highest number of articles (n = 37) come from the journal Sustainability,[1] which contains topics from energy, environmental science and social sciences in relation to sustainability and sustainable development. The journal *Society and Natural Resources*, which also published articles citing Berkes and Ross 2013 (n = 19), is another popular multidisciplinary journal covering research on multiple disciplines such as environmental science, development, and sociology and political science. For the rest of the articles, twelve subject areas are identified, including social sciences (n = 137), environmental science (n = 60), business, management and accounting (n = 29), agricultural and biological sciences (n = 17), medicine (n = 16), earth and planetary sciences (n = 9) and psychology (n = 8).

While social sciences and environmental science are common disciplines, diverse research fields are included (see Fig. 17.2). Social sciences: anthropology, development, geography, planning and development, health (social science), safety research, sociology and political science, etc.; Environmental science: ecology, global and planetary change, pollution, water science and technology, etc. Within those articles, a range of topics (e.g. climate change, ecosystem service, tourism, crisis management, disaster management, entrepreneurship, etc.) can also be observed. Moreover, it can also be noticed that a few articles come from very distinct fields beyond the original disciplines like engineering, computer science, arts and humanities, decision sciences and economics, econometrics and finance in recent years.

Lastly, we want to know how Berkes and Ross (2013) influences the thinking and practice of community resilience. For the analysis, we took a subset of the most highly cited articles across all years to make a more manageable sample of 10% of all articles. The search resulted in 55 articles after two further self-citations were removed. These articles each have 35 or more citations. These articles were searched for their citations of Berkes and Ross 2013, categorised and tabulated based on where and how they

[1] *Sustainability* is a competitively priced, open access journal with a broad subject matter intake and high number of articles per volume, which may explain the higher number of articles in that journal.

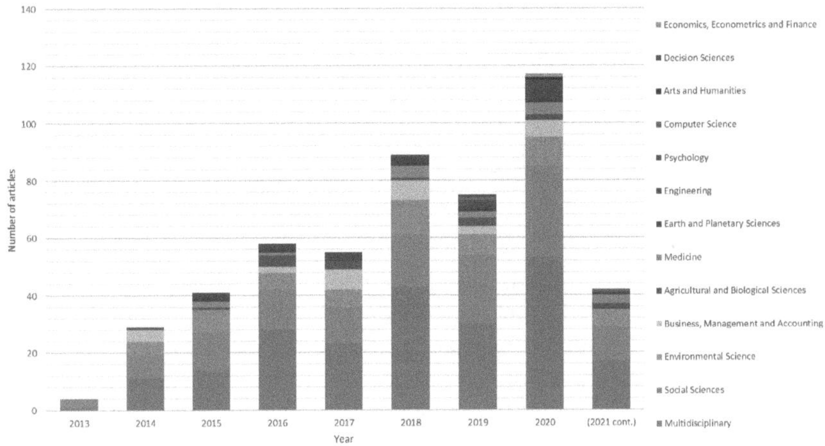


**Fig. 17.2** Subject areas of the journals reviewed

were utilised. Berkes and Ross 2013 can be found in every component of the articles—introduction (n = 10), conceptual framing (n = 30), methods (n = 3), results (n = 2), discussion (n = 16) and conclusion (n = 7). The articles were also manually reviewed to understand their contributions to the concept of community resilience in theory and practice. Noting that the highly cited papers were published mostly between 2014 and 2018 (n = 52), we took a further step and manually reviewed the articles that were published in 2019–2021 in understanding emerging themes in community resilience.

## Thinking and Practice of Community Resilience: ten years on

In this section, we first demonstrate how the propositions of Berkes and Ross 2013 have been supported by research in the field, drawing on the 55 highly cited articles. We then discuss how the propositions of Berkes and Ross 2013 have been enhanced or critiqued, as well as how these articles contribute to the discussion of community resilience. Thirdly, we present the emerging discussion and fields of research around community resilience resulting from a review of articles that were published in 2019–2021. Lastly, we reflect on the current understanding of community resilience and highlight the persistent challenges in applying community resilience as an analytical concept.

## How Berkes and Ross 2013 Was Cited and Used in the 10% Most Highly Cited Articles

The influence of Berkes and Ross 2013 can be found in different components of the 55 journal articles in advancing the understanding of community resilience in theory and practice. In some papers, Berkes and Ross 2013 is cited in the introduction section and used for establishing the context and highlighting the significance of the research topic, in particular identifying the knowledge gap(s) within the field of study. For example, articles highlighted how Berkes and Ross 2013 argue the need for better understanding of related concepts such as agency, structure, culture and power (Keating et al., 2017; Lyon & Parkins, 2013); as well as the interactions between community and regional scales (Singh-Peterson et al., 2014) within the concept of resilience.

The majority of articles draw on Berkes and Ross 2013's explanation of community resilience to support the conceptual framing of their research. Scholars emphasised that the integrated thinking that constitutes community resilience is an important component of resilience and sustainability more broadly (Julie L. Davidson et al., 2016; Khalili et al., 2015; Townshend et al., 2014). Considering the work of Berkes and Ross 2013 and other scholars, articles also highlight the importance of collective capacities and resources within resilience-building processes for communities when responding to challenges (Darnhofer, 2014; Julie L. Davidson et al., 2016). For example, authors cited Berkes and Ross 2013 to explain agency and self-organisation in building resilience for communities (Curtis et al., 2014; DeVerteuil & Golubchikov, 2016). Several articles cited the key strengths and characteristics of community resilience as identified by Berkes and Ross 2013 (Matarrita-Cascante et al., 2017), e.g. people–place relationships (Adams, 2016; Amundsen, 2015; Tschakert et al., 2017); social networks and engaged governance (Spies et al., 2014); values and beliefs (Tschakert et al., 2017).

While methodological challenges of conducting resilience research were mentioned in some articles (Darnhofer, 2014; Roberts & Townsend, 2016; Singh-Peterson et al., 2014), the indicators of community resilience proposed by Berkes and Ross 2013 appear to offer valid insights for empirical research. For example, Miles (2015) took the indicators of community resilience to develop a theoretical framework of well-being, identity, services and capitals for research in community disaster resilience. Similarly, Cretney (2016) showed the value of a community-led and defined framework for resilience by aggregating and reapplying the social factors from Berkes and Ross 2013 and other scholars. The application of Berkes and Ross 2013 in this empirical context advances resilience research by providing an evidence base for understanding the role and function of critical attributes of community resilience in practice (Maclean et al., 2014).

## Impact—The Contribution to Community Resilience from the Top-Cited Articles

In this section, the most highly cited articles are organised into three categories according to how they used the concepts proposed by Berkes and Ross 2013 and are discussed in terms of how they have contributed to a more integrated approach to understanding community resilience in a range of different contexts (Table 17.1). The categories identified are: (1) those articles that support the concepts and arguments of Berkes and Ross 2013 and have applied them directly in their research in new and different contexts, (2) those articles that cite Berkes and Ross 2013 and enhance understanding of community resilience by contextualising it within a new body of literature or adding new dimensions towards a more integrated approach; and (3) those articles that critique the concept of community resilience or highlight gaps in understanding or challenges in its application.

Of the articles that support Berkes and Ross 2013's propositions in terms of community resilience (n = 14), the main themes discussed were place, adaptive capacity and scale and how these were applied and related to community resilience in different research contexts. Regarding

**Table 17.1** Distribution of the top-cited articles by years and categories

| | *2013* | *2014* | *2015* | *2016* | *2017* | *2018* | *2019* | *2020* | *Total* |
|---|---|---|---|---|---|---|---|---|---|
| Support | | 2 | 4 | 4 | 4 | | | | 14 |
| Enhance | 1 | 3 | 4 | 5 | 5 | 3 | | 1 | 22 |
| Critique | 1 | 7 | 2 | 7 | 1 | 1 | | | 19 |

people–place relationships, Amundsen (2015) positioned place attachment as an essential entry point for engaging communities around the need to adapt in the face of climate change. Similarly, Tschakert et al. (2017) highlighted an explicit focus on local values and place-based decision-making processes to address climate threats. Berkes and Ross 2013's delineation of community resilience was also used in tourism research, where Moscardo et al. (2017) demonstrated how tourism could enhance adaptive capacity by contributing to the social capital and well-being of destination communities. In applying community resilience to fisheries research, Finkbeiner (2015) highlighted the importance of adaptive capacity for the resilience of small-scale fishing cooperatives in their ability to move across fishing strategies in response to changing environmental conditions. In crisis management research, Williams et al. (2017) suggested that community resilience attributes of community embeddedness (place attachment) and social networks could be activated to accelerate decision speed when preparing for and responding to a crisis. Spies et al. (2014) also found that activating social networks and using landscape visualisation tools accelerated communication and social learning for enhanced bushfire resilience. In terms of scale, Béné et al. (2015) supports Berkes and Ross 2013's position and argues for the importance of identifying a meaningful system boundary when talking about food security. Cumming et al. (2015) also highlighted the importance of cross-scale institutional linkages between top-down and bottom-up influences in the context of protected area management.

These articles are examples of the category of articles that support Berkes and Ross 2013's representation of community resilience and particularly the critical roles of place, adaptive capacity and scale.

Among the articles that take up the challenge laid down by Berkes and Ross 2013 and enhance the concept of community resilience towards a more integrated approach (n = 22), the concept of agency is a central focus. Several scholars have explored how agency can be influenced and developed, as well as how it relates to adaptive capacity. For example, Cheshire et al. (2015) added dimensions from psychology and trauma recovery to show that an integrated approach to community resilience also needs to consider exogenous forces on communities, not just endogenous characteristics such as local social capital and networks. Bristow and Healy (2014) introduced the dynamics of collective agency of communities (that also operate in their firms, organisations and governments), that create and perpetuate norms, culture, expectations, etc., and thus play a significant role in organising individual behaviours and influencing community resilience. From a strong systems perspective, Liu et al. (2017) stressed that a more integrated approach to community resilience would address the multiple interactions between intra-individual, interpersonal and socio-ecological variables. Also from a socio-ecological systems perspective, Dwiartama and Rosin (2014), using actor-network theory, viewed nonhumans as active agents that can enhance or hinder resilience pathways. By examining case studies of resilience in Indonesia's rice agriculture and the New Zealand kiwifruit industry, Dwiartama and Rosin (2014) demonstrated how human adaptive capacity is influenced by a heterogeneous network of agents that include nonhumans (such as crops and diseases) and the relationships that are developed with a wider set of influencing agents (e.g. climate, technology, markets).

Many of these articles have explored processes of adaptive capacity building in different empirical contexts. For example, Islam and Walkerden (2015) investigated the role of NGOs in building adaptive capacity through their work in establishing partnerships and networks that enhance social learning among households in Bangladesh for disaster resilience and recovery. Kontokosta and Malik (2018) expanded the concept of community resilience to apply to larger and more complex urban systems, but at nested scales of neighbourhoods. In their research of urban resilience in New York City, they suggested research on neighbourhood (community) resilience should also take into account the interrelationships with and resilience capacity of surrounding neighbourhoods.

Some authors have bravely attempted to address the operationalisation of scale. In relation to temporal scales in disaster management, Khalili et al. (2015) argued that community resilience must be understood over longer periods of time, as communities must be able to deal with changes that require a rapid and flexible response at all phases of disaster: pre-disaster, response and recovery. Quandt (2018) highlighted intra-community scales by illustrating the differences among groups of people within a community in an analysis of livelihood resilience in agricultural households in Kenya. Julie L. Davidson et al. (2016) identified cross-scale drivers of community resilience in policymaking and the influences of larger scale governance and planning. They highlighted how higher-level planning policies could hinder local strategies for community resilience. Furthering this argument, Ashkenazy et al. (2018) evaluated agricultural development policies and strategies for fourteen case study communities across Europe and beyond and found that policies and strategies to build community resilience were working against each other at different scales. These studies highlight the effects of scale and the need to understand community resilience as a fluid and dynamic, but also a necessarily bounded concept.

Culture has also been shown to play a vital role in an integrated concept of community resilience with arguments that community resilience should reflect the culture of local communities. In their study of forest-based communities in Canada, Lyon and Parkins (2013) showed the link between local cultural systems and adaptive capacity in terms of socially coordinated responses to change. Several scholars elaborated on culture as a key component of community resilience. For example, Cox and Hamlen (2015) developed an applied disaster resilience assessment index for use in rural and remote communities, underlining that community resilience has to reflect unique local culture, shifting demographics, and the specific capacities and vulnerabilities of each community. McCrea et al. (2014) demonstrated how the preservation and stewardship of traditional local knowledge and practices plays a substantial role in community resilience for climate change adaptation, particularly for subsistence-based, Indigenous, rural communities. Pascua et al. (2017) stressed the importance of including socio-cultural perspectives in decision-making for community resilience, and proposed using the cultural ecosystem services model. The premise being that understanding and incorporating cultural values and connections will facilitate community buy-in and support for management decisions in social-ecological systems.

While the concept of community resilience is certainly being used as a framework for understanding change, resilience and sustainability across a growing range of disciplines that deal with change, articles also provide a thorough critique of conceptualising and applying the concept (n = 19). For instance, several authors have identified some persistent challenges in the conceptualisation that limit its usefulness as a guiding concept for practice and policy. Alexander (2013) warned practitioners and scholars alike not to 'overwork' the term resilience, nor to expect that community resilience as a framing concept can necessarily provide greater insight or modelling capacity. Brown (2014) specifically noted how the multiple meanings and interpretations of resilience create tensions between normative and analytical stances on resilience. Simon and Randalls (2016) also highlighted how community resilience can be interpreted differently according to ontological conceptions within and between academic domains. All these arguments relate to attempts to quantify and measure community resilience in terms of its definable attributes.

Although Berkes and Ross 2013 did not propose specific methods for assessment in this paper, they did review and evaluate a range of methods in a subsequent paper, with recommendations for multi-method and participatory approaches (Ross & Berkes, 2014). Over the years, there has been increased interest in developing standardised tools for assessments and quantitative measures. Quinlan et al. (2016) argued that the amount of information needed to assess a community's resilience is much more than is useful for decision-making, and simpler measures should be sought. Singh-Peterson et al. (2014) compared two main forms of community resilience assessment tools for disaster management: community-based participatory assessment tools and top-down assessment tools. They concluded in favour of participatory approaches that can acknowledge a broader range of community values whereas top-down assessment can imply homogeneity across the study area. Thus converting the concept of community resilience into an assessment tool for communities to track their progress towards resilience appears to remain problematic.

Another criticism of the community resilience concept (mostly from the political sciences) is that it does not adequately consider or reflect local politics and power relations, particularly because community resilience initiatives often involve complex collaborations. From a public administration perspective, Duit (2016) noted how collaborative and participatory processes can be very difficult to initiate and sustain, and are highly sensitive to power asymmetries and elite capture. Keating et al. (2017) also argued for more explicit articulation of the roles of power and agency in community resilience, particularly in relation to understanding and measuring social capital for disaster risk management.

Several authors took insights from public administration and political science in attempts to integrate power relations with community resilience. For example, Estoque and Murayama (2014) suggested using local administrative or political units as logical boundaries for assessing social-ecological systems. Looking specifically at the role of institutions, Sjostedt (2015) emphasised the need to recognise the fundamentally political nature of institutions when thinking about community resilience, as the play out of politics can lead to path dependency and the reinforcement of certain views or the adoption of certain knowledge in addressing the environmental challenges we face today. Exploring relationships between power and knowledge further, Ingalls and Stedman (2016) recommended using discourse analysis as a valuable approach to community resilience thinking, to understand how power functions to structure environmental knowledge within social-ecological systems and to what effect.

There were also some questions raised in the articles around the role of place attachment in community resilience. While many articles showed place attachment as a key motivating factor for communities to engage and coordinate around community resilience strategies, in a study of mobility and resilience, Adams (2016) found place attachment also led to immobilisation if people do not want to move or change, that can significantly hinder resilience. In a similar line of argument, Townshend et al. (2014) in demonstrating a clear positive association between cohesion and resilience in post-disaster communities, warned that cohesion is a driver of some but not all resilience attributes, and that focusing too strongly on cohesion assumes that people want to work together. McCrea et al. (2014) noted some confusion between community resilience and community well-being, particularly when local strategies often have the dual objective. They highlight how measures of community resilience frequently overlap with measures of community well-being and that it can be unclear in an applied sense how the two concepts differ.

## Emerging Themes in Community Resilience (Articles Between 2019 and 2021)

This section explores the most recent articles citing Berkes and Ross 2013 and identifies new applications and emerging themes that contribute towards a more integrated approach to community resilience thinking and practice. The review focuses on articles published from 2019 to 2021.

### *Emerging Applications in Climate Change Responses and Energy Transition*

Building community resilience for climate change adaptation has been a topic of research since the Berkes and Ross paper published in 2013, however recent citations in this field show an increasing interest in collaborative and multi-scale approaches. Several scholars have explored communities' vulnerability, adaptive capacity and resilience in relation to climate-induced events such as wildfire (Higuera et al., 2019) and floods (Kim et al., 2022). There is also an increased interest in understanding communities not as a whole, but as nested subsets, and ways particular community groups can adapt to changes associated with climate change. Some examples include studies of gendered experiences of women to climate stress in South-west Bangladesh (Jordan, 2019), a fishing community in the US (Haugen et al., 2021) and small-scale farmers in Kenya (Elijah & Odiyo, 2020). The attributes of community resilience proposed by Berkes and Ross 2013 have also been applied to leadership and policymaking around responses to climate change (Crosweller & Tschakert, 2021), as well as planning and governance for low-carbon resilient cities integrating both climate mitigation and adaption approaches (He et al., 2020). Apart from recognising community development processes for adapting to the changing climate, recent work on community resilience points towards a growing movement of bottom-up community-based action across Europe intended to support larger-scale systems change for a zero-carbon future (Revell & Henderson, 2018).

The recent literature also discusses the impacts and opportunities for community resilience in the context of energy system transition and the development of renewable energy. One paper uses community resilience in an analysis of the recent experience of coal communities in the US to highlight the need to provide opportunities and resources to support communities navigating the social and economic impacts of energy transition (Roemer & Haggerty, 2021). Similarly, in a case study of microgrid adoption in the US, Ajaz (2019) proposes resilience thinking as one of the key drivers for achieving decentralised energy systems, as it facilitates deliberative solutions to dealing with changing conditions caused by climate change or other threats. Zurba and Bullock (2019) used community resilience to explore bioenergy partnerships with Indigenous business leaders in Canada, highlighting the critical roles communities play in supporting transitions to renewable resources.

### *Resilience in Relation to Community and Environmental Health (The COVID-19 Pandemic)*

Community Resilience has recently been applied in studies on how communities can prevent and cope with emerging health and environmental risks. Highlighting the linkages between social and ecological systems, and particularly between natural disasters and public health, Cairns et al. (2017) highlight the need to frame, explain and illustrate how public health and overall well-being are supported by the environment. Citing increasing concern about climate change impacts, Banwell et al. (2018) provide insights into the conceptual, technical and operational synergies between management paradigms for disaster risk reduction and climate change adaptation for health outcomes. More recently, in response to the COVID-19 pandemic, De Garine-Wichatitsky et al. (2021) draw on community resilience concepts from a socio-ecological systems perspective and highlight the need to understand, and adaptively manage, the complex inter-linkages between human health and bio-physical environments.

The recent studies also highlight the importance of resilience thinking to social determinants of health in terms of public administration and community resources and supports. Marks and Knassmüller (2021), using COVID-19 as an example, argue that community resilience approaches can help articulate the underlying dynamics in society and its administrative systems to allow for better incorporation of long-term perspectives in short-term strategies, particularly in times of disturbances and temporary perturbations. Stressing the importance of localised responses to global issues, Paarlberg et al. (2020), in their research of COVID-19 support funding across the US, explored the role of community-based philanthropic organisations in building community resilience during a crisis.

Other studies highlight links between individual and community resilience and positive health outcomes. For example, Winstanley (2020) demonstrate how mindfulness, self-care and resilience skills may be useful strategies to respond to overdose-related compassion fatigue at the individual- and community level in the US and Shevell and Denov (2021) stressed resilience-enabling systems at the family, community, national, global and intergenerational levels for war-affected children and youth in post-genocide Rwanda.

### *Larger Systems—Food Systems and International Development*

Community resilience has also been used to explain complexity in global food systems. For example, Stoll et al. (2021) used community resilience to examine the interconnections between local and regional food systems in the US and Canada and Elias and Marsh (2020) applied community resilience in studies of sustainable farming and sustainable food systems in California. The trend towards applying community resilience to global-scale issues is again observed in the paper by Haggard et al. (2019) which examines international development rhetoric across several emerging economies through a lens of community resilience.

Recent papers also apply community resilience to social justice issues and highlight tensions between different geographic scales. For example, Matin et al. (2018) propose the term 'equitable resilience' and highlight the need to consider in social and environmental justice along with power influences, in any proposed economic development. Ensor et al. (2021) build on this work and describe policy and practice for equitable resilience in Bangladesh. Community resilience has also been applied to the global peacebuilding context, where Joseph and Juncos (2020) explore the resilience of the EU and other peace-keeping institutions.

### *The Application of Technology in Improving Resilience*

A growing body of research has examined the role of new technologies in enhancing community resilience. Several authors have emphasised how information and communication technologies (ICT) can contribute to resilience for rural communities. For example, Roberts and Townsend (2016) show how Internet-enabled technologies impact creative work, which in turn build cultural capacity for community resilience in rural Scotland. Ashmore et al. (2017), in their study on community-led broadband initiatives in England, support that technology can contribute to social resilience by developing new spatial understandings of community identity and identifying new communities of interest, but also point to the existing uneven rural development. Recent studies explore how technology can be applied in crisis and disaster management. Comes et al. (2019) highlight ICT can help a community to generate a coherent understanding of a crisis and enable de facto decision-makers in the community to make well-informed decisions. Technological applications can also be found in the forms of simulation and gaming for enhancing community resilience against large-scale earthquakes in Japan (Toyoda, 2020), as well as an online tool for self-assessment of community resilience in Europe (Revell & Henderson, 2018). Hence, with the broader applications of technology, it is necessary to examine how community resilience could benefit from technology and the problems encountered with the more substantial role of technology in future research.

### *Focus on Developing Countries*

Special attention has been paid recently to understanding resilience in communities from emerging economies and developing countries, which were underrepresented in 'Western' research in the past. For example, with the increased urgency of climate change threats, several scholars have examined resilience in disaster management and climate change response for communities in the Pacific including Vanuatu (McDonnell, 2020) and the Solomon Islands (Basel et al., 2020). The propositions of Berkes and Ross 2013 also support research in capacity building for communities in African countries. Some examples include applying community resilience to examine the impact of trophy hunting tourism on local communities and wildlife in Namibia (Thomsen et al., 2022), enhancing the sustainability of tree growing by smallholder farmers in Madagascar (Kubo et al., 2022), and exploring ecosystem services based income-generating activities in Tanzania (Mngumi, 2019). In the context of urban resilience and change, Botezat et al. (2021) explore the effects of the largest post-communist urbanisation waves that took place in Romania in the early 2000s. Merilainen (2020) uses community resilience to analyse how disaster-affected people and neighbourhoods in Chile self-organised to develop a new discourse of 'robustness' that empowered recovery.

### *Urban Resilience*

While initially described as being more relevant to smaller and spatially defined communities, community resilience is now being increasingly applied in urban and spatially diffuse communities. As more than half of the world's population lives in urban areas, and the urban population is projected to continue increasing, community resilience is emerging as a key concept in dealing with issues within complex urban systems and in strategies for making cities more resilient and sustainable for the future. With a focus on planning and embedding resilience within policymaking, Wardekker et al. (2020) analyse urban flood risk management in Rotterdam and highlight the importance of incorporating local aspirations for resilience-building in urban plans. Drawing on cases in Spain, LopezDeAsiain and Díaz-García (2020) stress the importance of participatory approaches for building urban resilience. In a novel study, Valjakka (2019) illustrate how communities of urban-based online 'hackers' in Hong Kong contribute to cultural resilience, which is a direct challenge to socio-political and spatial-aesthetic dynamics of the urban 'public space'. In another example of community resilience in an urban context, Quaranta et al. (2019) demonstrate how an art-food hub in Italy is a central place of connection that can stimulate adaptation to change, enhancing local resilience and strengthening urban–rural links for the local communities.

While these studies all acknowledge the importance of applying resilience thinking to address complex urban issues, questions are also raised about whether community resilience as an analytic framework can capture the complexity of urban systems, particularly the diversity of people and political influences, diffuse boundaries and scales, cultures and sub-cultures, the range of vulnerabilities and motivations for change within cities.

### *Community Resilience Beyond the Socio-Ecological Interest*

The influence of Berkes and Ross 2013 can also be seen in studies beyond the initial interest in social-ecological systems. Community resilience has recently been applied in corporate and organisational settings. For example, Chapman et al. (2020) examine community resilience in relation to workplace teams across occupational settings, specifically the capacity of team members as a collective to resist, bounce back or recover from adversity. Lim et al. (2020) look into workforce resilience in the context of human resource development, highlighting resilience as an essential element needed to effectively deal with unexpected changes that arise in the current workplace.

Focusing on marginal groups, Miller and Bowen (2020) studied youth experiencing homelessness in New York and concluded the need for developmentally relevant services to enhance resilience. Tippens (2020), in an analysis of urban Congolese refugee communities in Kenya, illustrated how refugees use social capital to access and distribute resilience-promoting resources.

## Persistent Challenges in Applying Community Resilience

When Berkes and Ross 2013 was first published, their conceptualisation of community resilience was challenged (in the same issue) by Debra J. Davidson (2013, p. 21), as being an "unfinished project" with "a long way to go". Davidson (2013) questioned whether the 'community' was an appropriate unit of analysis for resilience and argued that resilience was essentially an outcome of individual agency. She argued that 'community resilience' did not adequately address power relationships and differences in adaptive capacities within communities, especially where trauma may be involved. Davidson (2013) also challenged community resilience as a concept more suited to stochastic events rather than longer-term macro-trends such as resources decline and climate change. Although Ross and Berkes (2013) provided a response to Davidson's (2013) challenge highlighting many points of agreement, some of Davidson's concerns have persisted over the years and are discussed below.

While the concept of community resilience has been applied to many different types of communities and the different types of changes that communities may experience, there are also some recurring themes about the challenges faced when applying community resilience as an analytical concept.

Of note is the persistent challenge of defining a meaningful boundary around the system or community of focus. Drawing a boundary requires making decisions about which factors and influences should be included in the study and which not. In complex, interconnected systems conclusions drawn about resilience are inevitably influenced by where and how the boundaries of the system are drawn.

Another observation is that while some authors have highlighted the need to consider the role of underlying dynamics in social-ecological systems, particularly those around power and influence relationships, many of the articles reviewed here describe a 'snapshot' or single point in time studies. Particularly in relation to studies on vulnerable communities, some authors have questioned if a community resilience approach adequately addresses underlying system dynamics and root causes of vulnerability (e.g. Cutter (2016)).

The more recent studies applying community resilience to social justice issues highlight inequalities in processes and pathways for resilience. They also raise ethical questions about the framing of community resilience as a shared aspiration, when there may be contested futures within a community, or privileged types of knowledge, communication and understanding. Indeed, several articles in this review have shown how subsets of communities can experience the same phenomenon differently.

While several articles described resilience as an embedded concept in planning and community development (particularly the more recent studies on urban and city planning), resilience in itself is not a clear policy objective. Localised strategies for building community resilience have been shown to be convoluted by a range of factors, including exogenous and scale factors that communities have little influence over.

By looking at how community resilience has been used in different contexts also raises the question of whether a more integrated concept of community resilience can be at once comprehensive enough to capture the different drivers and types of change within a community, yet specific enough to draw clear boundaries and inform practical strategies for building resilience.

## Conclusion

This chapter has reviewed 96 (55 of the most highly cited and 41 most recent) articles that cite Berkes and Ross 2013 to show the breadth and nature of the impact that it has had on the world. We conclude that it has indeed influenced thought and practice towards a more integrated approach to community resilience.

Over the decade since its publication, the Berkes and Ross paper has been used to support a wide range of research agendas, methods, analytical frameworks, and has informed discussions of research findings that have over time enriched our collective understanding of communities as complex systems learning from and adapting to change.

The concept of community resilience has become more integrated over time to include greater understanding and appreciation of power relations between actors, individual and collective agency, adaptive capacities and the role of institutions. It has an enhanced ethical dimension that makes approaches to community resilience more inclusive of vulnerable and marginal groups, and gives greater consideration of inequalities, unconscious bias and privilege among subsets of communities. This stronger ethical dimension enables the application of community resilience approaches to complex and contentious contemporary social, environmental and energy justice issues.

An initial focus on community resilience as a place-based concept and discussions on the role of place attachment as a motivating factor in resilience has been questioned over time and this has opened up a much broader contextualisation of community. Communities being researched using the community resilience framework are no longer place-based communities, but include complex supply chains, global food systems, corporate workforces and even online communities.

Similarly, there is increasing recognition of the role of culture, including the incorporation of traditional knowledge, but also in relation to formal institutions and the role of the arts in shaping and perpetuating culture, social learning and social norms that might enhance (or impede) resilience.

Despite a more integrated understanding and more widespread application, there are still some fundamental challenges in terms of operationalising the concept of community resilience as a tool for communities to use. While it is an attractive aspiration in community planning and development, it is not yet a clear policy objective with a consistent set of measurable targets and outcomes. There are challenges in defining the appropriate scale at which strategies for building resilience will be effective, where there is enough specificity to enable collective agency and collaboration yet enough scope to be inclusive of longer-term dynamics, and neighbouring or nested influences.

Nevertheless, it is clear that the article by Berkes and Ross has had a significant influence on the scholarship and practical application of community resilience in the last ten years. This chapter is a celebration of that contribution by showcasing the nature and extent of the impact it has had. It is also an expression of gratitude to Helen Ross for her mentoring and guidance over an even longer period of time.

## References


Adams, H. (2016). Why populations persist: Mobility, place attachment and climate change. *Population and Environment, 37*(4), 429–448. https://doi.org/10.1007/s11111-015-0246-3

Ajaz, W. (2019). Resilience, environmental concern, or energy democracy? A panel data analysis of microgrid adoption in the United States. *Energy Research & Social Science, 49*, 26–35. https://doi.org/10.1016/j.erss.2018.10.027

Alexander, D. E. (2013). Resilience and disaster risk reduction: An etymological journey. *Natural Hazards and Earth System Sciences, 13*(11), 2707–2716. https://doi.org/10.5194/nhess-13-2707-2013

Amundsen, H. (2015). Place attachment as a driver of adaptation in coastal communities in Northern Norway. *Local Environment, 20*(3), 257–276. https://doi.org/10.1080/13549839.2013.838751

Ashkenazy, A., Calvão Chebach, T., Knickel, K., Peter, S., Horowitz, B., & Offenbach, R. (2018). Operationalising resilience in farms and rural regions—Findings from fourteen case studies. *Journal of Rural Studies, 59*, 211–221. https://doi.org/10.1016/j.jrurstud.2017.07.008

Ashmore, F. H., Farrington, J. H., & Skerratt, S. (2017). Community-led broadband in rural digital infrastructure development: Implications for resilience. *Journal of Rural Studies, 54*, 408–425. https://doi.org/10.1016/j.jrurstud.2016.09.004

Banwell, N., Rutherford, S., Mackey, B., & Chu, C. (2018). Towards improved linkage of disaster risk reduction and climate change adaptation in health: A review. *International Journal of Environmental Research and Public Health, 15*(4), 793. https://doi.org/10.3390/ijerph15040793

Basel, B., Goby, G., & Johnson, J. (2020). Community-based adaptation to climate change in villages of Western province, Solomon Islands. *Marine Pollution Bulletin, 156*, 111266. https://doi.org/10.1016/j.marpolbul.2020.111266

Béné, C., Headey, D., Haddad, L., & von Grebmer, K. (2015). Is resilience a useful concept in the context of food security and nutrition programmes? Some conceptual and practical considerations. *Food Security, 8*(1), 123–138. https://doi.org/10.1007/s12571-015-0526-x

Berkes, F., Folke, C., & Colding, J. (1998). *Linking social and ecological systems: Management practices and social mechanisms for building resilience*. Cambridge University Press.

Berkes, F., & Ross, H. (2013). Community resilience: Toward an integrated approach. *Society & Natural Resources, 26*(1), 5–20. https://doi.org/10.1080/08941920.2012.736605

Botezat, A., David, M., Incaltarau, C., & Nijkamp, P. (2021). The illusion of urbanization: Impact of administrative reform on communities' resilience.

*International Regional Science Review, 44*(1), 33–84. https://doi.org/10.1177/0160017620964861

Bristow, G., & Healy, A. (2014). Regional resilience: An agency perspective. *Regional Studies, 48*(5), 923–935. https://doi.org/10.1080/00343404.2013.854879

Brown, K. (2014). Global environmental change I: A social turn for resilience? *Progress in Human Geography, 38*(1), 107–117. https://doi.org/10.1177/0309132513498837

Cairns, M. R., Cox, C. E., Zambrana, J., Flotemersch, J., Lan, A., Phillips, A., et al. (2017). Building multi-country collaboration on watershed management: Lessons on linking environment and public health from the Western Balkans. *Reviews on Environmental Health, 32*(1), 15–22. https://doi.org/10.1515/reveh-2016-0053

Chapman, M. T., Lines, R. L. J., Crane, M., Ducker, K. J., Ntoumanis, N., Peeling, P., et al. (2020). Team resilience: A scoping review of conceptual and empirical work. *Work and Stress, 34*(1), 57–81. https://doi.org/10.1080/02678373.2018.1529064

Cheshire, L., Esparcia, J., & Shucksmith, M. (2015). Community resilience, social capital and territorial governance. *Ager, 18*, 7–7.

Comes, T., Meesters, K., & Torjesen, S. (2019). Making sense of crises: The implications of information asymmetries for resilience and social justice in disaster-ridden communities. *Sustainable and Resilient Infrastructure, 4*(3), 124–136. https://doi.org/10.1080/23789689.2017.1405653

Cox, R. S., & Hamlen, M. (2015). Community disaster resilience and the rural resilience index. *The American Behavioral Scientist, 59*(2), 220–237. https://doi.org/10.1177/0002764214550297

Cretney, R. M. (2016). Local responses to disaster: The value of community led post disaster response action in a resilience framework. *Disaster Prevention and Management, 25*(1), 27–40. https://doi.org/10.1108/DPM-02-2015-0043

Crosweller, M., & Tschakert, P. (2021). Disaster management leadership and policy making: A critical examination of communitarian and individualistic understandings of resilience and vulnerability. *Climate Policy, 21*(2), 203–221. https://doi.org/10.1080/14693062.2020.1833825

Cumming, D. H. M., Cumming, G. S., Allen, C. R., Ban, N. C., Biggs, D., Biggs, H. C., et al. (2015). Understanding protected area resilience: A multi-scale, social-ecological approach. *Ecological Applications, 25*(2), 299–319. https://doi.org/10.1890/13-2113.1

Curtis, A., Ross, H., Marshall, G. R., Baldwin, C., Cavaye, J., Freeman, C., et al. (2014). The great experiment with devolved NRM governance: Lessons from community engagement in Australia and New Zealand since the 1980s.

*Australasian Journal of Environmental Management, 21*(2), 175–199. https://doi.org/10.1080/14486563.2014.935747
Cutter, S. L. (2016). Resilience to what? Resilience for whom? *Geographical Journal, 182*(2), 110–113. https://doi.org/10.1111/geoj.12174
Darnhofer, I. (2014). Resilience and why it matters for farm management. *European Review of Agricultural Economics, 41*(3), 461–484. https://doi.org/10.1093/erae/jbu012
Davidson, D. J. (2013). We still have a long way to go, and a short time to get there: A response to Fikret Berkes and Helen Ross. *Society & Natural Resources, 26*(1), 21–24. https://doi.org/10.1080/08941920.2012.749758
Davidson, J. L., Jacobson, C., Lyth, A., Dedekorkut-Howes, A., Baldwin, C. L., Ellison, J. C., et al. (2016). Interrogating resilience: Toward a typology to improve its operationalization. *Ecology and Society, 21*(2), 27. https://doi.org/10.5751/ES-08450-210227
De Garine-Wichatitsky, M., Binot, A., Ward, J., Caron, A., Perrotton, A., Ross, H., et al. (2021). "Health in" and "Health of" social-ecological systems: A practical framework for the management of healthy and resilient agricultural and natural ecosystems. *Frontiers in Public Health, 8*, 616328. https://doi.org/10.3389/fpubh.2020.616328
DeVerteuil, G., & Golubchikov, O. (2016). Can resilience be redeemed? Resilience as a metaphor for change, not against change. *City, 20*(1), 143–151. https://doi.org/10.1080/13604813.2015.1125714
Duit, A. (2016). Resilience thinking: Lessons for public administration. *Public Admin, 94*(2), 364–380. https://doi.org/10.1111/padm.12182
Dwiartama, A., & Rosin, C. (2014). Exploring agency beyond humans: The compatibility of Actor-Network Theory (ANT) and resilience thinking. *Ecology and Society, 19*(3), 28. https://doi.org/10.5751/ES-06805-190328
Elias, M., & Marsh, R. (2020). Innovations in agricultural and food systems sustainability in California. *Case Studies in the Environment, 4*(1), 1–14. https://doi.org/10.1525/cse.2019.002170
Elijah, V., & T., & Odiyo, J., O. (2020). Perception of environmental spillovers across scale in climate change adaptation planning: The case of small-scale farmers' irrigation strategies, Kenya. *Climate, 8*(1), 3. https://doi.org/10.3390/cli8010003
Ensor, J. E., Mohan, T., Forrester, J., Khisa, U. K., Karim, T., & Howley, P. (2021). Opening space for equity and justice in resilience: A subjective approach to household resilience assessment. *Global Environmental Change, 68*, 102251. https://doi.org/10.1016/j.gloenvcha.2021.102251
Estoque, R. C., & Murayama, Y. (2014). Social–ecological status index: A preliminary study of its structural composition and application. *Ecological Indicators, 43*, 183–194. https://doi.org/10.1016/j.ecolind.2014.02.031

Finkbeiner, E. M. (2015). The role of diversification in dynamic small-scale fisheries: Lessons from Baja California Sur, Mexico. *Global Environmental Change, 32*, 139–152. https://doi.org/10.1016/j.gloenvcha.2015.03.009
Haggard, R., Cafer, A., & Green, J. (2019). Community resilience: A meta-study of international development rhetoric in emerging economies. *Community Development, 50*(2), 160–180. https://doi.org/10.1080/15575330.2019.1574851
Haugen, B. I., Cramer, L. A., Waldbusser, G. G., & Conway, F. D. L. (2021). Resilience and adaptive capacity of Oregon's fishing community: Cumulative impacts of climate change and the graying of the fleet. *Marine Policy, 126*, 104424. https://doi.org/10.1016/j.marpol.2021.104424
He, X., Lin, M., Chen, T.-L., Liu, B., Tseng, P.-C., Cao, W., et al. (2020). Implementation plan for low-carbon resilient city towards sustainable development goals: Challenges and perspectives. *Aerosol and Air Quality Research, 20*(3), 444–464. https://doi.org/10.4209/aaqr.2019.11.0568
Higuera, P. E., Metcalf, A. L., Miller, C., Buma, B., McWethy, D. B., Metcalf, E. C., et al. (2019). Integrating subjective and objective dimensions of resilience in fire-prone landscapes. *BioScience, 69*(5), 379–388. https://doi.org/10.1093/biosci/biz030
Ingalls, M. L., & Stedman, R. C. (2016). The power problematic: Exploring the uncertain terrains of political ecology and the resilience framework. *Ecology and Society, 21*(1), 6–6. https://doi.org/10.5751/ES-08124-210106
Islam, R., & Walkerden, G. (2015). How do links between households and NGOs promote disaster resilience and recovery?: A case study of linking social networks on the Bangladeshi coast. *Natural Hazards, 78*(3), 1707–1727. https://doi.org/10.1007/s11069-015-1797-4
Jordan, J. C. (2019). Deconstructing resilience: Why gender and power matter in responding to climate stress in Bangladesh. *Climate and Development, 11*(2), 167–179. https://doi.org/10.1080/17565529.2018.1442790
Joseph, J., & Juncos, A. E. (2020). A promise not fulfilled: The (non) implementation of the resilience turn in EU peacebuilding. *Contemporary Security Policy, 41*(2), 287–310. https://doi.org/10.1080/13523260.2019.1703082
Keating, A., Campbell, K., Szoenyi, M., McQuistan, C., Nash, D., & Burer, M. (2017). Development and testing of a community flood resilience measurement tool. *Natural Hazards and Earth System Sciences, 17*(1), 77–101. https://doi.org/10.5194/nhess-17-77-2017
Khalili, S., Harre, M., & Morley, P. (2015). A temporal framework of social resilience indicators of communities to flood, case studies: Wagga wagga and Kempsey, NSW, Australia. *International Journal of Disaster Risk Reduction, 13*, 248–254. https://doi.org/10.1016/j.ijdrr.2015.06.009

Kim, H., Woosnam, K. M., Marcouiller, D. W., & Kim, H. (2022). Seeking anticipatory adaptation: Adaptive capacity and resilience to flood risk. *Environmental Hazards, 21*(1), 36–57. https://doi.org/10.1080/17477891.2021.1902783

Kontokosta, C. E., & Malik, A. (2018). The resilience to emergencies and disasters index: Applying big data to benchmark and validate neighborhood resilience capacity. *Sustainable Cities and Society, 36*, 272–285. https://doi.org/10.1016/j.scs.2017.10.025

Kubo, H., Andriamizaka, C., Yamanoshita, M., & Shiraishi, T. (2022). Enhancing the sustainability of tree growing by smallholder farmers in remote areas: An insight from Madagascar. *International Journal of Agricultural Sustainability, 20*(1), 42–53. https://doi.org/10.1080/14735903.2021.1914447

Lim, D. H., Hur, H., Ho, Y., Yoo, S., & Yoon, S. W. (2020). Workforce resilience: Integrative review for human resource development. *Performance Improvement Quarterly, 33*(1), 77–101. https://doi.org/10.1002/piq.21318

Liu, J. J. W., Reed, M., & Girard, T. A. (2017). Advancing resilience: An integrative, multi-system model of resilience. *Personality and Individual Differences, 111*, 111–118. https://doi.org/10.1016/j.paid.2017.02.007

LopezDeAsiain, M., & Díaz-García, V. (2020). The importance of the participatory dimension in urban resilience improvement processes. *Sustainability, 12*(18), 7305. https://doi.org/10.3390/SU12187305

Lyon, C., & Parkins, J. R. (2013). Toward a social theory of resilience: Social systems, cultural systems, and collective action in transitioning forest-based communities. *Rural Sociology, 78*(4), 528–549. https://doi.org/10.1111/ruso.12018

Maclean, K., Cuthill, M., & Ross, H. (2014). Six attributes of social resilience. *Journal of Environmental Planning and Management, 57*(1), 144–156. https://doi.org/10.1080/09640568.2013.763774

Marks, P., & Knassmüller, M. (2021). Bouncing around in a turbulent teaching environment: Teaching and researching resilience for public administration. *Teaching Public Administration, 41*(1). https://doi.org/10.1177/01447394211000407.

Matarrita-Cascante, D., Trejos, B., Qin, H., Joo, D., & Debner, S. (2017). Conceptualizing community resilience: Revisiting conceptual distinctions. *Community Development, 48*(1), 105–123. https://doi.org/10.1080/15575330.2016.1248458

Matin, N., Forrester, J., & Ensor, J. (2018). What is equitable resilience? *World Development, 109*, 197–205. https://doi.org/10.1016/j.worlddev.2018.04.020

McCrea, R., Walton, A., & Leonard, R. (2014). A conceptual framework for investigating community wellbeing and resilience. *Rural Society, 23*(3), 270–282. https://doi.org/10.1080/10371656.2014.11082070

McDonnell, S. (2020). Other dark sides of resilience: Politics and power in community-based efforts to strengthen resilience. *Anthropological Forum, 30*(1–2), 55–72. https://doi.org/10.1080/00664677.2019.1647828

Merilainen, E. (2020). The dual discourse of urban resilience: Robust city and self-organised neighbourhoods. *Disasters, 44*(1), 125–151. https://doi.org/10.1111/disa.12367

Miles, S. B. (2015). Foundations of community disaster resilience: Well-being, identity, services, and capitals. *Environmental Hazards, 14*(2), 103–121. https://doi.org/10.1080/17477891.2014.999018

Miller, B., & Bowen, E. (2020). I know where the rest of my life is going: Attitudinal and behavioral dimensions of resilience for homeless emerging adults. *Journal of Social Service Research, 46*(4), 553–570. https://doi.org/10.1080/01488376.2019.1607647

Mngumi, L. E. (2019). Socio-ecological resilience to climate change effects in peri-urban areas: Insights from the Pugu and Kazimzumbwi forest reserves of Dar es Salaam, Tanzania. *GeoJournal, 86*(1), 339–355. https://doi.org/10.1007/s10708-019-10071-9

Moscardo, G., Konovalov, E., Murphy, L., McGehee, N. G., & Schurmann, A. (2017). Linking tourism to social capital in destination communities. *Journal of Destination Marketing & Management, 6*(4), 286–295. https://doi.org/10.1016/j.jdmm.2017.10.001

Paarlberg, L. E., LePere-Schloop, M., Walk, M., Ai, J., & Ming, Y. (2020). Activating community resilience: The emergence of COVID-19 funds across the United States. *Nonprofit and Voluntary Sector Quarterly, 49*(6), 1119–1128. https://doi.org/10.1177/0899764020968155

Pascua, P., McMillen, H., Ticktin, T., Vaughan, M., & Winter, K. B. (2017). Beyond services: A process and framework to incorporate cultural, genealogical, place-based, and indigenous relationships in ecosystem service assessments. *Ecosystem Services, 26*, 465–475. https://doi.org/10.1016/j.ecoser.2017.03.012

Quandt, A. (2018). Measuring livelihood resilience: The Household Livelihood Resilience Approach (HLRA). *World Development, 107*, 253–263. https://doi.org/10.1016/j.worlddev.2018.02.024

Quaranta, G., Dalia, C., Salvati, L., & Salvia, R. (2019). Building resilience: An art-food hub to connect local communities. *Sustainability, 11*(24), 1–15. https://doi.org/10.3390/su11247169

Quinlan, A. E., Berbés-Blázquez, M., Haider, L. J., & Peterson, G. D. (2016). Measuring and assessing resilience: Broadening understanding through multiple disciplinary perspectives. *The Journal of Applied Ecology, 53*(3), 677–687. https://doi.org/10.1111/1365-2664.12550

Revell, P., & Henderson, C. (2018). Operationalising a framework for understanding community resilience in Europe. *Regional Environmental Change, 19*(4), 967–979. https://doi.org/10.1007/s10113-018-1390-y

Roberts, E., & Townsend, L. (2016). The contribution of the creative economy to the resilience of rural communities: Exploring cultural and digital capital. *Sociol Ruralis, 56*(2), 197–219. https://doi.org/10.1111/soru.12075

Roemer, K. F., & Haggerty, J. H. (2021). Coal communities and the U.S. energy transition: A policy corridors assessment. *Energy Policy*, *151*, 112112. https://doi.org/10.1016/j.enpol.2020.112112.

Ross, H., & Berkes, F. (2014). Research approaches for understanding, enhancing, and monitoring community resilience. *Society & Natural Resources, 27*(8), 787–804. https://doi.org/10.1080/08941920.2014.905668

Shevell, M. C., & Denov, M. S. (2021). A multidimensional model of resilience: Family, community, national, global and intergenerational resilience. *Child Abuse & Neglect, 119*(Pt 2), 105035–105035. https://doi.org/10.1016/j.chiabu.2021.105035

Simon, S., & Randalls, S. (2016). Geography, ontological politics and the resilient future. *Dialogues in Human Geography, 6*(1), 3–18. https://doi.org/10.1177/2043820615624047

Singh-Peterson, L., Salmon, P., Goode, N., & Gallina, J. (2014). Translation and evaluation of the baseline resilience indicators for communities on the sunshine coast, Queensland Australia. *International Journal of Disaster Risk Reduction, 10*, 116–126. https://doi.org/10.1016/j.ijdrr.2014.07.004

Sjostedt, M. (2015). Resilience revisited: Taking institutional theory seriously. *Ecology and Society, 20*(4), 23. https://doi.org/10.5751/ES-08034-200423

Spies, T. A., White, E. M., Kline, J. D., Paige Fischer, A., Ager, A., Bailey, J., et al. (2014). Examining fire-prone forest landscapes as coupled human and natural systems. *Ecology and Society, 19*(3), 9. https://doi.org/10.5751/ES-06584-190309

Stoll, J. S., Harrison, H., L., Sousa, E. D., Callaway, D., Collier, M., Harrell, K., et al. (2021). Alternative seafood networks during COVID-19: Implications for resilience and sustainability. *Frontiers in Sustainable Food Systems, 5*. https://doi.org/10.3389/fsufs.2021.614368

Thomsen, J. M., Lendelvo, S., Coe, K., & Rispel, M. (2022). Community perspectives of empowerment from trophy hunting tourism in Namibia's Bwabwata National Park. *Journal of Sustainable Tourism, 30*(1), 223–239. https://doi.org/10.1080/09669582.2021.1874394

Tippens, J. A. (2020). Urban Congolese refugees' social capital and community resilience during a period of political violence in Kenya: A qualitative study.

*Journal of Immigrant & Refugee Studies, 18*(1), 42–59. https://doi.org/10.1080/15562948.2019.1569744
Townshend, I., Awosoga, O., Kulig, J., & Fan, H. (2014). Social cohesion and resilience across communities that have experienced a disaster. *Natural Hazards, 76*(2), 913–938. https://doi.org/10.1007/s11069-014-1526-4
Toyoda, Y. (2020). A framework of simulation and gaming for enhancing community resilience against large-scale earthquakes: Application for achievements in Japan. *Simulation & Gaming, 51*(2), 180–211. https://doi.org/10.1177/1046878119899424
Tschakert, P., Barnett, J., Ellis, N., Lawrence, C., Tuana, N., New, M., et al. (2017). Climate change and loss, as if people mattered: Values, places, and experiences. *Wiley Interdisciplinary Reviews. Climate Change, 8*(5), n/a. https://doi.org/10.1002/wcc.476.
Valjakka, M. (2019). Urban hacking: The versatile forms of cultural resilience in Hong Kong. *Urban Design International, 25*(2), 152–164. https://doi.org/10.1057/s41289-019-00079-5
Wardekker, A., Wilk, B., Brown, V., Uittenbroek, C., Mees, H., Driessen, P., et al. (2020). A diagnostic tool for supporting policymaking on urban resilience. *Cities, 101*, 102691. https://doi.org/10.1016/j.cities.2020.102691
Williams, T. A., Gruber, D. A., Sutcliffe, K. M., Shepherd, D. A., & Zhao, E. Y. (2017). Organizational response to adversity: Fusing crisis management and resilience research streams. *Academy of Management Annals, 11*(2), 733–769.
Winstanley, E. L. (2020). The bell tolls for thee & thine: Compassion fatigue & the overdose epidemic. *International Journal of Drug Policy, 85*, 102796–102796. https://doi.org/10.1016/j.drugpo.2020.102796
Zurba, M., & Bullock, R. (2019). Bioenergy development and the implications for the social wellbeing of Indigenous peoples in Canada. *Ambio, 49*(1), 299–309. https://doi.org/10.1007/s13280-019-01166-1

CHAPTER 18

# Tracing the Impact and Contribution of the "Six Attributes of Social Resilience"

*Kirsten Maclean and Bradd Witt*

## Introduction

> … there is an urgent need for researchers to shift focus away from examining the nature of resilience to accelerating learning about fostering resilience in practice. (Fazey et al., 2018, p. 30)

The concept of resilience continues to attract attention both in theory and practice, and in a range of disciplinary domains. This can be linked to a progression in international developing thinking that has moved

K. Maclean (✉)
CSIRO Environment, Dutton Park, QLD, Australia
e-mail: kirsten.maclean@csiro.au

B. Witt
School of the Environment, The University of Queensland, Brisbane, QLD, Australia
e-mail: bwitt@uq.edu.au



C. Baldwin and S. van Bommel (eds.), *Rural Development for Sustainable Social-ecological Systems*,
Palgrave Studies in Environmental Sustainability,
https://doi.org/10.1007/978-3-031-34225-7_18

to embrace sustainability, and the need for balance between the social, economic and environmental aspects of development. More recent theory has considered the complex and dynamic links between and across these aspects of development, and a focus on systems theory and behaviour, and how to manage for the resilience and adaptive capacity of linked social-ecological systems in the face of change. In practice, and for example within political fora, the concepts of resilience and social resilience (and the linked concepts of adaptive capacity and vulnerability) are increasingly used to inform decision-making in domains such as 'climate resilience and adaptation' (e.g. Climate & Disaster: Technical Reports, 2020; Dale et al., 2015, 2016; Gawith et al., 2016; Kim et al., 2020), and disaster management (e.g. Townshend et al., 2015) and adaptation, relocations and migration of people in relation to climate change (Clarke et al., 2019) and large-scale projects such as dams (Gong et al., 2020).

Resilience thinking has developed separately in specific bodies of literature (Maclean et al., 2017). It has evolved in the literature focussed on social-ecological systems (e.g. Adger, 2000; Gunderson & Holling, 2002; Gunderson et al., 1995, 2010; Holling, 1973, 1996, 2001; Holling et al., 2001; Westley, 2002), and that published within developmental psychology (e.g. Buikstra et al., 2010 Garmezy, 1974; Hegney et al., 2007; Kaplan, 2005; Masten & Garmezy, 1985) and mental health to do with the resilience of individuals (e.g. Berkes & Ross, 2013; Buikstra et al., 2010). Although thinking on social resilience in both these fields has developed separately for several decades, there has been little cross-fertilisation or even mutual awareness (Berkes & Ross, 2013; Maclean et al., 2017). Drawing on these literatures, a team of Australian researchers (including Ross, Maclean, Cuthill, Witt and Jansen) have focussed their individual and collective scholarship on considering what constitutes resilience at diverse scales. Their research has woven together theory and empirical work to consider the social aspects of individual

---

K. Maclean
The Fenner School of Environment & Society, The Australian National University, Canberra, Australia

B. Witt
School of Agriculture and Food Sustainability, The University of Queensland, Brisbane, Australia

resilience (e.g. Maclean et al., 2014), community resilience (e.g. Ross and Berkes, 2013) and, as is the focus of this chapter, regional resilience (e.g. Cuthill et al., 2008; Maclean et al., 2013, 2014, 2017, 2018; Ross et al., 2010).

Our work, located conceptually across the literature discussed above, recognises the nested nature of social and social-ecological systems and considers the mutual influences that exist within these systems from the individual to the global scale (Cuthill et al., 2008). We define 'social resilience' as a way "to describe how individuals, communities and societies adapt, transform and potentially become stronger when faced with environmental, social, economic or political challenges" (Cuthill et al., 2008, p. 146). We focus on peoples' strengths and the social disadvantages they may wish to overcome. This definition, that draws from research into social-ecological systems and from the social and health sciences, acknowledges the complex interplay between "social, cultural, spiritual, economic and environmental systems" (Cuthill et al., 2008, p. 146), and in particular the synergistic relationship between people and the environments in which they live.

In this chapter, we draw on published research (Cuthill et al., 2008; Maclean et al., 2013, 2014, 2017; Ross et al., 2010) from a study undertaken to develop a framework to monitor and report on social resilience at the regional scale in far north Queensland, Australia. The study was conducted in partnership with a set of representatives from government agencies, non-government organisations and Aboriginal Traditional Owners from the Wet Tropics region, who sought to improve their understanding of the social dimensions of resilience, to enhance their planning and capacity building interventions. Six case studies were conducted in the region, including a total of 71 interviews, and was focussed on specific 'change events' that impacted on the livelihoods of those living in the place-based region. They included: dairy deregulation on the Atherton Tablelands; changes to the water allocation process in the upper zone of the Barron River catchment; declaration of the Wet Tropics World Heritage area; rapid expansion in the Cairns coastal zone; the third crown of thorns starfish outbreak in the near shore area of Cairns; and the institutional role of Girringun Aboriginal Corporation in enhancing social resilience. Interviews and case studies explored how the relevant industry, organisation or group was able to 'get through the change period'; the skills they thought their community would need to be proactive in the face of future changes; and the kinds of future changes and challenges they predicted (see Ross et al., 2010 for specific details of the studies).

This work has informed the Social-ecological Systems literature in five main ways.

First, we drew on the literature highlighted above to show how each could learn from the other towards a more integrated and enriched understanding of both social-ecological systems and social resilience (Maclean et al., 2017).

Second, we illustrated how to describe a region as a linked, and nested, social-ecological system and provided a succinct description of the Wet Tropics region as such a system (see Maclean et al., 2013; Ross et al., 2010).

Third, we drew parallels and insights between literature on social-ecological systems and adaptive management and used these to better understand and describe the adaptive governance strategies used by an Aboriginal organisation from the Wet Tropics (Girringun Aboriginal Corporation) to enhance its social-ecological system (see Maclean et al., 2013).

Fourth, we derived six attributes of social resilience—(1) knowledge skills and learning; (2) community networks; (3) people-place connections; (4) community infrastructure; (5) diverse and innovative economy; (6) engaged governance (see Maclean et al., 2014).

Finally, we drew insights from our theoretical and empirical work to develop, design and complete a 'Social Resilience Monitoring and Reporting Framework' (see Ross et al., 2010) that could be used by regional managers to inform their management of their social-ecological system, in three potential ways. As articulated in Ross et al. (2010) and Maclean et al. (2014, p. 153), these potential management options were described to include:

> 1. being aware of social resilience context, without intention to intervene ('acknowledge it'); 2. making constructive use of social resilience characteristics in management actions ('use it'); or 3. meeting environmental and other mandates in ways that simultaneously enhance social resilience ('grow it').

The aim of the present chapter is to consider how our work on the six attributes of social resilience (Maclean et al., 2014) has informed the scholarship on the social dimensions of resilience of social-ecological systems, and thus interrogate a portion of the ongoing legacy of the work of Professor Helen Ross, and her colleagues, to this important and growing field of research. The chapter is structured in the following way: first we present the methodology used to better understand the extent to which this work has informed the literature. Next, we present the results of our interrogation, and end with a discussion about the current and ongoing contributions of our work to the body of literature.

## Method

We used two specific methods to collect data to analyse how our work has informed and continues to inform the theoretical and empirical literature and the management of social resilience of social-ecological systems. First, to gather data about the academic impact of our work, we reviewed the number of citations of Maclean et al. (2014) that appeared in Scopus until 12th May, 2021. We extracted the reference data, abstract, keywords and PDF of all these papers into Endnote ($n = 73$). Some papers were inaccessible and left out of the analysis ($n = 10$). In total, we systematically interrogated 63 papers to consider the extent to which the authors used the 6 attributes to inform their work. Our analysis focused on whether they: (1) simply cited MacLean et al., 2014 to augment their introduction or literature review; (2) critiqued the 6 attributes; or (3), used the 6 attributes, by either, integrating and adapting, or extending the 6 attributes in their work. We also noted: (1) the topic of the work; (2) the geographic location where the work was conducted; and, if the item was published in a journal (as opposed to in an edited book) and (3) the journal in which they published their work.

Second, to visualise the focus of the literature that either cited Maclean et al. (2014) in passing or applied or extended this work, we gathered the keywords downloaded from Scopus database and displayed them as a word cloud created using TagCrowd.com. We omitted the terms Social, Community and Resilience as these appeared on almost every list of keywords. The resulting cloud only uses terms that appear two or more times. The Scopus database often gathers more than the normal number of approximately six keywords associated with peer-reviewed literature and book chapters.

## Results and Discussion

In this section, we present a description and analysis of the results (as presented in Tables 18.1 and 18.2, Appendix A and Appendix B). To understand how the six attributes of social resilience (Maclean et al., 2014) has informed, and continues to inform scholarship on social resilience, it is first useful to consider the topics of the journals in which authors have chosen to publish their work as this can indicate the breadth of impact on academic disciplinary scholarship. It is also useful to analyse the geographic location where authors conducted their empirical research as this provides insights into the influence of our work on geographic regions. Further, we can develop a better understanding of the theoretical impact of our work by considering the extent to which authors have used, adapted or extended the 6 attributes (Maclean et al., 2014). We end the chapter with a discussion on the kinds of impact this work appears to have had, to date, on the growing field of research.

### *Journals and Disciplinary Focus of the Citing Research*

Five book chapters were included in the data set, each with a core focus on social resilience, including two chapters focussed on the Australian context, one in Turkey and one in the developing world and economies. We identified that the set of papers ($n = 58$) were published in a mix of journals covering diverse disciplines and subject areas. The most popular journals in which authors chose to publish their work were: Sustainability ($n = 8$); Community Development ($n = 6$), Ecology and Society ($n = 4$); Area ($n = 2$), Geoforum ($n = 2$), Journal of Environmental Planning and Management ($n = 2$), Rural Society ($n = 2$) and Land Use Policy ($n = 2$). Disciplinary-focussed journals included those that publish Geography related research (Geoforum, Area, Antipode, Human Ecology, Tourism Geographies), Sociology (Sociologia Rurualis) and those that deal with social-ecological systems (e.g. Ecology and Society).

A large subsection of the journals publish work on social and natural resource management (e.g. Local Environment, Society and Natural Resources, Natural Resources Forum, Reviews in Aquaculture, Ocean and Coastal Management, Rangelands Journal), management of 'hazards' (e.g. Plant Disease, Natural Hazards, Natural Disasters) and journals relating to natural resource policy and management (e.g. Land Use Policy,

**Table 18.1** The number of empirical studies by geographic region, and by country

| *Geographic region* | *Country* | *Count* | *Geographic region* | *Country* | *Count* |
|---|---|---|---|---|---|
| **Australia, New Zealand, and the Pacific (Total)** | | **17** | **South America (Total)** | | **5** |
| | Australia | 11 | | Brazil | 2 |
| | Queensland | 8 | | Bolivia | 1 |
| | New Zealand | 5 | | Panama | 1 |
| | Fiji | 1 | | Suriname | 1 |
| **Asia (Total)** | | 7 | **North America (Total)** | | **4** |
| | Bangladesh | 2 | | Canada | 3 |
| | Cambodia | 1 | | USA | 1 |
| | China—1 | 1 | | | |
| | Hong Kong | 1 | | | |
| | India | 1 | | | |
| | Vietnam | 1 | | | |
| **Europe and the United Kingdom (Total)** | | 7 | **Africa (Total)** | | **3** |
| | United Kingdom | 3 | | Senegal | 1 |
| | Italy | 2 | | South Africa | 1 |
| | Denmark | 1 | | Sudan | 1 |
| | Finland | 1 | | | |
| | Ireland | 1 | | | |
| | Spain | 1 | | | |
| | Turkey | 1 | | | |
| | 'European study' | 1 | | | |

**Table 18.2** Total number of references analysed as either using the Maclean et al. (2014) as a 'simple citation', 'critique' or 'used, adapted and/or extended'

| *Simple citation to support authors' own theoretical and empirical assertions* | *Critiqued the 6 attributes* | *Used, adapted and/or extended the research in Maclean et al. (2014) including the 6 attributes* |
|---|---|---|
| 37 | 0 | 26 |

Journal of Natural Resources Policy Research, Journal of Environmental Planning and Management, Marine Policy).

Another group of journals include those related to rural life (e.g. Australian Journal of Rural Health, Journal of Rural Studies, Rural

Development, Rural Society), change and development (e.g. Journal of Organisational change, Technological Forecasting and Social Change, Community Development, Regional Environmental Change) and urban research (Urban Research and Practice: City, Territory and Architecture: Sustainability Cities and Society). This diversity reflects the comment made at the outset of the chapter: that resilience has grown to span a diverse range of research and practice. Perhaps of greater interest is to consider the more disparate journals where relevant papers were published. These journals included the International Journal of Green Economics; a journal by the name of Auditing; and Tourism Planning and Development.

### *Geographic Location of Citing Research*

Our analysis considered the geographic location of the empirical research. The results show the use of the Maclean et al. (2014) work by authors who have conducted empirical research in a total of 26 countries. This indicates the relative impact of the Maclean et al. (2014) over many geographies. Twelve of the items from the 63 were desktop studies and literature reviews, as such this work was not conducted in a specific geographic location, and thus we did not record a geographic location for these items. Of the remaining items ($n = 51$), the empirical research was conducted in diverse locations as outlined in Table 18.1. Given Maclean et al. (2014) drew on empirical research conducted in Australia (specifically the Wet Tropics region of Queensland), it is not surprising that the largest number of empirical studies that drew on the paper, were conducted in Australia, with 8 of those being conducted in Queensland.

### *How Authors Used Maclean et al. (2014) in Their Publication*

In this section, we discuss how authors used the Maclean et al. (2014) study in their own work. First, we present the overall results of our interrogation (see Table 18.2). Next, we interrogate how the authors chose to use Maclean et al. (2014) as either a simple citation (see Appendix A) or whether they engaged with the work in a deeper way (e.g. critiqued, used, adapted and/or extended the 6 attributes, see Appendix B). A common way that we classified papers in these groups was (1) as a 'simple citation' if a paper included Maclean et al. (2014) with a string of other references to support the theoretical and empirical assertions made by the

**Fig. 18.1** Word cloud of the keywords from the papers that were only simple citations

authors; (2) one that critiqued the paper; (3) a paper that used Maclean et al. (2014), if the authors reproduced any aspect of the paper, for example, the definition of social resilience and/or the six attributes or a part thereof; (4) one that adapted Maclean et al. (2014) if authors used the six attributes in a framework and/or to analyse their own empirical data; (5) extended, if authors used the six attributes as a central organising concept in their paper, and applied to a different case study.

As is shown in Tables 18.1 and 18.2, 37 authorship teams used the Maclean et al. (2014) as a simple citation and the remaining 26 authorship teams used, adapted or extended the research shared in Maclean et al. (2014). None of the collected papers critiqued Maclean et al. (2014) or the 6 attributes that are presented therein. Figure 18.1 shows the word cloud of the keywords from the 37 papers that were only simple citations. Note that we omitted the terms Social, Community and Resilience as these appeared on almost every list of keywords.

### *Simple Citation*

As indicated in Table 18.2 and Appendix A, over half of the citing papers (37 out of a total of 63 items) were identified as having used the Maclean et al. (2014) as a 'simple citation'. Our analysis identified that this includes authors citing the reference in the introduction and/or literature review as part of a series of other papers on the topic of social resilience ($n = 22$). Our analysis shows that of the remaining publications,

Maclean et al. (2014) is cited in a variety of ways. Firstly, to support the authors' definition of 'cultural resilience' (see Saxena, 2020), 'individual adaptive capacity' (see Durocher et al., 2016), 'people-place attachment' (see Guo et al., 2018) and 'engaged governance' (see Altay Kaya, 2019). Secondly, Maclean et al. (2014) is also cited to support the authors' assertion that social resilience is proactive rather than reactive (see Mahon & Hyyryläinen, 2019), that leadership is essential for the development of shared visions (see Hung, 2017), and that community networks and social capital help actors to seek new opportunities during times of change (see Wesselow & Mashele, 2019). Third, several authors cited the Maclean et al. (2014) to highlight how a focus on sets of attributes (or 'components') is necessary to support communities undergoing times of change, including: to upscale responses to climate change (see Kim et al., 2020), to support urban communities to recover from disturbances (see Torabi et al., 2021) and to build resilience to disasters (see Townshend et al., 2015). One desktop review cited Maclean et al. (2014) as a case study about ways to monitor social resilience through sets of attributes (see Ross & Berkes, 2014) and a further study cited it with regard to the need for methods to measure community and social resilience (see Brown, 2016). One paper, Ollerenshaw et al. (2016), incorrectly cited Maclean et al. (2014) to supplement their argument about emergency responses to build community resilience. Given the topic and focus of their analysis, the role of local government to build resilience to support communities during a (natural disaster) biosecurity event, they likely meant to cite Maclean et al. (2018).

### *Used, Adapted and/or Extended Maclean et al. (2014)*

As indicated in Table 18.2 and Appendix B, less than half (26 out of a total of 63 items) were identified as having used ($n$ = 13), adapted ($n$ = 12) and/or extended ($n$ = 1) the research presented in Maclean et al. (2014). Figure 18.2 shows a word cloud of the keywords from the 26 papers that used, adapted and/or extended the research in Maclean et al. (2014). Only words appearing 2 or more times are shown. Note that we omitted the terms Social, Community and Resilience. Here we discuss each of these in turn.

**Fig. 18.2** Word cloud of the keywords from the papers that used, adapted and/ or extended the research in Maclean et al. (2014)

### *Used the Research Presented in Maclean et al. (2014) Including the Six Attributes*

Nine of the 13 items in this category, directly used the definition of social resilience that was originally presented in Cuthill et al. (2008) and used to frame the empirical research (and the derived six attributes of social resilience), presented in Maclean et al. (2014). One item (see Leite et al., 2019) paraphrased this definition of social resilience in their paper. This use of Maclean et al. (2014) is notable, given that the majority of these papers presents empirical research into social resilience, that is framed by the definition coined in Maclean et al. (2014) and Cuthill et al. (2008). Three further items used either one or all the six attributes of social resilience, shared in Maclean et al. (2014) as central to the theoretical framing of the empirical work or paper. For example, Dos Reis (2015, p. 330) includes the six attributes in their theoretical framework that is developed to assess "whether and how a policy is leading towards sustainability in a broad, deep and transformational sense". Bueno and Schiavetti (2019) cite one of the six attributes (knowledge) to support their choice of indicator development in their research into marine protect areas and the impact on fishers; and Cavaye and Ross (2019) also cite one of the six attributes (engaged governance) in their desktop study as important to community resilience and development.

### *Adapted the Research Presented in Maclean et al. (2014) Including the Six Attributes*

Another set of authors adapted the 6 attributes to inform the design of their own theoretical framework for social resilience ($n$ = 3) and to analyse their data ($n = 1$, Uddin et al., 2020) or to analyse their empirical data ($n = 3$). Authors who adapted the 6 attributes to inform their theoretical framework for social resilience include the following. Dale et al. (2016) applied social resilience concepts and indicators for an assessment of social resilience, to support climate adaptation, in north Queensland, Australia. They concluded that "a purist approach to measuring and monitoring social resilience is neither practical nor feasible" (Dale et al., 2016: page). Kulig and Botey (2016) applied social resilience 'components' to highlight how a community responded to a wildfire in Canada, their work showed links between individual and community resilience. Vaneeckhaute et al. (2019) investigated the existing resiliency of a community impacted by the disruptions of a mine site in Suriname. They applied social resilience concepts to study "how systems respond (i.e. cope, adapt for transform) in the face of exogenous stressors and disruptions" (Vaneeckhaute et al. 2019, p. 735). The adaptation of the six attributes from Maclean et al. (2014) in these contexts shows now this work has informed theoretical frameworks that have been applied across individual, community and regional scales.

A further way in which authors adapted the research presented in Maclean et al. (2014) was by using the 6 attributes in their own framing of how to measure social resilience. For example, in their desktop study, Matarrita-Cascante et al. (2017, p. 105) adapted the six attributes to inform their "conceptual clarification to the term community resiliency" and conclude that "community resilience provides a better framework for understanding community change and how it is addressed". Fielke et al. (2018) use the six attributes to inform a resilience framework to interrogate four case studies of community and social resilience in New Zealand. Kaye-Blake et al. (2019, p. 161) adapted the six attributes into their own set of 'resilience indicators' in their investigation into "the possibility of using official statistics to measure community resilience" in four rural communities in New Zealand.

A final set of authors adapted the six attributes to inform their data collection methods and/or to analyse their empirical data. This indicates the impact of the Maclean et al. (2014) research as a key theoretical framework instrumental to be used as the basis of data collection, to analyse data and interrogate the results of empirical studies conducted in Europe, South America and Australia. Fabbricatti et al. (2020, p. 1) used the six attributes to inform their survey instrument, whereby they collected data on several case studies focussed on "heritage-driven practices in Italian inner peripheral areas", with the prediction that "challenges and highlights the potential that culture and heritage can develop for community resilience".

Three final authorship teams use the six attributes to analyse their data of their empirical studies. For example, Kelly et al. (2019) use the six attributes to interrogate the results of their study into the impacts of drought on rural towns in central western Queensland, and also use them to inform their recommendations with regard to building a resilient regional community. Singh-Peterson and Underhill (2017, p. 124) integrate the six attributes into their "multi-scalar, mixed methods framework" to assess rural communities' capacity for resilience, adaptation and transformation. They apply this methodological framework via a Queensland case study, to demonstrate how local actors can use it to derive "an applicable suite of locally relevant indicators" (p. 124). Finally, Chelleri et al. (2016, p. 2229) use the six attributes to help interpret the results of their empirical research into resilience and vulnerability of areas in transition, using the case study of the Southern Bolivian Altiplano from a "remote rural area of subsistence farming to a global leader in quinoa production and exportation".

### *Extended the Research Presented in Maclean et al. (2014) Including the Six Attributes*

Only one authorship team (Maclean et al., 2018) was identified to have 'extended' the Maclean et al. (2014) research in their own work. Important to note that the lead author of this publication is Maclean—also the lead author of the Maclean et al. (2014) paper. The authors extended Maclean et al. (2014) in three specific ways. First, they use the attributes as a basis to their data analysis; second, they use the 6 attributes as foundational to a framework they derived; finally, they apply the 6 attributes to a new context—biosecurity risk management. The focus of this work is to inform government responses to the management of biosecurity incursion events. The view is to design responses to build and enhance social resilience through the various stages of an incursion event, and thus support the affected community members of any given social-ecological system to adapt and potentially become stronger in the fact of future incursion events. Notably, the authors articulate how

> …policies and programs must acknowledge the actual and potential diversity of reactions to a biological invasion within any socio-ecological system to ensure a diverse approach to building adaptive capacity […] Programs could work to build the capacities of unengaged or disengaged actors, while at the same time support other groups to enhance the capacities that already exist. (Maclean et al., 2018, p. 103)

Importantly, the research presented in this paper will be used as the basis for a case study to inform a chapter (led by Patricia Howard, forthcoming) in the Global Assessment of Invasive Alien Species, for the Intergovernmental Science-Policy Platform on Biodiversity and Ecosystem Services (IPBES).

### *Overview of the Impacts*

We opened this paper with a quotation from Fazey et al. (2018) that

> …there is an urgent need for researchers to shift focus away from examining the nature of resilience to accelerating learning about fostering resilience in practice.

It is clear that Maclean et al. (2014) has been used in the literature not just as an endless rehashing of debates over exactly how to measure and define resilience in some sort of "purist" fashion (see Dale et al., 2016 p. 37), but rather the work has provided a widely cited foundation for applied research into a range of issues of resilience and development around the world.

Originating in Cuthill et al. (2008, p. 146) and reproduced in Maclean et al. (2014), as well as Maclean et al. (2017), the six attributes of social resilience were defined as:

> one way to describe how individuals, communities and societies adapt, transform, and potentially become stronger when faced with environmental, social, economic or political challenges. This working definition […] appreciates the complex interplay between social, cultural, spiritual, economic and environmental systems. In particular, it recognises the synergistic relationship between people and the environments in which they live. Our conceptualisation of social resilience focuses on people's strengths, in addition to the social disadvantages they may wish to overcome.

As we have seen, this definition has been applied in a wide range of disciplines and practice-based contexts demonstrating their applicability across individual, community and regional scales. In addition, the work has been used widely and adapted to measure social resilience. Specifically, the work has been used in studies as broad as architecture, Natural Resources Management, rural studies, disasters, and others and in diverse geographic setting.

In line with the opening quotation of Fazey et al. (2018), it is noteworthy that none of the literature reviewed for this chapter engaged in further critique or redefinition of the six attributes and this is encouraging as it demonstrates we have moved on sufficiently that people are comfortable with the attributes that people are applying it (see below). Important to note here, however, is that we do not assert that the Maclean et al. (2014) article is the definitive research on attributes of social resilience. This is certainly not the case. Berkes and Ross (2013), for example, augment these six attributes with three further ones: 'values and beliefs, leadership and positive outlooks' as central to social resilience; Walsh's (2014) work in the Bolivian Highlands (that did not cite Maclean et al., 2014) articulates other factors as being central to resilience including maintaining social heterogeneity; learning to live with (and learning from) change; creating space for building trust, building practices that favour the common good; and being able to assess power. In this chapter, we posit rather that the dimensions of social resilience articulated in Maclean et al. (2014) are regarded by those who have cited them in their work (see Appendices A and B) as effectively bedded down and empirically measurable aspects of resilience. As we have shown above, these dimensions and definition are being applied to inform practical studies and research. These points above reinforce the view that the time of theorising and critiquing in an academic sense the dimension/attributes of social resilience are largely behind us, and that the applications of the concepts in research is well established.

## Conclusion

As we finalise the drafting of this chapter, the Maclean et al. (2014) manuscript sits at a noteworthy "Field-Weighted citation impact" of 4.01 and in the 95th percentile of papers in the Scopus database. As we have seen from the above review of citing literature, the research impact of this body of work is very clear. And this, in its own right, is a significant contribution to the field. However, it raises the common question for academic work, about the ability to evaluate translation, impact, relevance and uptake into practice (Ravenscroft et al., 2017).

From the finest scale of a local town or village attempting to build and maintain resilience, to regions adapting to economic, social or climatic change, or national-scale efforts, it can be difficult to know how much the body of academic work in this space directly (or indirectly) informs practice and actions. It can at times be difficult to know how local natural resource managers, NGO, planners and so on have used this research directly or indirectly. However, the great diversity of applications shown in this chapter does suggest a wide adoption of the concepts and definitions of the six attributes of social resilience. In addition to this, the Technical Report (Ross et al., 2010), on which much of this was based, has enjoyed very wide readership as indicated by the number of reads in the ResearchGate online researcher platform (over 1,600 reads), and the high citation activity in Google Scholar (69 citations by Jan 2022). Much of the readership of this work via ResearchGate is by non-academic actors and is indicative of a wider public and community interest in the applications of resilience thinking into practice. The insights from this chapter, along with the broader body of work of Prof. Helen Ross and colleagues, do clearly show a significant contribution to research and practice in resilience thinking and practice in social-ecological systems.

## Appendix A Table Showing the 37 Papers/Book Chapters That Used Maclean et al. (2014) as a 'Simple Citation'

| *Lead author* | *Year* | *Journal/book title* | *Title of paper/chapter* | *Location of work* |
|---|---|---|---|---|
| Altay | 2019 | *Urban Book Series* | Integrating the resilience perspective into the Turkish planning system: Issues and challenges | Turkey |

(continued)

(continued)

| *Lead author* | *Year* | *Journal/book title* | *Title of paper/chapter* | *Location of work* |
|---|---|---|---|---|
| Apgar et al. | 2015 | *Ecology and Society* | Understanding adaptation and transformation through indigenous practice: The case of the Guna of Panama | Panama |
| Baldwin | 2020 | *Natural Hazards and Disaster Justice: Challenges for Australia and Its Neighbours* | Book Chapter: Justice, resilience and participatory processes | Australia (QLD) |
| Bricknell et al. | 2021 | *Reviews in Aquaculture* | Resilience of cold-water aquaculture: a review of likely scenarios as climate changes in the Gulf of Maine | USA |
| Broegaard et al. | 2020 | *Tourism Geographies* | Rural destination development contributions by outdoor tourism actors: A Bornholm case study | Denmark |
| Brown | 2016 | *Resilience, development and global change* | Book | |
| Bunch et al. | 2020 | *Ecology and Society* | Quantifying community resilience in South Sudan: The feed project (fortifying equality and economic diversification) | Sudan |
| Cradock-Henry et al. | 2018 | *Sustainability (Switzerland)* | Transformations for resilient rural futures: The case of Kaikōura, Aotearoa-New Zealand | New Zealand |
| Cramer et al. | 2018 | *Marine Policy* | Graying of the fleet: Perceived impacts on coastal resilience and local policy | USA |
| Dale et al. | 2015 | *Journal of Natural Resources Policy Research* | Can social resilience inform SA/SIA for adaptive planning for climate change in vulnerable regions? | Literature review |

(continued)

(continued)

| *Lead author* | *Year* | *Journal/book title* | *Title of paper/chapter* | *Location of work* |
|---|---|---|---|---|
| Durocher et al. | 2016 | *Auditing* | Waves of Global standardization: Small practitioners' resilience and intra-professional fragmentation within the accounting profession | Canada |
| Evans et al. | 2020 | *Plant Disease* | Pathogen incursions—Integrating technical expertise in a socio-political context | Australia (TAS) |
| Fazey et al. | 2018 | *Current Opinion in Environmental Sustainability* | Community resilience for a 1.5 °C world | Literature review |
| Gerrard | 2018 | *Area* | Developing an index of community competence in flood response for flood-affected rural parishes on the Somerset Levels and Moors using composite and spatial datasets | UK |
| Green et al. | 2017 | *Community Development* | The Local Agrifood System Sustainability/ Resilience Index (SRI): Constructing a data tool applied to counties in the southern United States | USA |
| Guo et al. | 2018 | *Sustainability* | Catalyst or barrier? The influence of place attachment on perceived community resilience in tourism destinations | China |
| Hung | 2017 | *Land Use Policy* | Formation of new property rights on government land through informal co-management: case studies on countryside guerilla gardening | Hong Kong |

(continued)

(continued)

| *Lead author* | *Year* | *Journal/book title* | *Title of paper/chapter* | *Location of work* |
|---|---|---|---|---|
| Johnson et al. | 2014 | *Human Ecology Review* | Qualitative Indicators of SR in smale-scale Fishing communities: an emphasis on perceptions and practice | USA |
| Karman | 2020 | *Journal of Organizational Change* | Flexibility, coping capacity and resilience of organizations: between synergy and support | Literature Review |
| Kim et al. | 2020 | *Geoforum* | Coordinated planning effort as multilevel climate governance: Insights from coastal resilience and climate adaptation | USA |
| Maclean et al. | 2017 | *Journal of Environmental Planning and Management* | Converging disciplinary understandings of social aspects of resilience | Australia (QLD) |
| Mahon & Hyyryläinen | 2019 | *Sociologia Ruruails* | Rural Arts Festivals as Contributors to Rural Development and Resilience | Ireland and Finland |
| McElduff & Howitt | 2018 | *Area* | Fostering coastal community resilience: Mobilising people–place relationships | Ireland |
| Mozumder et al. | 2018 | *Sustainability (Switzerland)* | Enhancing social resilience of the coastal fishing communities: A case study of hilsa (*Tenualosa ilisha* H.) fishery in Bangladesh | Bangladesh |
| Natarajan et al. | 2019 | *Antipode* | Debt-Bonded Brick Kiln Workers and Their Intent to Return: Towards a Labour Geography of Smallholder Farming Persistence in Cambodia | Cambodia |
| Ollerenshaw et al. | 2016 | *Rural Society* | Beyond the call of duty: the integral role of rural local government in emergency management | Australia (VIC) |

(continued)

(continued)

| *Lead author* | *Year* | *Journal/book title* | *Title of paper/chapter* | *Location of work* |
|---|---|---|---|---|
| Powell et al. | 2018 | *Tourism Planning and Development* | Examining Community Resilience to Assist in Sustainable Tourism Development Planning in Dong Van Karst Plateau Geopark, Vietnam | Vietnam |
| Rakhimova | 2018 | *Natural Resources Forum* | Aging with financial insecurity: social resilience and adaptation in urban areas of the United States | USA |
| Redshaw et al. | 2018 | *Australian Journal of Rural Health* | Assessing the impact of vulnerability on perceptions of social cohesion in the context of community resilience to disaster in the Blue Mountains | Australia (NSW) |
| Ribeiro & Shapira | 2019 | *Technological Forecasting and Social Change* | Anticipating governance challenges in synthetic biology: Insights from biosynthetic menthol | Desk-top study |
| Ross and Berkes | 2014 | *Society and Natural Resources* | Research Approaches for Understanding, Enhancing, and Monitoring Community Resilience | Desktop study |
| Saxena | 2020 | *Sustainability (Switzerland)* | Community self-organisation from a social-ecological perspective: 'Burlang Yatra' and revival of millets in Odisha (India) | India |
| Torabi et al. | 2021 | *Urban Research and Practice* | A framework for using the concept of urban resilience in responding to climate-related disasters | Australia (QLD) |
| Townshend et al. | 2015 | *Natural Hazards* | Social cohesion and resilience across communities that have experienced a disaster | Canada |

(continued)

(continued)

| *Lead author* | *Year* | *Journal/book title* | *Title of paper/chapter* | *Location of work* |
|---|---|---|---|---|
| Vaneeckhaute et al. | 2017 | *Community Development* | Community resilience 2.0: Toward a comprehensive conception of community-level resilience | Desktop study |
| Wesselow & Mashele | 2019 | *Human Ecology* | "Who Needs Money if You Got Hands, if You Got Plants" Forming Community Resilience in Two Urban Gardening Networks in South Africa | South Africa |

# Appendix B Table Showing 26 Papers/Book Chapters Sorted According to Whether They Used, Adapted and/or Extended the Research Presented in Maclean et al. (2014)

| *Author* | *Year* | *Journal/book* | *Title* | *Geographic location of work* | *How used* |
|---|---|---|---|---|---|
| Baldwin et al. | 2017 | *Journal of Rural Studies* | Love of the land: Social-ecological connectivity of rural landholders | Australia (QLD) | Social Resilience (SR) definition |
| Bec et al. | 2018 | *Sustainability* | Resilient and sustainable communities | Australia (QLD) | SR definition |
| Bottazzi et al. | 2018 | *Sustainability* | Measuring subjective flood resilience in Suburban Dakar: A before-after evaluation of the "Live with Water" project | Senegal | SR definition |
| Bousquet & Mathevet | 2019 | *Resilience in Social, Cultural and Political Spheres* | Chapter: Cultural resilience as the resilience of a distinctness. Distinctness from what? For what? | Spain | SR definition |
| Quaranta et al. | 2019 | *Sustainability (Switzerland)* | Building resilience: An art-food hub to connect local communities | Italy | SR definition |
| Finkbeiner et al. | 2017 | *Conservation for the Anthropocene Ocean: Interdisciplinary Science in Support of Nature and People* | Chapter: Social Resilience in the Anthropocene Ocean | Literature review | SR definition |
| Gawith et al. | 2016 | *Journal of Environmental Planning and Management* | Does community resilience mitigate loss and damage from climate related disasters? Evidence based on survey data | Fiji | SR definition |
| Teen et al. | 2020 | *Local Environment* | Disruption as opportunity for transformation? Insights from water supply contamination in Havelock North, New Zealand | New Zealand | SR definition |

(continued)

(continued)

| *Author* | *Year* | *Journal/book* | *Title* | *Geographic location of work* | *How used* |
|---|---|---|---|---|---|
| Lariman et al. | 2020 | *Sustainability (Switzerland)* | Neighbourhood Social Resilience (NSR): Definition, conceptualisation, and measurement scale development | New Zealand | SR definition |
| Leite et al. | 2019 | *Ecology and Society* | Interactions between individual, household, and fishing community resilience in southeast Brazil | Brazil | Paraphrased SR definition |
| Cavaye & Ross | 2019 | *Community Development* | Community resilience and community development: What mutual opportunities arise from interactions between the two concepts? | Desktop study | Used one of the six attributes (engaged governance) |
| Bueno & Schiavetti | 2019 | *Ocean and Coastal Management* | The influence of fisherman scale in the resilience of socio-ecological systems: An analysis using Q methodology | Brazil | Used one of the six attributes (knowledge) |
| Dos Reis, T. N. P | 2015 | *International Journal of Green Economics* | A policy analysis framework for green political economy | Theoretical framework | Used to define the six attributes of resilience |
| Kulig & Botey | 2016 | *Natural Disasters* | Facing a wildfire: What did we learn about individual and community resilience? | Canada | Adapted to inform social resilience 'components' |
| Dale, A. P. et al. | 2016 | *Climate Adaptation Governance in Cities and Regions: Theoretical Fundamentals and Practical Evidence* | Applying social resilience concepts and indicators to support climate adaptation in tropical North Queensland, Australia | Australia (QLD) | Adapted into social resilience attributes |

(continued)

(continued)

| *Author* | *Year* | *Journal/book* | *Title* | *Geographic location of work* | *How used* |
|---|---|---|---|---|---|
| Vaneeckhaute et al. | 2019 | *Community Development* | Community capitals of a Paramaca Maroon village in pictures: A Photovoice study on community resilience in the context of large-scale gold mining | Suriname | Adapted to inform theoretical framework |
| Uddin et al. | 2020 | *Journal of Environmental Management* | Community resilience to cyclone and storm surge disasters: Evidence from coastal communities of Bangladesh | Bangladesh | Adapted three attributes to inform framework & analyse data |
| Singh-Peterson & Underhill | 2017 | *Community Development* | A multi-scalar, mixed methods framework for assessing rural communities' capacity for resilience, adaptation, and transformation | Australia (QLD) | Adapted to analyse case study data |
| Faulkner et al. | 2018 | *Ecology and Society* | Analyzing community resilience as an emergent property of dynamic social-ecological systems | UK | Adapted to inform integrated understanding of resilience |
| Matarrita-Cascante et al. | 2017 | *Community Development* | Conceptualizing community resilience: Revisiting conceptual distinctions | Desktop study | Adapted to inform derived community resilience 'factors' for community |
| Fielke et al. | 2018 | *Land Use Policy* | Learning from resilience research: Findings from four projects in New Zealand | New Zealand | Adapted to inform framework (dimensions of resilience) |

(continued)

(continued)

| *Author* | *Year* | *Journal/book* | *Title* | *Geographic location of work* | *How used* |
|---|---|---|---|---|---|
| Kayne-Blake et al. | 2019 | *Rural Society* | Testing indicators of resilience for rural communities | New Zealand | Adapted to inform resilience indicators |
| Kelly & Phelps | 2019 | *Rangeland Journal* | Looking beyond the D.U.S.T.-building resilient rangeland communities | Australia (QLD) | Adapted for data analysis & inform recommendations |
| Fabbricatti et al. | 2020 | *City, Territory and Architecture* | Heritage Community Resilience: towards new approaches for urban resilience and sustainability | Italy | Adapted for survey instrument |
| Chelleri et al. | 2016 | *Regional Environmental Change* | Does community resilience decrease social–ecological vulnerability? Adaptation pathways trade-off in the Bolivian Altiplano | Southern Bolivian Altiplano | Adapted to interpret their data |
| Maclean et al. | 2018 | *Geoforum* | Building social resilience to biological invasions. A case study of Panama Tropical Race 4 in the Australian Banana Industry | Australia (QLD) | Extended to inform a novel conceptual framework for biosecurity incursion decision-making |

**Acknowledgements** Naturally, we would like to express our gratitude to Prof. Helen Ross, as well as Michael Cuthill and Danni Jansen (all co-authors of various papers cited in this chapter). We thank Professor Claudia Baldwin and Dr. Severine van Bommel, for their invitation to join this Liber Amicorum that celebrates the career, to date, of our colleague and mentor Professor Helen Ross. We extend our thanks to Dr. Stuart Whitten (CSIRO) and an anonymous reviewer, whose feedback on an earlier version has resulted in a much-improved chapter. We appreciate the input and support from the many organisations and individuals who contributed to the original study: Terrain NRM Ltd, the Wet Tropics Management Authority, Aboriginal Rainforest Council (disbanded after 2008), Girringun Aboriginal Corporation, Great Barrier Reef Marine Park Authority and Queensland Department of Communities. We also thank all the 71 participants in the original case studies for their valuable time and insights at the time of the research. Finally, we gratefully acknowledge the funding support from the Marine and Tropical Sciences Research Facility (project MTSRF 4.9.7) and The University of Queensland that funded the original project research.

## References

Adger, W. N. (2000). Social and ecological resilience: Are they related? *Progress in Human Geography, 23*, 347–364.

Altay Kaya, D. (2019. Integrating the resilience perspective into the Turkish planning system: Issues and challenges. In. N. Uzun, Ö. Burcu Özdemir Sarı, S. S. Özdemir (Eds.), *Urban book series* (pp. 213–233). Springer.

Apgar, M. J., Allen, W., Moore, K., & Ataria, J. (2015). Understanding adaptation and transformation through indigenous practice: The case of the Guna of Panama. *Ecology and Society, 20*(1). https://doi.org/10.5751/ES-07314-200145

Baldwin, C. (2020). Justice, resilience and participatory processes. In A. Lukasiewicz & C. Baldwin (Eds.), *Natural Hazards and disaster justice: Challenges for Australia and its neighbours* (pp. 279–298). Springer Singapore.

Baldwin, C., Smith, T., & Jacobson, C. (2017). Love of the land: Social-ecological connectivity of rural landholders. *Journal of Rural Studies, 51*, 37–52. https://doi.org/10.1016/j.jrurstud.2017.01.012

Bec, A., Moyle, B., & Moyle, C. L. (2018). Resilient and sustainable communities. *Sustainability (Switzerland), 10*(12). https://doi.org/10.3390/su10124810

Berkes, F., & Ross, H. (2013). Community resilience: Toward an integrated approach. *Society & Natural Resources, 26*(1), 5–20. https://doi.org/10.1080/08941920.2012.736605

Bottazzi, P., Winkler, M. S., Boillat, S., Diagne, A., Maman Chabi Sika, M., Kpangon, A., Faye, S., & Speranza, C. I. (2018). Measuring subjective flood resilience in Suburban Dakar: A before-after evaluation of the "Live with Water" project. *Sustainability (Switzerland), 10*(7). https://doi.org/10.3390/su10072135

Bousquet, F., & Mathevet, R. (2019). Cultural resilience as the resilience of a distinctness. Distinctness from what? For what? In R. Benjamin, E. Martin, & N. Marie (Eds.), *Resilience in social, cultural and political spheres* (pp. 305–321). Springer Fachmedien Wiesbaden.

Bricknell, I. R., Birkel, S. D., Brawley, S. H., Van Kirk, T., Hamlin, H. J., Capistrant-Fossa, K., & Moeykens, S. (2021). Resilience of cold water aquaculture: A review of likely scenarios as climate changes in the Gulf of Maine. *Reviews in Aquaculture, 13*(1), 460–503. https://doi.org/10.1111/raq.12483

Broegaard, R. B. (2020). Rural destination development contributions by outdoor tourism actors: A Bornholm case study. *Tourism Geographies*. https://doi.org/10.1080/14616688.2020.1795708

Brown, K. (2016). *Resilience, development and global change*. Taylor & Francis.

Bueno, P. F., & Schiavetti, A. (2019). The influence of fisherman scale in the resilience of socio-ecological systems: An analysis using Q methodology. *Ocean and Coastal Management, 169*, 214–224. https://doi.org/10.1016/j.ocecoaman.2018.12.008

Buikstra, E., Ross, H., Baker, P., Hegney, D. G., King, C., Mclachlan, K., & Rogers-Clark, C. (2010). The components of resilience—Perceptions of an Australian Rural Community. *Journal of Community Psychology, 38*(8), 975–991. https://doi.org/10.1002/jcop.20409

Bunch, M. J., Pathan, S., Battaglia, A. G., Greer-Wootten, B., Mascoll, A., Russell, T., & Folkema, J. (2020). Quantifying community resilience in South Sudan: The feed project (fortifying equality and economic diversification). *Ecology and Society, 25*(2), 1-43. https://doi.org/10.5751/ES-11450-250212

Cavaye, J., & Ross, H. (2019). Community resilience and community development: What mutual opportunities arise from interactions between the two concepts? *Community Development, 50*(2), 181–200. https://doi.org/10.1080/15575330.2019.1572634

Chelleri, L., Minucci, G., & Skrimizea, E. (2016). Does community resilience decrease social–ecological vulnerability? Adaptation pathways trade-off in the Bolivian Altiplano. *Regional Environmental Change, 16*(8), 2229–2241. https://doi.org/10.1007/s10113-016-1046-8

Clarke, T., McNamara, K. E., Clissold, R., & Nunn, P. D. (2019). Community-based adaptation to climate change: Lessons from Tanna Island, Vanuatu. *Island Studies Journal, 14* (1) 59–80. https://doi.org/10.24043/isj.80

Climate and Disaster: Technical Reports. (2020). *CSIRO, Australia*. https://www.csiro.au/en/research/natural-disasters/bushfires/report-climate-disaster-resilience

Cradock-Henry, N. A., Fountain, J., & Buelow, F. (2018). Transformations for resilient rural futures: The case of Kaikōura, Aotearoa-New Zealand. *Sustainability (Switzerland), 10*(6). https://doi.org/10.3390/su10061952

Cramer, L. A., Flathers, C., Caracciolo, D., Russell, S. M., & Conway, F. (2018). Graying of the fleet: Perceived impacts on coastal resilience and local policy. *Marine Policy, 96*, 27–35. https://doi.org/10.1016/j.marpol.2018.07.012

Cuthill, M., Ross, H., Maclean, K., Owens, K., Witt, B., & King, C. (2008). Reporting social outcomes of development: An analysis of diverse approaches. *The International Journal of Interdisciplinary Social Science, 3*, 145–158.

Dale, A., Vella, K. J., & Cottrell, A. (2015). Can social resilience inform SA/SIA for adaptive planning for climate change in vulnerable regions? *Journal of Natural Resources Policy Research, 7*(1), 93–104. https://doi.org/10.1080/19390459.2014.963371

Dale, A. P., Vella, K., Potts, R., Voyce, B., Stevenson, B., Cottrell, A., King, D., Babacan, H., Boon, H., Gooch, M., & Pert, P. (2016). Applying social resilience concepts and indicators to support climate adaptation in tropical North Queensland, Australia. In *Climate adaptation governance in cities and regions: Theoretical fundamentals and practical evidence* (pp. 23–43). Wiley Blackwell.

Dos Reis, T. N. P. (2015). A policy analysis framework for green political economy. *International Journal of Green Economics, 9*(3–4), 330–353. https://doi.org/10.1504/IJGE.2015.075197

Durocher, S., Gendron, Y., & Picard, C. F. (2016). Waves of Global standardization: Small practitioners' resilience and intra-professional fragmentation within the accounting profession. *Auditing, 35*(1), 65–88. https://doi.org/10.2308/ajpt-51303

Evans, K. J., Scott, J. B., & Barry, K. M. (2020). Pathogen incursions—Integrating technical expertise in a socio-political context. *Plant Disease, 104*(12), 3097–3109. https://doi.org/10.1094/PDIS-04-20-0812-FE

Fabbricatti, K., Boissenin, L., & Citoni, M. (2020). Heritage Community Resilience: Towards new approaches for urban resilience and sustainability. *City, Territory and Architecture, 7*(1). https://doi.org/10.1186/s40410-020-00126-7

Faulkner, L., Brown, K., & Quinn, T. (2018). Analyzing community resilience as an emergent property of dynamic social-ecological systems. *Ecology and Society, 23*(1). https://doi.org/10.5751/ES-09784-230124

Fazey, I., Carmen, E., Chapin III, F. S., Ross, H., Rao-Williams, J., Lyon, C., Connon, I. L. C., Searle, B. A., & Knox, K. (2018). Community resilience for a 1.5 °C world. *Current Opinion in Environmental Sustainability, 31*, 30-40. https://doi.org/10.1016/j.cosust.2017.12.006

Fielke, S. J., Kaye-Blake, W., Mackay, A., Smith, W., Rendel, J., & Dominati, E. (2018). Learning from resilience research: Findings from four projects in New Zealand. *Land Use Policy, 70*, 322–333. https://doi.org/10.1016/j.landusepol.2017.10.041

Finkbeiner, E. M., Oleson, K. L. L., & Kittinger, J. N. (2017). Social resilience in the Anthropocene Ocean. In P. Levin & M. Poe (Eds.), *Conservation for the Anthropocene Ocean: Interdisciplinary science in support of nature and people* (pp. 89–106). Elsevier Inc.

Garmezy, N. (Ed.). (1974). *The study of competence in children at risk for severe psychopathology*. Wiley.

Gawith, D., Daigneault, A., & Brown, P. (2016). Does community resilience mitigate loss and damage from climaterelated disasters? Evidence based on survey data. *Journal of Environmental Planning and Management, 59*(12), 2102–2123. https://doi.org/10.1080/09640568.2015.1126241

Gerrard, R. E. C. (2018). Developing an index of community competence in flood response for flood-affected rural parishes on the Somerset Levels and Moors using composite and spatial datasets. *Area, 50*(3), 344–352. https://doi.org/10.1111/area.12416

GHD. (2013). *Multiple benefits of Landcare and natural resource management* (Report for the Australian Landcare Council—Multiple Benefits of Landcare, 21/21673). https://www.agriculture.gov.au/sites/default/files/style%20library/images/daff/__data/assets/pdffile/0009/2386926/multiple-benefits-of-landcare-and-natural-resource-management-report.pdf

Gong, Y., Zhang, R. Yao, K., Liu, B., & Wang, F. (2020). A Livelihood resilience measurement framework for dam-induced displacement and resettlement. *Water, 12*, 3191. https://doi.org/10.3390/w12113191

Green, J. J., Worstell, J., & Canarios, C. (2017). The Local agrifood system Sustainability/Resilience Index (SRI): Constructing a data tool applied to counties in the southern United States. *Community Development, 48*(5), 697–710. https://doi.org/10.1080/15575330.2017.1370001

Gunderson, L. H., Light, S. S., & Holling, C. S. (1995). Lessons from the Everglades. *BioScience, 45*, S66–S73. http://links.jstor.org/sici?sici=0006-3568%281995%2945%3CS66%3ALFTE%3E2.0.CO%3B2-6

Gunderson, L. H., & Holling, C. S. (Eds.). (2002). *Panarchy: Understanding transformations in human and natural systems*. Island Press.

Gunderson, L. H., Allen, C. R., & Holling, C. S. (Eds.). (2010). *Foundations of ecological resilience*. Island Press.

Guo, Y., Zhang, J., Zhang, Y., & Zheng, C. (2018). Catalyst or barrier? The influence of place attachment on perceived community resilience in tourism destinations. *Sustainability (Switzerland), 10*(7).https://doi.org/10.3390/su10072347

Hegney, D. G., Buikstra, E., Baker, P., Rogers-Clark, C., Pearce, S., Ross, H., King, C., & Watson-Luke, A. (2007). Individual resilience in rural people: A Queensland study, Australia. *Rural and Remote Health, 7*, 620. http://www.rrh.org.au/publishedarticles/article_print_620.pdf.

Holling, C. S. (1973). Resilience and stability of ecological systems. *Annual Review of Ecology and Systematics, 4*, 1–23. http://links.jstor.org/sici?sici=0066-4162%281973%294%3C1%3ARASOES%3E2.0.CO%3B2-M

Holling, C. S. (1996). Surprise for science, resilience for ecosystems, and incentives for people. *Ecological Applications, 6*, 733–735.

Holling, C. S. (2001). Understanding the complexity of economic, ecological, and social systems. *Ecosystems, 4*, 390–405.

Holling, C. S., Gunderson, L. H., & Peterson, G. (2001). Sustainability and panarchies. In L. H. Gunderson & C. S. Holling (Ed.), *Panarchy: Understanding transformations in human and natural systems* (pp. 63–102). Island Press

Hung, H. (2017). Formation of new property rights on government land through informal co-management: Case studies on countryside guerilla gardening. *Land Use Policy, 63*, 381–393. https://doi.org/10.1016/j.landusepol.2017.01.024

Johnson, T. R., Henry, A. M., & Thompson, C. (2014). Qualitative indicators of social resilience in small-scale fishing communities: An emphasis on perceptions and practice. *Human Ecology Review, 20*(2), 97–115. https://doi.org/10.22459/her.20.02.2014.05

Kaplan, H. (2005). Understanding the concept of resilience In S. Goldstein & R. B. Brooks (Eds.), *Handbook of resilience in children* (pp. 39–47). Springer.

Karman, A. (2020). Flexibility, coping capacity and resilience of organizations: Between synergy and support. *Journal of Organizational Change Management, 33*(5), 883–907. https://doi.org/10.1108/JOCM-10-2019-0305

Kaye-Blake, W., Stirrat, K., Smith, M., & Fielke, S. (2019). Testing indicators of resilience for rural communities. *Rural Society, 28*(2), 161–179. https://doi.org/10.1080/10371656.2019.1658285

Kelly, D., & Phelps, D. (2019). Looking beyond the D.U.S.T.-building resilient rangeland communities. *Rangeland Journal, 4*(3), 233–250. https://doi.org/10.1071/RJ18047

Kim, H., Marcouiller, D. W., & Woosnam, K. M. (2020). Coordinated planning effort as multilevel climate governance: Insights from coastal resilience and climate adaptation. *Geoforum, 114*, 77–88. https://doi.org/10.1016/j.geoforum.2020.05.023

Kulig, J., & Botey, A. P. (2016). Facing a wildfire: What did we learn about individual and community resilience? *Natural Hazards, 82*(3), 1919–1929. https://doi.org/10.1007/s11069-016-2277-1

Larimian, T., Sadeghi, A., Palaiologou, G., & Schmidt, R., III. (2020). Neighbourhood Social Resilience (NSR): Definition, conceptualisation, and measurement scale development. *Sustainability (Switzerland), 12*(16). https://doi.org/10.3390/SU12166363

Leite, M., Ross, H., & Berkes, F. (2019). Interactions between individual, household, and fishing community resilience in southeast Brazil. *Ecology and Society, 24*(3). https://doi.org/10.5751/ES-10910-240302

Maclean, K., Cuthill, M., & Ross, H. (2014). Six attributes of social resilience. *Journal of Environmental Planning and Management, 57*(1), 144–156. https://doi.org/10.1080/09640568.2013.763774

Maclean, K., Farbotko, C., Mankad, A., Robinson, C. J., Curnock, M., Collins, K., & McAllister, R. R. J. (2018). Building social resilience to biological invasions. A case study. *Geoforum, 97*, 95–105. https://doi.org/10.1016/j.geoforum.2018.10.018

Maclean, K., Ross, H., Cuthill, M., & Rist, P. (2013). Healthy country, healthy people: An Australian Aboriginal organisation's adaptive governance to enhance its social–ecological system. *Geoforum, 45*, 94–105. https://doi.org/10.1016/j.geoforum.2012.10.005

Maclean, K., Ross, H., Cuthill, M., & Witt, B. (2017). Converging disciplinary understandings of social aspects of resilience. *Journal of Environmental Planning and Management, 60*(3), 519–537. https://doi.org/10.1080/09640568.2016.1162706

Mahon, M., & Hyyryläinen, T. (2019). Rural arts festivals as contributors to rural development and resilience. *Sociologia Ruralis, 59*(4), 612–635. https://doi.org/10.1111/soru.12231

Masten, A., & Garmezy, N. (1985). Risk, vulnerability, and protective factors in developmental psychopathology. In B. Lahey & A. Kazdin (Eds.), *Advances in clinical child psychology* (pp. 1–52). Plenum Press.

Matarrita-Cascante, D., Trejos, B., Qin, H., Joo, D., & Debner, S. (2017). Conceptualizing community resilience: Revisiting conceptual distinctions. *Community Development, 48*(1), 105–123. https://doi.org/10.1080/15575330.2016.1248458

McElduff, L., & Ritchie, H. (2018). Fostering coastal community resilience: Mobilising people-place relationships. *Area, 50*(2), 186–194. https://doi.org/10.1111/area.12419

Mozumder, M. M. H., Wahab, A., Sarkki, S., Schneider, P., & Islam, M. M. (2018). Enhancing social resilience of the coastal fishing communities: A case study of hilsa (*Tenualosa ilisha* H.) fishery in Bangladesh. *Sustainability (Switzerland), 10*(10). https://doi.org/10.3390/su10103501

Natarajan, N., Parsons, L., & Brickell, K. (2019). Debt-bonded brick kiln workers and their intent to return: Towards a labour geography of smallholder farming persistence in Cambodia. *Antipode, 51*(5), 1581–1599. https://doi.org/10.1111/anti.12564

Ollerenshaw, A., Graymore, M., & McDonald, K. (2016). Beyond the call of duty: The integral role of rural local government in emergency management. *Rural Society, 25*(3), 185–203. https://doi.org/10.1080/10371656.2016.1255476

Powell, R. B., Green, T. F., Holladay, P. J., Krafte, K. E., Duda, M., Nguyen, M. T., Spencer, J. H., & Das, P. (2018). Examining community resilience to assist in sustainable tourism development planning in Dong Van Karst Plateau Geopark, Vietnam. *Tourism Planning and Development, 15*(4), 436-457. doi:https://doi.org/10.1080/21568316.2017.1338202

Quaranta, G., Dalia, C., Salvati, L., & Salvia, R. (2019). Building resilience: An art-food hub to connect local communities. *Sustainability (Switzerland), 11*(24), 1–15. https://doi.org/10.3390/su11247169

Rakhimova, N. (2018). Aging with financial insecurity: Social resilience and adaptation in urban areas of the United States. *Natural Resources Forum, 42*(4), 227–242. https://doi.org/10.1111/1477-8947.12160

Ravenscroft, J., Liakata, M., Clare, A., & Duma, D. (2017). Measuring scientific impact beyond academia: An assessment of existing impact metrics and proposed improvements. *PLOS One, 12*(3), e0173152. https://doi.org/10.1371/journal.pone.0173152

Redshaw, S., Ingham, V., McCutcheon, M., Hicks, J., & Burmeister, O. (2018). Assessing the impact of vulnerability on perceptions of social cohesion in the context of community resilience to disaster in the Blue Mountains. *Australian Journal of Rural Health, 26*(1), 14–19. https://doi.org/10.1111/ajr.12377

Ribeiro, B., & Shapira, P. (2019). Anticipating governance challenges in synthetic biology: Insights from biosynthetic menthol. *Technological Forecasting and Social Change, 139*, 311–320. https://doi.org/10.1016/j.techfore.2018.11.020

Ross, H., Cuthill, M., Maclean, K., Jansen, D., & Witt, B. (2010). *Understanding, enhancing and managing for social resilience at the regional scale: Opportunities in North Queensland*. Retrieved from Cairns https://www.rrrc.org.au/wp-content/uploads/2014/06/497-UQ-Ross-H-et-al-2010-Social-Resilience-at-the-Regional-Scale.pdf

Ross, H., & Berkes, F. (2014). Research approaches for understanding, enhancing, and monitoring community resilience. *Society and Natural Resources, 27*(8), 787–804. https://doi.org/10.1080/08941920.2014.905668

Saxena, L. P. (2020). Community self-organisation from a social-ecological perspective: 'Burlang Yatra' and revival of millets in Odisha (India). *Sustainability (Switzerland), 12*(5), 1–27. https://doi.org/10.3390/su12051867

Singh-Peterson, L., & Underhill, S. J. R. (2017). A multi-scalar, mixed methods framework for assessing rural communities' capacity for resilience, adaptation, and transformation. *Community Development, 48*(1), 124–140. https://doi.org/10.1080/15575330.2016.1250103

Teen, R., Roberts, L., & Challies, E. (2020). Disruption as opportunity for transformation? Insights from water supply contamination in Havelock North, New Zealand. *Local Environment, 25*(10), 810–827. https://doi.org/10.1080/13549839.2020.1839030

Torabi, E., Dedekorkut-Howes, A., & Howes, M. (2021). A framework for using the concept of urban resilience in responding to climate-related disasters. *Urban Research and Practice*. https://doi.org/10.1080/17535069.2020.1846771

Townshend, I., Awosoga, O., Kulig, J., & Fan, H. Y. (2015). Social cohesion and resilience across communities that have experienced a disaster. *Natural Hazards, 76*(2), 913–938. https://doi.org/10.1007/s11069-014-1526-4

Uddin, M. S., Haque, C. E., Walker, D., & Choudhury, M. U. I. (2020). Community resilience to cyclone and storm surge disasters: Evidence from coastal communities of Bangladesh. *Journal of Environmental Management, 264*. https://doi.org/10.1016/j.jenvman.2020.110457

Vaneeckhaute, L. E., Vanwing, T., Jacquet, W., Abelshausen, B., & Meurs, P. (2017). Community resilience 2.0: Toward a comprehensive conception of community-level resilience. *Community Development, 48*(5), 735–751. https://doi.org/10.1080/15575330.2017.1369443

Vaneeckhaute, L. E., Vanwing, T., Meurs, P., Abelshausen, B., & Jacquet, W. (2019). Community capitals of a Paramaca Maroon village in pictures: A Photovoice study on community resilience in the context of large-scale gold mining. *Community Development Journal, 54*(2), 233–253. https://doi.org/10.1093/cdj/bsx036

Walsh, S. (2014). *Trojan-horse aid: Seeds of resistance and resilience in the Bolivian highlands and beyond*. McGill-Queen's University Press.

Wesselow, M., & Mashele, N. J. (2019). "Who Needs Money if You Got Hands, if You Got Plants" forming community resilience in two urban gardening networks in South Africa. *Human Ecology, 47*(6), 855–864. https://doi.org/10.1007/s10745-019-00116-5

Westley, F. (2002). The devil in the dynamics: Adaptive management on the front lines. In L. H. Gunderson & C. S. Holling (Eds.), *Panarchy: Understanding transformations in human and natural systems* (pp. 333–360). Island Press.

CHAPTER 19

# Understanding Resilience in the Mining Sector

*Ursula Harman and Ian Thomson*

## Introduction

The San Cristobal Mine is located approximately five hundred kilometers south of the city of La Paz at an elevation of close to 4000 meters in the sparsely populated semi-desert of the southern Altiplano of Bolivia. The mine is owned and operated by Minera San Cristobal S.A (MSC) a Bolivian company and wholly-owned subsidiary of Japan-based Sumitomo Corporation. MSC commenced commercial operations in August 2007 as an open-pit operation with an anticipated 17-year mine life based on known mineral reserves. Its flotation plant processes fifty-two

U. Harman (✉)
School of Agriculture and Food Sciences, The University of Queensland, Brisbane, QLD, Australia
e-mail: u.harman@pucp.pe

I. Thomson
Shinglespit Consultants Inc., Hornby Island, BC, Canada
e-mail: ian@shinglespit.com



C. Baldwin and S. van Bommel (eds.), *Rural Development for Sustainable Social-ecological Systems*, Palgrave Studies in Environmental Sustainability,
https://doi.org/10.1007/978-3-031-34225-7_19

thousand tons per day of mineralized material producing lead-zinc-silver-concentrates that are shipped internationally via the port of Mejillones on the coast of Chile.

To allow for the development of the mine, the community (an historic UNESCO site) of San Cristobal had to be relocated and substantial land acquired from the community. Further land was acquired from the nearby communities of Culpina K for a tailings facility and Vila Vila for potential mine expansion. The population of San Cristobal was relocated to 'Nuevo San Cristobal' in 1999 under a comprehensive resettlement agreement with the mining company that included a clause stating:

> the Company will give priority for contract of workers of active age to members of the population of San Cristobal ......... (Hernández & Fernández Moscoso, 2014).

The agreement for the land acquisition from Culpina K also contained this clause (Hernández & Fernández Moscoso, 2014), and the company made a similar commitment to Vila Vila (Valverde & Martinez, 2008).

Prior to road construction in 2004 to allow the passage of heavy transport required to build the mine, the communities in closest proximity to the proposed San Cristobal Mine were isolated, de-populated and living a traditional campesino lifestyle. Economic activity depended heavily on quinoa production and remittance monies from family members who had migrated out to find work in cities within Bolivia and in mines in Chile and Argentina. Llama husbandry, traditionally the responsibility of women, provided a further source of income and food supply, which together with the seasonal production of vegetables provided the principal source of food security. Annual incomes were low and gender roles were well defined (Knight Piesold Consulting, 2001).

To fulfill the written commitment for employment, the Company established a three-shift regime that maximized the number of jobs available. A 2018 census shows the impact of this policy. In San Cristobal, 47% of the economically active members of the population were either direct employees of MSC or in the employ of contractors, including a significant number of women. In Culpina K, it was 56% and in Vila Vila, 52% (Cumbre de Sahama, 2019).

In practice, almost every family living in San Cristobal, Culpina K and Vila Vila has at least one family member working for the Company or one of its contractors. In addition, a commercial and service sector has emerged that includes shops of various kinds, restaurants, hotels, trades (building, electrician, plumber, electronics, etc.), school and college teachers and domestic help, which benefit from the trickle-down effect of well-paid jobs at the mine. Indeed, there has been a net in-migration to all three of the communities of people, many with family connections, in search of employment. There has also been a significant 'baby boom' encouraged by the fiscal security of regular employment. In 1998, the population of San Cristobal was 450 individuals in 140 houses with 130 pupils in the school. By 2018, the population had grown to an estimated 2000 individuals with 1000 pupils in the schools (Boutilier & Thomson, 2018).

Conscious of the finite life of the mine, the Company initiated a coordinated process of closure planning in 2017 and in 2019 commissioned a study of the socio-economic status of the three communities and the possible impacts of closure. The principal finding from this investigation was that the population was 'highly dependent on the mine for its economic security' and vulnerable to economic collapse when the mine ends production (Cumbre de Sahama, 2019). However, from the beginning of commercial operations, the Company had worked closely with community leaders to support a number of initiatives to develop economic activities that would continue after the mine closed down. Furthermore, by 2019 the individuals who led the communities through first contact with the mining company, exploration, construction and the first ten years of operation of the San Cristobal Mine had been replaced by a new, aggressively forward-looking younger generation of leaders.

At the same time, Company personnel were also aware that a number of families had actively invested in businesses within the community and also in the urban centers of Uyuni, Oruro, Potosi, La Paz and Santa Cruz, diversifying their potential sources of future income and economic security for themselves and their children. This, largely anecdotal evidence, suggested that describing the population as 'highly dependent' on the mine and hence 'vulnerable' was perhaps an oversimplification of the situation on the ground. It is within this context that Minera San Cristobal commissioned a study of community resilience—the potential within the population to withstand the impacts of mine closure—that is the subject of this chapter.

## Resilience in Mining

The noun 'resilience' is derived from the Latin and was first used in the mid-seventeenth century to describe the property of materials, typically metals, to recoil or rebound to their original shape after being physically stressed or deformed. In the early nineteenth century, the word was subsequently applied to individuals and groups of people. A further application was initiated by Holling in his seminal paper, 'Resilience and the Stability of Ecological Systems' (Holling, 1973). In this, Holling postulated that, 'in a system that is subject to external shocks and disturbances and 'continually confronted by the unexpected', the 'persistence of the relationships' within the system is crucial in achieving stability and resilience. Holling, followed by others, defined resilience as having two components, thus:

1. the amount of time required for an ecosystem to return to a state of equilibrium following a perturbation/disturbance; and,
2. the capacity of a system to adapt to and recover from disturbance such that the system retains essentially the same function, or, in the case of very resilient systems, experience enhancement.

Subsequently, resilience, as defined by Holling (above), has been applied in numerous contexts, including social, and particularly studies of communities that undergo disturbance or change, often with an emphasis on sudden shocks or crises. Berkes and Ross (2013) noted that, increasingly from 2000, the concept of resilience has been applied to communities in terms of social-ecological systems. Societies that are dependent, directly and indirectly, on ecological and environmental resources, including those related to natural resources such as mining, clearly meet this linkage.

Within the mining industry, the most frequent reference to resilience is the degree to which a mine remains financially viable in the face of disturbances such as fluctuating metal prices, supply chain challenges and environmental factors such as weather, water supplies and natural disasters. It should be noted that the communities dependent on a mine also experience these disturbances. However, mines ultimately exhaust the resource that is being exploited leading to what may be the ultimate disturbance, the termination of extraction and closure of the mine.

The application of social resilience 'thinking' in the mining industry is relatively recent. The industry led Mines Minerals and Sustainable Development project has scant reference to the topic in its final report (MMSD, 2002), noting only that community resilience is an attribute supportive of sustainable development objectives. A particular turning point may have been the publication of the United Nations Sustainable Development Goals in 2015 in which Goal 11 highlights the importance of resilience to sustainable cities and human settlements. Currently, the International Council on Mines and Metals is emphasizing the importance of mining companies working to strengthen community resilience to deal with disruptions such as the present COVID-19 epidemic, and, particularly, the challenges created by mine closure (ICMM, 2021a, b).

To date, published studies on resilience of mining-related communities are dominated by the examination of current conditions at and around the mine and how these may be modified to enhance resilience. For example, Gibson (2008) examined the situation with the Dene people of northern Canada and their relationship with the diamond mines, noting that community and cultural resilience could be linked to the vision of self-determination held by the Dene and the relationships of reciprocity that held the community together. In further studies, Wasylycia-Leis et al. (2014) looked at circumstances around the iron ore mines of Itabira, Brazil; Kuir-Ayius (2016) community resilience and health at mine affected communities in New Guinea; Asuah and Ankoye (2016) the gold mines of Obuasi, Ghana; Kanakis (2018) metal mines in Australia; Vaneeckhaute et al. (2019) gold mines and the tribal Maroon society in Suriname; and, Rela et al. (2020) nickel mines in Indonesia.

Across this broad range of geographic local, mine type, culture, language, religion and political regime, there is considerable consistency in the studies as to the nature of community resilience. Common factors include community cohesion, social capital, leadership, action and adaptation, involvement of government and other parties and quality of communication between community and mining company.

More relevant to the present investigation are published studies of the interrelationship between mine closure and community resilience; all of which are necessarily retrospective assessments. For example, Skeard (2014, 2015) looked at outcomes in the community of Buchans, Newfoundland, Canada, formerly a 'company town', following closure of the mine; Stefănescu and Constantin (2017) studied the situation in the Apuseni Mountians of Romania following the closure of mines that had operated for many years under the previous communist regime; and, Mfune et al. (2020) examined the consequences of closure of a giant lead-zinc mine on the community of Kabwe, Zambia. These examples demonstrate that 'resilience thinking' (the application of resilience frameworks and methodologies) can provide policy and practice recommendations to develop measures and interventions to mitigate negative outcomes from mine closure.

Yang et al. (2018a, 2018b) argue convincingly that environmental sustainability is inseparable from the socio-economic development of mining communities. This becomes especially important in indigenous communities with a traditional mode of agricultural existence and dependence upon nature (Pandey, 2020).

While the coupling, interdependence and co-evolution of social and natural systems in resource-dependent communities are crucial; Berkes and Ross (2013) introduce other relevant concepts from social science and psychology to develop an integrated approach to resilience (see Fig. 19.1). This framework, one of the most cited in the resilience literature, is here applied in a mining context for the first time and offers a methodology to understand the strengths, weaknesses and flexibilities of a community that experience this form of rapid and dramatic change. It also acknowledges the "complex interplay between social, cultural, spiritual, economic and environmental systems, and recognizes the synergistic relationship between people and the environments in which they live and derive livelihoods" (Ross et al., 2010, p. 2).

This conceptual framework was chosen for the present study not only because it provides insights that other concepts of resilience may not include, but also because this integrated approach emphasizes practice and theory as equally relevant (Berkes & Ross, 2013; Ross & Berkes, 2014). By understanding resilience as a field for emphasizing and developing strengths rather than redressing weaknesses, it enables analysis of communities' adaptive capacity, self-organization and agency and assists in practical ways by monitoring and evaluating resilience through processes such as community development, community-based planning and social learning (Berkes & Ross, 2013; Ross & Berkes, 2014).

**Fig. 19.1** Characteristics of resilient communities (Berkes & Ross, 2013)

## Methodology

The methodological design of the resilience study in MSC was based on multiple cases. The rural communities of San Cristóbal, Culpina K and Vila Vila were selected as case studies because they are the sites directly impacted by the mine operations. The multiple case study design allows the results to be analyzed case by case using the same variables or aspects, as well as the same research techniques for data collection, and to discuss, collectively, differences and similarities across the three cases (Hernández et al., 2014).

Data collection consisted of group and individual interviews, which took place between the 2nd and 18th of March, 2020. A total of 65 residents (male and female) participated in the group interviews in Culpina K, 40 in Vila Vila and 41 in San Cristobal, with one key informant also interviewed individually, making an overall of 147 participants in the study. The topics discussed in each interview were designed to reveal resilience characteristics, as summarized in Table 19.1.

Since the primary information was collected just before the COVID-19 pandemic, this study does not reflect any changes that may have

**Table 19.1** Discussion topics in the interviews

| *Resilience characteristics* | *Discussion topics* | *Resilience characteristics* | *Discussion topics* |
|---|---|---|---|
| Agency | • Ability to take action in the face of a problem<br>• Confidence to solve their own problems | Self-organizing | • Organization to tackle problems<br>• Meetings between community members to take control of the situation |
| Positive outlook | • Optimism in the future<br>• Perseverance | Leadership | • Community leaders representing different groups<br>• Different types of leadership that promote planning, flexibility and adaptation |
| Diversified and innovative economy | • Different economic activities<br>• Entrepreneurship: new and improved ideas<br>• Connected local economy<br>• Interdependence between local businesses | Engaged Governance | • Participation in decision-making institutions in the community<br>• Interactions between different organizations planning the future<br>• Collaboration between different actors<br>• Collaboration with other institutions with the community |
| Knowledge, skills and learning | • Transferable technical knowledge<br>• Interest in learning new things<br>• Skills to adapt to change | Social networks | • Knowledge of everyone in the community<br>• Help among community members in times of difficulty<br>• Collaboration with other groups outside the community |
| Community Infrastructure | • Interactive spaces that meet people's needs and activities<br>• Access to basic services<br>• Access to facilities that facilitate economic activities | People place relationships | • Opportunities and limitations that the environment offers<br>• Cultural connection to the land<br>• Commitment to protecting the environment<br>• Sustainable lifestyles |
| Values and Beliefs | • Identity values<br>• Spirituality<br>• Rituals and myths | | |

arisen from the mine's subsequent period in caretaker mode or any other COVID-19 impacts in the participating communities.

## Communities' Brief Background

The case study communities are located in the region known as Nor (i.e., 'North') Lípez, 50 kilometers south of the great Salar de Uyuni within the Bolivian Altiplano and belong to the Municipality of Colcha K (see Fig. 19.2). Nor Lípez is a sparsely populated semi-desert region situated between 3,600 and 3,800 meters (more than two miles) above sea level where sources of potable water are few and far between with rains fall, sometimes as snow, between October and April supporting the seasonal cultivation of food crops—quinoa, potatoes and other vegetables (Boutilier & Thomson, 2018).

Before the establishment of MSC, San Cristobal, Culpina K and Vila Vila communities were geographically isolated, impoverished and lived a traditional campesino lifestyle based on a subsistence economy. According to Boutilier and Thomson (2018), the small-scale agriculture was complemented by the men taking seasonal work in construction or other physical labor during the summer and younger, unmarried women going to the cities to work as housemaids. Other sources of money were family remittances and the odd sale of a llama for meat. With respect of infrastructure, the only way in and out from the communities was via a poorly maintained network of dirt roads, which became impassable during the wet season. There were basic, rural school and health facilities. Uyuni was the nearest commercial city.

Despite colonialization, capitalist investment by mining firms in the 1980s and 1990s, and the substitution of barter with money in the rest of the economy, the three communities continued to rely heavily on cooperative and traditional communal institutions. As indigenous communities, the land was communally owned with use rights allocated to families; reciprocal duties bound families together; and, community membership provided support and security in exchange for involvement in communal activities. Knowledge of and reverence for the gods and spirits that inhabited certain locations within the community's territory further bolstered collective identity (Boutilier & Thomson, 2018).

Currently, the main economic activity of the communities' families is mining, having agriculture and livestock as secondary and third activities, respectively. In the three communities, the most frequent occupations

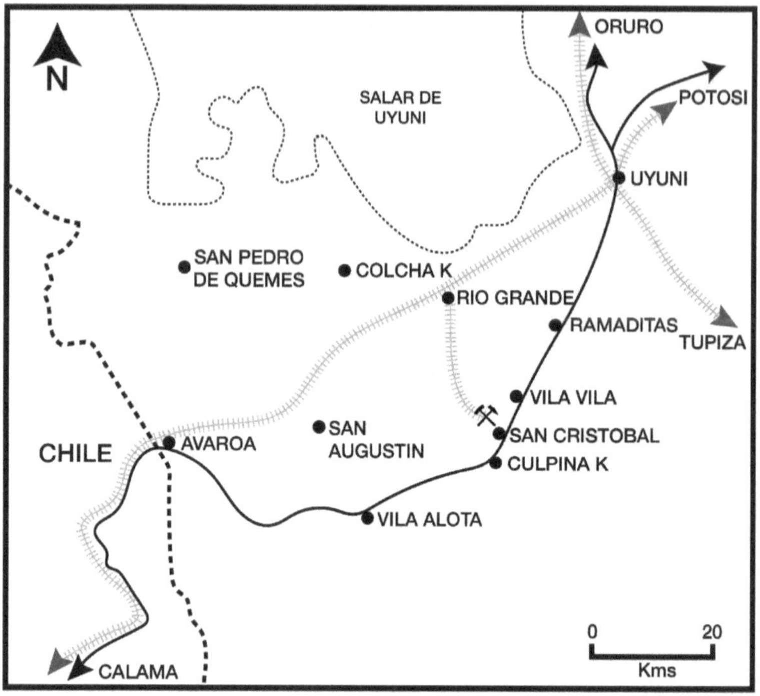


**Fig. 19.2** Minera San Cristóbal's region of influence (Boutilier & Thomson, 2018)

among families linked to MSC are heavy machinery operator, plant operator and automotive mechanic. Another common denominator among the three communities is that, in addition to agriculture or livestock, most families own small commercial service businesses (Cumbre de Sahama, 2019).

## Communities' Resilience Analysis

This section examines the resilience of San Cristobal, Culpina K and Vila Vila with respect to the future closure of MSC. The analysis focuses on the resilience characteristics and their interrelationships at the community, group and individual level.

### *San Cristobal*

San Cristobal stands out for its capacity in agency and self-organizing from first contact with MSC 1996, with community leaders negotiating the successful relocation of the town and comprehensive benefits for the community, including creation of the San Cristobal Foundation as a vehicle for economic development. Currently, San Cristobal has managed to maintain its agency and self-organizing as a strength in which community members display a proactive attitude and look to the sustainability of their community enterprises and farmers' associations, even though they currently have difficulties in initiating and realizing projects.

The formation of the Consultative Committee, an initiative of community leaders from San Cristobal, Culpina K and Vila Vila at the time of mine construction (Boutilier & Thomson, 2018) as an inclusive governance space for the future of the three communities, has allowed important projects to develop in San Cristobal, such as a llama meat processing plant, a new hospital, a tourist center, a higher education institute and a technological college. Members of the Committee are convinced that sustainability depends on the community itself, so they are devising new ways to show the results of their projects in order to encourage community participation in their own organizations. They are also integrating community traditions into social activities in order to show the progress of their projects.

There is also a high potential to generate an interconnected economy involving the various existing producer organizations, enterprises and new initiatives that entrepreneurs want to promote. An interconnected economy is possible because there are young leaders in positions of authority in the community, who yearn to maintain a living and dynamic community after the closure of the MSC. Likewise, young and middle-aged women leaders who are members of farmers' associations believe that their projects can generate jobs and open up many additional opportunities for the community.

However, the establishment and initial operation of these initiatives depends on support from MSC and do not yet represent a source of work for young professionals in the community. In addition, many of the companies created by the San Cristobal Foundation are suppliers to MSC; as such, the mining company is the most important customer and often the only customer.

This dependency raises concerns in the community because if they fail to generate alternative job opportunities; the population will choose to migrate, leaving an infrastructure without people. Having this in mind, local governance institutions are working on strategies to encourage participation and engagement by members of the respective producer organizations. Knowing the progress of their organizations generates trust in the community and a positive outlook on the future closure of MSC. Community members believe their organizations can be effective and sustainable: but success depends on their own effort and dedication to overcome challenges.

A weak aspect of the San Cristobal community is the limited capacity of its basic utilities, resulting in insufficient water and electricity services for the population and ongoing economic projects. The shortage of water for human consumption and agricultural activities is in the process of resolution through the joint action of the community, MSC and the Municipality of Colcha K. At the time of the study, electricity was still administered by MSC because the resettlement agreement had established provision of water and electricity as the responsibility of the company, which over time has reinforced the identity of the community as an annex of the company rather than an independent entity responsible for making its own decisions. However, it is anticipated that the community will be connected to the national power grid before the end of 2021. This will remove a significant vulnerability to operation of the llama meat processing plant, the new hospital and mechanical laboratories of the technical institute (still to be inaugurated), which require a stable energy supply to operate medical equipment and machinery.

Dependence on MSC is also economic, as most of the enterprises generated by the San Cristobal Foundation are suppliers to the mining company. Furthermore, the llama processing plant, new hospital, mechanical laboratories of the technical institute and the San Cristóbal Hotel are not fully operational, and thus not yet a source of employment. As a consequence, there is immediate concern within the community that their youth will leave seeking employment elsewhere and that in the longer-term San Cristobal will become a ghost town, as has happened with other post-mining towns. This concern also reflects the reality that many community members are investing in properties and/or businesses in other cities and will probably be the first group to leave San Cristobal at the closure of MSC.

Due to a lack of personal free time, there is low commitment and participation by community members in the farmers' associations and the Consultative Committee. This slows down and delays current projects in tourism, health, education and agribusiness. In addition, low participation has handicapped incorporation of water and energy services into planning for the future of San Cristobal. The limited time of mine workers to pursue the affairs of their community has also transformed the character of the region's traditional festivities, which were important spaces for social cohesion prior to the arrival of the mine and continuous employment.

### *Culpina K*

Culpina K's agency and self-organizing are currently focused on ensuring implementation of the community-MSC agreements, specifically those related to a water supply project and preventing environmental pollution. This limits the community's agency and self-organizing around the sustainability of livestock and agriculture, which are economic activities necessary for Culpina K to sustain itself without MSC. As a result, there are few discussions about the future of the community, despite the experience of a MSC workers' strike in 2019, which showed what the community would look like without an active enterprise that employs much of the population. Also, the perception that the community is more of a company residential camp prevents Culpina K from seeing itself as an independent community with a prosperous future and its own natural, human and cultural capitals.

On the other hand, the presence of various producers' associations, communal enterprises and organizations that manage basic services (water, energy, health and education) are clear examples of self-organizing and agency oriented toward the diversification of its economy. Nevertheless, all of these organizations maintain a strong economic dependence on MSC. MSC underpins the local economy not only through direct employment of residents but also as the dominant and often the only customer of community enterprises and producer associations.

A foundation has been established to build a diversified economy, with a quinoa processing and packaging plant and other organizations aimed at revitalizing tourism. However, community participation in these associations and enterprises is limited for a number of reasons that include MSC workers lack of time to engage in future-oriented institutions, and community members' demotivation because their effort has not generated rewards so far. Negativity is further reinforced via a patten in which senior leaders restrain proposals from young people.

Another reason for the limited involvement of the community in projects and associations is the loss of community cohesion, mainly due to population growth, and personal fulfillment from mining employment. This affects the ability of community organizations to function and progress while they are not generating economic gains in line with the expectations of their members. This slow progress generates mistrust in the associations and community enterprises—the quinoa plant is an example of this. Likewise, the five-year plans are perceived as very long, in which results are not usually noticed by the community.

In practice, the implementation of projects promoted by subcommittees of the Consultative Committee depends on the management of MSC, which is responsible for providing training and linking the community with NGOs, foundations and international cooperation. The strong link between the quinoa sub-committee and MSC as the intermediary in the process of technology transfer is useful because the community lacks the skills and contacts to carry out projects. However, this relationship simultaneously limits the agency, independence and self-organizing of the community with regard to finding markets for quinoa and the opening of the quinoa plant.

The termination of operations at MSC is an important issue for the community because people are actively planning alternative futures: to stay or leave Culpina K. The community is concerned about those young people who already consider migrating because they have invested in properties and/or businesses in other cities, have specialized knowledge in mining that is not transferable to the rural agricultural context or do not have the necessary resources to work in agriculture or livestock, such as land, animals and machinery. In addition, water scarcity now affects their everyday people-place relationships.

In this context, it is crucial to continue to strengthen the resilience characteristics of Culpina K so that the core population, who wants to stay where they are, continue their development. This majority of Culpina K residents has a sense of certainty that there is some future without MSC, albeit with difficulties and with less income. A commission to review the community-MSC agreements, which is a clear example of community self-organizing, gives them some peace of mind and hope to solve the problems of water and potential environmental pollution. As these issues come under control, focus can shift to strengthening strategies for the sustainability of their economy.

For the producers' associations and community enterprises, the cultural connection with the land through agriculture and livestock provides security because they are traditional and ancestral activities, and the residents of Culpina K have the resources to implement them successfully. The community organizations are motivated to work and build strategies for the sustainability of a local economy without mining, such as advertising their attractions at fairs and tourist agencies, investing in the genetic improvement of llamas and exporting quinoa internationally. To achieve these goals, Culpina K has female entrepreneurial leaders who push their associations to succeed. Young leaders in positions of authority in the community are crucial to the decision-making processes regarding the direction of the community, and they are in a position to take action to promote the community's resilience. These same younger people are key players to strengthen the collaboration with the Municipality of Colcha K and other communities, mainly for the implementation of a reliable, permanent water supply system.

The technical knowledge, interest, ability to learn and acquire skills to adapt to change are assets for all Culpina K community members, especially for new economic ventures. MSC workers know they can move to other mining projects where their skills will make them highly employable. Members of the current Water and Energy Committees are aware that they can transfer the skills learned in other companies into their community positions. Those MSC workers with traditional quinoa and llama breeding skills are confident that they can return to or continue these occupations, enhanced by the opportunities presented by the quinoa plant.

Inclusive institutions such as communal assemblies and the Consultative Committee allow participatory and collaborative governance of the economy and basic community services, as well as the integration of young leaders into community positions. This new generation of community leaders, along with female leaders and young entrepreneurs, has technical knowledge, skills for change and learning capacity that can enable the growth of the local economy. These are probably the main drivers for a significant part of the community to remain in Culpina K after the closure of MSC.

### *Vila Vila*

The agency and self-organizing capacities of the Vila Vila community have created community enterprises and farmers' associations that contribute to the growth of its economy and have enabled autonomy in water and energy services. However, the wages of MSC workers are the most important resource underpinning the local economy, despite others economic activities such as aggregate extraction, quinoa production, llama and vicuna rearing and textiles. Hence, MSC boosts the community's economy mainly through the income of company workers and as the most important client of the two communal enterprises. In addition, there is a lack of attention by the community to preparing for the future closure of MSC, reinforced by limited participation of members in projects and also by demands for greater employment of community members by MSC.

Since there is, as yet, no diversified and robust economy that can generate employment for young professionals in the community, there is a concern that they will leave on the closure of MSC. This preoccupation arises from the fact that many community members have properties and/or businesses in other cities or do not have local resources associated with agriculture and livestock, such as land, animals and machinery. Notably, a concentration of land in the hands of a few families leaves the youngest Vila Vila residents landless.

In Vila Vila, the youth find that they cannot exercise the traditional rights that come with community membership and in the long term; they will migrate in search of employment when MSC closes. This situation arises from competition for resources and social differences, which have not only broken the equitable institutions that formally underpinned community life, but also now require community input to deal with the undesired effects of population and economic growth. For example, there is a leadership crisis in the main women's association, whose success and economic gains have led to conflicts among their members. This has created consternation within the community because they do not yet know how to solve the problem. Such problems break with traditional values such as reciprocity and solidarity that were the foundations of Vila Vila prior to the arrival of MSC.

Water scarcity and community flooding are direct consequences of unplanned population growth. In addition to the irregularity of rains due to climate change, the excessive use of water has already affected sustainable lifestyles as residents can no longer grow the quantities of vegetables as before. This will have an impact on their agricultural activities, especially when the community returns to them as a livelihood when the mine company closes.

The weak resilience features in Vila Vila offer an explanation of why the community does not feel ready to face the closure of MSC. Therefore, it is important to continue to strengthen resilience characteristics in the community. For example, the agency and self-organizing capacities of the community have facilitated the creation of basic services committees with favorable results such as connection to the public electricity network, becoming independent from the service provided by the MSC. In addition, the women's association has become a benchmark of success, despite the current problem of leadership.

It is clear to the community that they must continue to strengthen their farmers' associations in order to generate employment opportunities. Accordingly, the membership has been thinking about strategies for the sustainability of their organizations in textiles, aggregates and llama husbandry. The aim of these strategies is to generate new markets and integrate their projects into the value chain of tourism. Because Vila Vila considers tourism as an economic activity that already generates benefits in other communities, producer organizations seek to guide their learning capacity and skills toward this sector, such as selling their textile products to tourists.

A return to agriculture and livestock after the closure of the MSC gives some confidence about the future. In particular, community members who are part of farmers' associations have plans to open a new market in the region in association with the llama meat processing plant in San Cristobal and the quinoa processing plant in Culpina K. To address the problems of water supply and management, community leaders have been discussing the idea of a water infrastructure project in collaboration with the Municipality of Colcha K.

Inter-institutional collaboration is important for the governance and autonomy of Vila Vila's basic services. This is strengthened by the presence of young leaders in positions of authority in the community. Middle-aged leaders who have a long-term perspective work collaboratively with larger bodies of government to solve their problems. This type of leadership has also facilitated mutual support in the community and with the municipality when flooding occurs.

## The Implications of Resilience in Sustainable Mining

Considering that sustainable mining entails socio-economic wellbeing, successful environmental outcomes and community engagement during the mine's life cycle and long after it is closed (Laurence, 2011; Mfune et al., 2020), the current study found that the framework of resilience proposed by Berkes and Ross (2013) is fundamental in providing an overview of viable livelihood options and how to achieve them outside of the mining company. Understanding the agency (being proactive, feeling empowered) and self-organization of communities toward a diverse and innovative local economy has to interact with other crucial characteristics, such as the effectiveness of community organizations, social relationships, inclusive institutions, connection to the environment, knowledge, skills, leadership, values and beliefs (Berkes & Ross, 2013).

As seen in Table 19.2, the three communities show some strengths that can contribute to their resilience, some weaknesses and some areas of high concern that differ between them. Comparison of the findings with those of other studies confirms that community residents are clearly unprepared for the socio-economic shifts that come with mine closure because they have little previous knowledge of the potential effects of mine closure or, indeed, a closure strategy (Mfune et al., 2020; Wasylycia-Leis et al., 2014).

In resilience, however, it is important to work from the strengths that can assist communities, and their members, in planning, monitoring and evaluating sustainability plans. According to Berkes and Ross (2013), in the present case, adaptive capacity, self-organization and agency can be facilitated by community members themselves operating in collaboration with MSC as an external change agent, to take action based on a resilience style model through processes such as community development and community-based planning.

In this sense, the company and communities together develop objectives that will represent what they need to do to improve community resilience. From these objectives, both parties could develop indicators to represent the key dimensions to be monitored and develop visual or numerical ways of representing progress on these indicators, which should align with the company's existing monitoring strategies and forms of representing data.

**Table 19.2** Summary of resilience analysis in the three communities

| *Characteristics* | *San Cristobal* | *Culpina K* | *Vila Vila* |
|---|---|---|---|
| Agency | Sustainability of their economy | Fulfillment of agreements | Economy dynamization |
| *Self-organizing* | Economic projects | Reviewing commission | Basic services |
| Diversified and innovative economy | MSC as the main customer | MSC as the main customer | MSC as the main customer |
| Positive outlook | Confidence in the future | Pessimist but with hope | Concern about migration |
| *Engaged governance* | Planning spaces for the future | Little participation and commitment in planning spaces | Unequal access to land |
| *Leadership* | Different types of leaders working together | Young leaders vs. old leaders | Crisis in one of the associations |
| Knowledge, skills and learning | New ideas and entrepreneurship | New ideas and entrepreneurship | New ideas and entrepreneurship |
| *Social networks* | Loss of cohesion | Loss of cohesion | Mutual support when flooding |
| Community infrastructure | Dependence on MSC for basic services | Dependence on MSC for basic services | Independence from MSC for basic services |
| *People-place relationships* | Loss of their sustainable lifestyles because of perceived water shortage | Loss of their sustainable lifestyles because of perceived water shortage and contamination | Loss of their sustainable lifestyles because of perceived water shortage |
| Values and beliefs | Erosion of community values toward individualistic | Erosion of community values toward individualistic | Erosion of community values toward individualistic |

Given the relative lack of participation and commitment in planning and discussion spaces for the future during the operational phase of the mine life, the findings of this study confirm that agency at community level is currently not strong. At the same time, representative committees of the communities have demonstrated self-organizing capacities and achievements around which the company and communities could re-focus their partnership, empowering communities to better plan and organize their futures.

Community dependence on the mine as an employment and basic services provider is a significant weakness for socio-economic wellbeing after mine closure. This is already recognized by the communities in their concerns about possible population out-migration because there will be fewer job opportunities when the mine closes. A preoccupation with ending up as a ghost town, as has happened with numerous other mining communities in Bolivia, is a common social problem created by mining activities (Githiria & Onifade, 2020).

However, the communities have strengths and potential for developing a diverse and innovative economy based on their new ideas for economic ventures and new post-mining enterprises. These strengths include entrepreneurial women, skills and learning capacities in different economic activities. In this sense, the monitoring and evaluation strategies from the company and communities could center on accelerating the development of new community enterprises by assessing and communicating their progress and reinforcing the interconnections and mutual financial benefits between economic activities.

Consistent with the literature, the present study found that an aspect of high concern for mining communities is related to environmental outcomes. As long as uncertainties as to mining impacts on the communities' natural resources remain unattended, concerns about the future increase, especially when people depend on the quality of water and land for agriculture and livestock. This finding broadly supports the work of other resilience studies in mining communities linking ecology and socio-economic aspects (Pandey, 2020; Yang et al., 2018a, 2018b).

In agreement with Pandey (2020), the findings of the study show that those communities that originally relied on traditional agriculture feared that mining operations would deplete their natural resources, self-sustaining lifestyles and agricultural economy, but there was also interest in mine-driven growth opportunities. Therefore, resilience for a sustainable mining involves balancing agricultural production, environmental conservation and industrial developments (Pandey, 2020; Yang et al., 2018a).

The cultural connection to the land gives communities some confidence in a return to agriculture and livestock in a post-mining scenario. Because of this commitment to their customary livelihoods, long-term residents wish to remain in their respective communities, despite the difficulties they will have to confront once the mine closes. In order to take advantage of communities' positive outlooks and social networks for self-organizing, it is important to enhance community' cohesion and unity, which have been gradually weakened by more individualistic values created by continuous employment and high incomes.

Resilience for a sustainable outcome also entails working along with the different types of leaders (young, old, women and entrepreneurs) because they are key actors in reinforcing community engagement and leading collaboration with their government, as a source of support in the transition to a post-mining society and economy. Nevertheless, the limited relationships between communities and their local government match those observed in earlier studies (Mfune et al., 2020; Wasylycia-Leis et al., 2014). Also, a need is seen for strategic planning at the municipal level, especially when it comes to allocating mine revenues to long-term sustainability planning regarding water infrastructure, sanitation, economic diversification, waste control and health.

Since resilience-building projects demand an engaged governance, not only local governments play a key role, but also community institutions. Besides the main goal of planning a desired future, these institutions in collaboration with government face other important challenges such as facilitating policy instruments to manage social inequalities created by the competition for community resources, generating skills and capacities to lead, monitor and evaluate technical projects for a diversified and innovative economy when it is no longer possible to rely on expertise from the mine.

## Conclusions

This study set out to understand the resilience of individuals, groups and communities in the three villages closest to the San Cristobal mine (MSC) utilizing qualitative research, based on group and individual interviews. The results confirm that the three communities are highly dependent on the mining company and, consequently, vulnerable to mine closure. Nevertheless, the study revealed that individual families, groups and the three communities have potential for resilience, in which their positive outlook, knowledge, skills and learning, social networks and support and people-place relationship are strengths to build from.

Furthermore, the resilience approach proves useful in explaining why communities' agency and self-organization, economy, governance within the communities, with MSC and the municipality, and community infrastructure are characteristics that need improvement in order to reduce this dependency.

The study also reveals a tension between individual/family and collective resilience. This is shown in the decisions to join in investing in community-based initiatives or business ventures elsewhere in Bolivia, or to participate in both. Such trade-offs reflect an ambiguity toward the future and a wish to be in a position to leave or strengthen the local economy so that it is possible to make staying an attractive option for themselves and others.

For Minera San Cristobal, application of the resilience framework converted a patchwork of information and understanding into a coherent analytical process. While recognizing that, today, the mine is the predominant economic factor in the lives of the local population, the resilience assessment gives clarity to both the current situation and future potential of the communities to cope with the social and economic stresses created by mine closure. The framework permits comparative assessment of strengths and weaknesses within and between the three communities.

It has further allowed both the company and the communities themselves to identify where and how new or intensified effort is required to enhance resilience ahead of mine closure. As such, it provides an opportunity for company and community to collaborate constructively in addressing the challenges and opportunities of closure previously seen largely through the overwhelmingly negative lens of today's economic dependency predicating an inevitable future collapse.

This study has significant implications for understanding the relationship between resilience and sustainability in the mining sector. The results support the premise that the concept of sustainable mining should not be limited to consequences on the environment. Because mine closures are ultimately inevitable and those that occur unexpectedly can have serious, life-threatening consequences for mine-dependent communities (Mfune et al., 2020), mining sustainability evaluations need to pay greater attention to the latter stages of mining operations where supporting the creation of resilient communities that can deal with the consequences of mine decline and closure is critical.

Finally, further research would be of great help in exploring the resilience of developing nations with mineral riches within a sustainable development framework. Is mining creating and leaving only negative social and environmental impacts or boosting resilient and sustainable countries and/or regions? Responding to this question will provide crucial insights for designing and implementing key cross-sectorial policies that allow the intersection between mining and other important areas such as education, health, science, technology and innovation, environment and labor and social development.

**Acknowledgements** This chapter is derived from a resilience study conducted for Minera San Cristobal and is published with the company's permission. Prof. Helen Ross was team leader, supplied the resilience framework and guided the entire process. Ursula Harman, as field researcher, collected most of the data. Ian Thompson first thought of the study and guided it in his capacity as consultant to Minera San Cristobal.

## References

Asuah, A. Y., & Ankoye, P. A. (2016). The resilience of mining communities in Obuasi, as Anglogold Ashanti shifts position. *Journal of Poverty, Investment and Development, 22*, 80–90.

Berkes, F., & Ross, H. (2013). Community resilience: Toward an integrated approach. *Society & Natural Resources, 26*(1), 5–20. https://doi.org/10.1080/08941920.2012.736605

Boutilier, R. G., & Thomson, I. (2018). *The social license: The story of the San Cristobal mine*. Routledge.

Cumbre de Sahama. (2019). *Estudio Socio-Económico Productivo y Ambiental de Adaptacion a los Cambios/Resilencia en Comunidades del Área de Influencia Directa de MSC (Minera San Cristobal Internal Document).*

Gibson, V. V. (2008). *Negotiated spaces: Work, home and relationships in the Dene diamond economy*. University of British Columbia Vancouver.

Githiria, J. M., & Onifade, M. (2020). The impact of mining on sustainable practices and the traditional culture of developing countries. *Journal of Environmental Studies and Sciences, 10*(4), 394–410. https://doi.org/10.1007/s13412-020-00613-w

Hernández, B. M., & Fernández Moscoso, M. (2014). Investigación 1: ¿Es posible construir beneficios mutuos entre comunidades y empresas mineras?: El caso San Cristóbal. In B. M. Hernández, M. Fernández Moscoso, M. Yapu, D. Romero Romay, P. Rocha Portugal, G. Damonte, et al. (Eds.), *América Latina hacia la Inclusión Social: Avances, aprendizajes y desafíos*. ASIES.

Hernández, R., Fernández, C., & Baptista, P. (2014). *Metodología de la investigación. Sexta edición*. Editorial McGraw-Hill.

Holling, C. S. (1973). Resilience and Stability of Ecological Systems. *Annual Review of Ecology and Systematics, 4*, 1–23. http://www.jstor.org.ezproxy.library.uq.edu.au/stable/2096802

ICMM. (2021a). *Community resilience*. https://www.icmm.com/en-gb/social-performance/minings-contribution/contributing-to-communities/community-resilience. Accessed 14 June 2021.

ICMM. (2021b). *Skills for our common future initiative*. https://www.icmm.com/en-gb/social-performance/skills-for-our-common-future. Accessed 14 June 2021.

Kanakis, K. (2018). *Mining community resilience explored through sustainable community development and perceptions of community wellbeing*. James Cook University.

Knight Piesold Consulting. (2001). *Apex Silver Mines Corporation, Proyecto San Cristóbal, Plan de Gestión Social, Planes y Programas (Minera San Cristobal Internal Document)*. Denver.

Kuir-Ayius, D. D. (2016). *Building community resilience in mine impacted communities: A study on delivery of health services in Papua New Guinea: A thesis presented in fulfilment of the requirements for the degree of Doctor of Philosophy in Development Studies at Massey University, Palmerston North*. Massey University.

Laurence, D. (2011). Establishing a sustainable mining operation: An overview. *Journal of Cleaner Production, 19*(2), 278–284. https://doi.org/10.1016/j.jclepro.2010.08.019

Mfune, O., Kunda-Wamuwi, C. F., Chansa-Kabali, T., Chisola, M. N., & Manchisi, J. (2020). The legacy of mine closure in Kabwe, Zambia: What can resilience thinking offer to the mining sustainability discourse? In *Sustainability challenges in Sub-Saharan Africa II* (pp. 167–188). Springer.

Mining Minerals and Sustainable Development (MMSD) Project. (2002). *Breaking new ground: The report of the Mining, Minerals, and Sustainable Development Project*. Earthscan Publications for International Institute for Environment and Development and World Business Council for Sustainable Development.

Pandey, R. (2020). Mining, agriculture change and resilience: Reflections from indigenous knowledge in the anthropocene. In K. Bauddh, S. Kumar, R. Pratap Singh, & J. Korstad (Eds.), *Ecological and practical applications for sustainable agriculture* (pp. 395–410). Springer Singapore. https://doi.org/10.1007/978-981-15-3372-3_18

Rela, I. Z., Awang, A. H., Ramli, Z., Taufik, Y., Sum, S. M., & Muhammad, M. (2020). Effect of corporate social responsibility on community resilience: Empirical evidence in the Nickel Mining Industry in Southeast Sulawesi, Indonesia. *Sustainability (Basel, Switzerland), 12*(4), 1395. https://doi.org/10.3390/su12041395

Ross, H., & Berkes, F. (2014). Research approaches for understanding, enhancing, and monitoring community resilience. *Society & Natural Resources, 27*(8), 787–804. https://doi.org/10.1080/08941920.2014.905668

Ross, H., Cuthill, M., Maclean, K., Jansen, D., & Witt, B. (2010). *Understanding, enhancing and managing for social resilience at the regional scale: opportunities in North Queensland*. The University of Queensland.

Skeard, J. (2014). *"Hope springs eternal": A study of company and community in the former mining town of Buchans*. Memorial University of Newfoundland.

Skeard, J. (2015). Come hell or high water: Identity and resilience in a mining town. *London Journal of Canadian Studies*. https://doi.org/10.14324/111.444.ljcs.2015v30.006

Stefănescu, L., & Constantin, V. (2017). Resilience of Mining Communities to Long-Term Environmental Stress in the Apuseni Mountains Area (NW of Romania). In A. Ozunu, I. A. Nistor, D. C. Petrescu, P. Burny, & R. M. Petrescu-Mag (Eds.), *Resilient society. Environment and human action series* (pp. 199–213).

Valverde, A. H., & Martinez, J. (2008) *Sistematización, Análisis, y Perspectivas de la Experiencia San Cristóbal (1996–2008) (internal document)*. Minera San Cristóbal.

Vaneeckhaute, L. E., Vanwing, T., Meurs, P., Abelshausen, B., & Jacquet, W. (2019). Community capitals of a Paramaca Maroon village in pictures: A Photovoice study on community resilience in the context of large-scale gold mining. *Community Development Journal, 54*(2), 233–253. https://doi.org/10.1093/cdj/bsx036

Wasylycia-Leis, J., Fitzpatrick, P., & Fonseca, A. (2014). Mining communities from a resilience perspective: Managing disturbance and vulnerability in Itabira, Brazil. *Environmental Management, 53*(3), 481–495. https://doi.org/10.1007/s00267-014-0230-1

Yang, Y., Erskine, P. D., Zhang, S., Wang, Y., Bian, Z., & Lei, S. (2018a). Effects of underground mining on vegetation and environmental patterns in a semi-arid watershed with implications for resilience management. *Environmental Earth Sciences, 77*(17), 1. https://doi.org/10.1007/s12665-018-7796-5

Yang, Y., Li, Y., Chen, F., Zhang, S., & Hou, H. (2018b). Regime shift and redevelopment of a mining area's socio-ecological system under resilience thinking: A case study in Shanxi Province, China. *Environment, Development and Sustainability, 21*(5), 2577–2598. https://doi.org/10.1007/s10668-018-0139-6

CHAPTER 20

# Mental Models and Values: Application to Natural Resource Management

*Natalie A. Jones and Claudia Baldwin*

## Introduction

Management of natural resources involves understanding the biophysical qualities and dynamics of environmental systems, as well as making decisions according to the priorities and needs (both physical and social) of people who use and value them. This often requires making tradeoffs between competing and conflicting interests. The way we manage natural resources, including their use and protection, are dependent on

N. A. Jones (✉)
School of the Environment, The University of Queensland, St Lucia, QLD, Australia
e-mail: n.jones3@uq.edu.au

C. Baldwin
Sustainability Research Centre, School of Law and Society, University of the Sunshine Coast, Sunshine Coast, QLD, Australia
e-mail: CBaldwin@usc.edu.au; claudiab@westnet.com.au



C. Baldwin and S. van Bommel (eds.), *Rural Development for Sustainable Social-ecological Systems*,
Palgrave Studies in Environmental Sustainability,
https://doi.org/10.1007/978-3-031-34225-7_20

how people think about, make sense of and interact with their environment. For Berkes et al., (1998, p. 2), "resource management is people management". Their social-ecological systems framework, designed to study local resource management systems, posits human thought, specifically local knowledge and understanding, as the key linkage between the environment and management practice (Berkes et al., 1998; Colding & Barthel, 2019). This framework thus situates the social science discipline, with its approaches and tools for studying how people understand, make sense of and give meaning to the world, at the core of natural resource management.

Over the past two decades, since Berkes et al., first proposed their framework in 1998, social scientists working in the field of environmental and natural resource management have focused on expanding conceptualisations and methods for providing a more nuanced understanding of human cognition as a link between environmental systems and people's management practices. This has involved drawing on insights from the cognitive science and psychology literature to explain how people perceive and form mental representations of the world, which they use to make decisions and act.

We, the authors of this chapter, have worked with Helen Ross to advance the theory and practice of two particular cognitive constructs: mental models and values. In the following sections, we define the meaning of these concepts as used within a natural resource management context, explain why they are useful, offering examples to illustrate their application, and finally, provide an overview of the methods that can be used to study mental models and values. Our aim is to inspire further research on the cognitive dimensions that link ecological systems with management regimes to advance our understanding of social-ecological systems research.

## Mental Models and Values and Their Application in Natural Resource Management Research

Mental models and values are both theoretical cognitive constructs that explain how people mentally represent information and knowledge about the world. They involve however different modes of thinking which influence decision-making processes in distinct ways.

### *Mental Models*

A mental model is a representation of external reality that exists within a person's mind. It provides a framework of concepts and relationships that are organised in way that reflects how the individual perceives the world.

Theoretical underpinnings of the mental model construct[1] have been developed within a number of fields, beginning with psychologist Kenneth Craik in the late 1930s. As outlined in Jones et al., (2011), Craik's work focused on providing an explanation of how people internally represent an aspect on the world within their mind. He coined the term 'mental model' to refer to a simplified representation of a given phenomenon that an individual creates within their mind for the purpose of anticipating events, reasoning and making decisions. He explains how these internal models can be run like a simulation within one's mind before acting in the real world (Craik, 1943).

Another key field that has advanced the theory of mental models is system dynamics (Jones et al., 2011). Researchers in this field strive to understand how people can learn and make better decisions in relation to managing complex systems, such as environmental systems. Acknowledging that humans are limited in their cognitive capacity to comprehend complexity, researchers draw on evidence demonstrating that people's cognitive conceptualisations of the world are 'simplified' representations and are often erroneous (Doyle & Ford, 1998). These observations are further supported by Herbert Simon's principle of 'bounded rationality', whereby rationality of human decision-making is limited by their cognitive thinking capacity. Systems dynamics researchers use mental model research to make explicit limitations in the human capacity to accurately represent a complex system. Their work focuses on revealing the types of concepts and relationships that people typically represent well, while also shedding light on the errors people make which help explain 'misunderstanding' in people's knowledge of a given phenomenon (Jones et al., 2011).

[1] Theoretical underpinnings of the mental model construct were explored in a PhD thesis by Jones, N. A. (2012). *Eliciting mental models of natural resource systems: A procedural comparison* (Unpublished doctoral dissertation). University of Queensland, under the supervision of Prof. Helen Ross, Dr. Timothy Lynam and Prof. Pascal Perez. This section provides a summary of that work, also presented in Jones, N. A., Ross, H., Lynam, T., Perez, P., & Leitch, A. (2011). Mental models: An interdisciplinary synthesis of theory and methods. *Ecology and Society, 16*(1). http://www.jstor.org/stable/26268859.

The theory of confirmation bias explains why these errors and misunderstandings in people's mental models may be difficult to overcome and why these misconceptions of the world can persist over time. Confirmation bias is the tendency to use pre-existing beliefs to filter and interpret incoming information (Allahverdyan & Galstyan, 2014). People are more likely to take in information that aligns with their understanding of the world and reject that which does not. New information can therefore serve to reinforce a person's existing mental model or may be rejected.

The mental model construct provides a dynamic perspective to reveal cause-and-effect understanding which represents how an individual perceives a system to function (Jones et al., 2014). By eliciting a person's internal conceptualisation of how the world works, we can better understand their reasoning processes which influence their decisions and interactions with the world.

Elicited mental models—i.e., those described through interviews or other methods—reveal (i) the concepts that feature in a person's representation of a given phenomenon and (ii) a structure representing how those concepts are organised within a person's mind, by depicting cause-and-effect relationships, for example, or associations more generally.

These insights can be useful in different ways in the context of natural resource management. As outlined in Jones et al. (2011), elicited mental models can improve overall understanding of an environmental system by integrating different elicited mental models to provide an ensemble view of a phenomenon. For example, by combining mental models from local community members, scientific experts, industry and government representatives, a more comprehensive understanding of a given environmental system can be developed to improve decision-making. Insights can also be used to identify where there is alignment and divergence in a given group of stakeholders' understanding of an issue: which concepts and relationships within a set of elicited mental models are shared, and which are not? Shared understanding provides common ground on which to build dialogue over how resources should be used and managed and can be a useful starting point in situations where there is known conflict. Gaps and errors in people's understanding can also be revealed through elicited mental models. By building on the concepts relating to a given phenomenon that stakeholders do use, awareness raising and knowledge building activities can be targeted to stakeholders' understanding of an issue.

While mental model studies typically reveal what concepts feature within people's understanding of an issue, they typically do not provide a qualitative measure of the extent to which an individual or group attributes importance and significance to that concept.

### *Values*

Values have been studied within a variety of fields including psychology, philosophy, economics, sociology and political science (Dietz et al., 2005), each presenting a unique body of theory explaining what values are and their function in guiding people's interaction with the world. A useful conceptualisation of values for social science research in natural resource management is that offered by Reser and Bentrupperbäumer (2005). Based on a review of social science and psychology literature, they deduce that there is consensus within these disciplines that "values are and represent important individual and collective investments and judgments about what in this world and in this life is truly important, worthwhile, and meaningful" (Reser & Bentrupperbäumer, 2005, p. 128). Values are thus held by individuals and can be shared within groups. They cognitively represent significance and meaning of both intangible and tangible aspects of the world.

Jones et al., (2016b) provide an overview of the different ways in which the construct has been conceived within the environmental management literature. They explain how values can represent the importance of intangible ideas or principles which signify desirable modes of interacting with the world, such as trust and loyalty, referred to as 'held values' (Brown, 1984). They also describe values that are ascribed to tangible features of the world, which are known in the natural resource management literature as 'assigned values' (Lockwood, 1999; Seymour et al., 2010). These values are studied in respect to specific places, terrestrial and marine environments (Biedenweg et al., 2019; Jones et al., 2016a; Schulz et al., 2017; Witt et al., 2019), as well as particular flora and fauna species (Engel et al., 2021; Reyes-Arroyo et al., 2021). 'Relational' values are also gaining prominence within the natural resource management realm (Ross et al., 2018). These values emerge from the relationship people have with an aspect of the world and involve the act of preferencing (Brown, 1984). The work of social ecologist Stephen Kellert established a typology of ten values that define the types of relationships people have with the natural world which are embedded within the biological make-up of humans

(Kellert, 2008). Ross et al., (2018) described how Kellert's typology is particularly useful for explaining human–environment interactions from a social-ecological systems perspective.

By representing perceptions of significance and worth, a study of people's values is useful for guiding decision-making, particularly when dealing with conflicts and trade-offs between social, environmental and other benefits (Baldwin & Ross, 2012; Jones et al., 2016a).

Mental models therefore represent rational and reason-based understandings of the world, while values represent the perceptions of significance and worth, existing more at the level of feeling and emotion.

## Application of These Constructs

In this section, we present an overview of research we have conducted with colleagues that use mental models or values to gain insight into how people think about and interact with their environment. These snapshots explain how methods were used in a variety of natural resource management challenges, such as water management, agriculture, fisheries and coastal development. They demonstrate how such research was applied for different purposes, such as understanding drivers of behaviour, how to better communicate, or as input to management decision-making and conflict resolution.

### *Mental Models*

Jones' PhD research (Jones et al., 2014), under the supervision of Prof. Helen Ross, focused on advancing application of the mental model construct within the field of natural resource management, particularly in terms of capturing people's understanding of complex environmental systems. Different elicitation procedures (i.e., methods) were compared to determine what effect those procedures may have on the mental model expressed, in this case participants' mental models of how riparian vegetation affects water quality. The research involved comparing an oral interview with a method involving a drawing task with oral commentary (see Fig. 20.1). These interviews were conducted at either a creek location (situated interview), where visual cues were available, or in the interviewee's home (non-situated interview).

Results show that the location of the interview had an effect, whereby the situated interview elicited more elaborate and specific mental models,

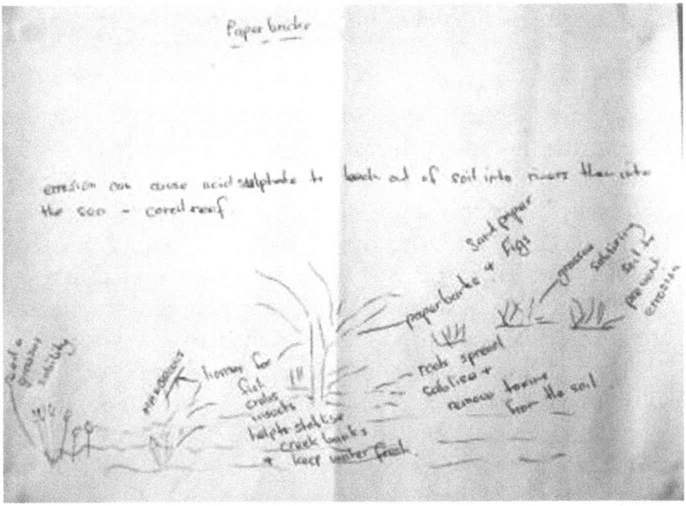

**Fig. 20.1** Elicited mental model using a diagramming technique combined with an oral-interview (Jones, 2012)

compared to the non-situated interview. No effect was identified in terms of the elicitation task used (oral versus oral-diagrammatic). A key finding of the research for natural resource management is that people's conceptual understanding of the world can change depending on the context in which that conceptual understanding is used or probed.

A study conducted by Guerrero et al., (2021) sought to create a systems perspective of the causal dynamics driving sustainability in the soy sector (Brazil) by reducing deforestation. It involved examining similarities and differences in the mental models of supply chain actors to establish common ground for building dialogue to reach collective agreement on pathways for sustainability. While the Jones et al. study discussed previously focused on biophysical dynamics of an environmental system, the scope of Guerrero et al.'s. research was broader to include an exploration of environmental, social, economic and political factors relating to a given natural resource management issue, i.e. deforestation within the soy industry. In addition to determining shared understanding of the issue, the study also identified a range of challenges faced by different actors within the soy supply chain which affect their capacity to commit to deforestation agreements. These findings will be used by organisations seeking to improve sustainability in the soy sector to better engage with soy supply chain actors.

A recent study was conducted by Dipu et al., (2022) which used mental models to identify the factors driving and inhibiting uptake of regenerative agriculture in south east Queensland (Australia). Rather than focus on identifying alignment and divergence in stakeholders' understanding, the purpose of this study was to provide an integrated, ensemble view of the phenomena under study. That is, participants' elicited mental models were integrated to provide a holistic picture of causal dynamics that are driving and constraining scaling of regenerative agriculture as a pathway to sustainable food production. Similar to the work of Guerrero et al., (2021), this research also elicited environmental, social, economic and political cause-and-effect relationships. The findings were used to identify opportunities to support regenerative agriculture in the region and overcome barriers faced, for example, by building new collaborations to facilitate the sharing of knowledge and information.

## *Values*

In this section, we describe two methods for studying values: (i) photovoice (ii) semi-structured interviews and (iii) participatory mapping. These methods have been used to reveal underlying values, namely relational values, for different purposes such as conflict resolution, climate change awareness, engaging the community about a controversial issue, insight into motivation for land stewardship and identifying new opportunities for managing waterways.

Photovoice involves participants taking photos of topics to be discussed in a workshop or individual interviews. It can be a comfortable way of engaging people who might be reticent to share their thoughts and feelings. The content of photos and script is analysed in combination. The analysis is usually fed back to participants to verify findings, as well as used to portray visually how others perceive a complex issue.

Baldwin applied photovoice for her thesis on consensus-building in water allocation and use (Baldwin, 2008) for which Prof. Ross was her PhD supervisor (Baldwin & Ross, 2012). Underlying values of irrigators, graziers, conservationists and government officers were revealed to shed light on factors contributing to social, environmental and economic sustainability, providing a basis for finding common ground on which to build consensus about sharing water from a catchment.

In an honours student project (Grant et al., 2015), photovoice was one of four techniques used to facilitate information exchange about climate risk in a Sunshine Coast canal estate community, all revealing insight into participants' values. Much high-value residential canal estate property in Southeast Queensland (Australia) is located in a low-lying coastal area and is predicted to be impacted by climate change, with specific threats including sea level rise, storm surge and severe weather events. The study surfaced residents' perceptions and understanding of vulnerability to climate change in their community. It compared the effectiveness of four participatory visual methods for information exchange, prompting dialogue about climate change risk: risk maps showing flood hazard (available on Council's website), participatory digital mapping by residents of known vulnerable areas (see Fig. 20.2), photovoice using resident-derived photos of local flood experience; and 3D climate change scenarios.

Participants preferred the 3D visualisation for its realistic futuristic representation of flood heights in a street-view and ability to digitally focus in on particular sites such as their own property. However, all of these techniques revealed an underlying concern about the length of time they would continue living there, and a future legacy they might leave the next generation. Such interactive engagement approaches could benefit

**Fig. 20.2** Participatory digital mapping—each participant adds a different colour to illustrate areas of vulnerability

State and local governments, the property market and communities to communicate and engage in dialogue around climate change issues in coastal areas, particularly within canal estates where long-term responses need to be considered.

Baldwin et al.,'s research project (2019) aimed to identify and share values of selected areas of the southeast Queensland (SEQ) wild catch commercial fishing industry which would assist in gaining ongoing social acceptance—a Social Licence to Operate (SLO), by the community and key stakeholders. SLO is needed to maintain access to the fishery resource, for example, by convincing regulators and interest groups that fishers engage in a sustainable harvest and do not negatively impact the environment through habitat destruction, by-catch or pollution. It can enhance market confidence and therefore industry profitability.

Photovoice interviews were used, followed by focus groups with commercial fishers, to reveal and confirm their values. Key stakeholders from government and conservation groups were interviewed to understand concerns or criticisms of the fishing industry. The photovoice method was selected for its ability to demonstrate and communicate values that might alleviate concerns raised. Fishers took photos to show how they fish, demonstrate their concerns, values and attitudes and indicate their capacity to contribute to engagement.

Analysis of interviews, photos and focus groups revealed four key themes (Baldwin et al., 2019):

1. Pride and market appeal of quality food product;
2. A sustainable fishery that is innovative, demonstrates compliance and cares for the environment;
3. Evidence of knowledge, skills and professionalism required for operation, e.g., navigation, weather, product handling, regulations; safety, boat handling, mechanics and maintenance; and crew management; and
4. Contribution to family and broader community, with involvement of many multi-generational family businesses, to provide direct and indirect employment in the region and play a role in community life.

Fishers were concerned about the community's lack of awareness of the fishing industry and rules that must be followed, and misrepresentation of the industry in media. The photographic images that displayed fishers'

**Fig. 20.3** Waste oil recycling (left) and saving a humpback whale...we were there...the Mum waiting just out of the breakers (right)

values were used in photo-stories posted in local libraries, on the industry association website and key stakeholder presentations. Theme two was illustrated by photos about sustainable stock research, marine garbage pickup, recycling and relaunching a stranded whale (see Fig. 20.3).

Interviews combined with a photovoice focus group revealed the social—ecological connectivity of rural landholders (Baldwin et al., 2017). The depth of landholder attachment to, and 'love of the land' was a key motivation for land and riparian stewardship, as opposed to concern about broader downstream impacts, social responsibility or connectedness (see Fig. 20.4). The catchment management and Landcare organisations found this information useful as a guide to build landholder commitment to, and programmes for riverine rehabilitation and maintenance.

Semi-structured interviews combined with a mapping exercise were used to identify how people value waterways in south east Queensland, focusing on Moreton Bay and the catchments that flow into it (Jones et al., 2016a; Pinner et al., 2019; Witt et al., 2019). Both Indigenous and non-Indigenous perspectives were captured in this study. The research aimed at identifying the diverse ways people interact with and derive benefit from marine and freshwater environments. Gaining an appreciation for how people value waterways can help ensure management efforts sustain the positive relationships people have with rivers, creeks and coastal areas. Relational values were elicited using semi-structured interviews whereby participants were asked to indicate on a

**Fig. 20.4** On the left: "it's our passion, we love our animals, we love the land". On the right: passing on knowledge to the next generation

map of the southeast Queensland region the waterways that are important and meaningful to them, and explain why. Stephen Kellert's typology of nature-based values (Kellert, 2008) was used as a coding frame to analyse the interview transcripts and identify the types of relationships that emerge through people's interaction with waterways. The study found that people's relationships with waterways in the study region are diverse with many shared values. The findings of the study were used by a regional natural resource management body to design a report card instrument for locating different benefits people derive from waterways in southeast Queensland, and monitoring how these change over time.

## Discussion

### *Implications of Using These Constructs for NRM*

Mental models and values offer two distinct lenses through which to examine how people understand, make sense of and derive benefit from the environment. By providing different perspectives on the cognition underpinning people's decisions and actions, they can be studied to meet different research and practice purposes.

Natural resource management involves (i) understanding how environmental systems function; and (ii) implementing decisions on how those systems should be used and managed. Over the past few decades there has been growing acknowledgement of the limitations of using a 'command and control' approach to natural resource management which involves a technical-fix from the top-down (Cox, 2016). The approach is guided by the assumption that the resource issue—or 'problem'—to be managed is "well-bounded, clearly defined, relatively simple, and generally linear with respect to cause and effect" (Holling & Meffe, 1996, p. 329). This approach is criticised by those who recognise that natural resource systems are both complex and uncertain, and that simple technical solutions that disregard social and ecological conditions can be detrimental (Cox, 2016). An adaptive management paradigm has emerged as an alternative approach, particularly within the social-ecological systems literature. This approach to management involves managers dealing with complexity and uncertainty, including social and ecological conditions, in an ongoing manner as they learn and adapt to changes in the system. The study of both mental models and values can offer critical insights for managing natural resources in an adaptive way.

In terms of understanding the system, mental models research provides a way of integrating local and experiential knowledge, and other perspectives, as additional forms of data that can be synthesised to provide a more comprehensive view, or explanation, of how a given system functions. Gaining an understanding of how local people perceive a system can also be used to reveal widely shared misunderstandings about system dynamics, such as misconceptions about how groundwater systems function which can result in over-use. This can provide resource managers with an opportunity to tailor engagement materials to the local context to overcome identified misunderstandings and knowledge gaps. Furthermore, incorporating local perspectives and other viewpoints can increase the legitimacy of knowledge produced for decision-making and can empower local people to play an active role in management of natural resources (Bélisle et al., 2018).

While integrating diverse perspectives can improve our understanding of a given system, it is important to acknowledge that as stated by Stirzaker et al., (2011, p. 3) "many of the ecological problems we face are as much a controversy over values as disputes about cause and effect". Good decisions in an adaptive management context therefore do not only depend on how accurately the system is represented; they also depend upon the processes through which trade-offs are made (Failing et al., 2013). Adaptive management is essentially a social learning process that prompts involved stakeholders to question worldviews or values that may constrain management efforts (Pahl-Wostl, 2009). By making the mental models and values of diverse stakeholders visible, they can be more readily discussed and negotiated. Lebel et al., (2010) describe social learning as the learning that occurs within groups that can lead to new knowledge, relationships, trust and collective action, and according to Cundill et al., (2012) can potentially lead to a shift in values. If all stakeholders feel their values have been acknowledged and adequately considered through a social learning process, they may be more capable of suggesting appropriate solutions and be more likely to accept the legitimacy of the decision-making process.

It is crucial that social science methods used to capture, document and integrate different sources of knowledge and values are selected according to the research purpose and context to ensure legitimacy and trust in both process and outcomes.

### *Critical Analysis of Benefits and Limitations of Various Methods*

As mental models and values are typically studied to understand the unique perspectives of a given group of people, qualitative approaches are commonly used, aligning with a constructivist or interpretivist paradigm. From this standpoint, social reality is shaped by lived experiences and social contexts. The following section discusses the benefits and limitations of methods to study mental models and values to understand diverse social realities as well as the social context in which they are embedded. The social context refers to the types of social settings where social interaction takes place and where meaning is created and shared (Given, 2008).

*Mental models*: Mental models are often elicited using oral interviews. This involves asking participants to explain their understanding of a given phenomenon. Elicitation interviews are typically 'semi-structured' whereby a set of key questions are asked, yet the wording and ordering of these questions can vary depending on responses received from the participant. This is beneficial as it enables the participant to form and express their own views on the topic and describe their unique social reality, rather than selecting from a pre-defined list of answers that may not hold any relevance to their perceptions of the world. Semi-structured interviews also involve probing questions to enable the interviewer to examine the participant's mental model in greater depth, typically by posing 'how' and 'why' questions (known as a laddering technique). These interviews are either conducted *in-situ* (i.e., in the same location of the phenomenon being studied) or *non-situ* (i.e., in a location away from the phenomenon studied).

Mental model interview transcripts are analysed to identify concepts and often the relationships between those concepts (e.g., causal relationships) that depict how concepts are organised or structured within a participant's mind. Rather than eliciting mental models orally, the participant may be asked to draw or visually represent their understanding of a phenomenon, using box and arrow diagrams for example (Blewett et al., 2021). These elicitation techniques may be conducted with individuals or with groups and can involve use of pen and paper or computer-based modelling software (Nyaki et al., 2014).

Results of mental model elicitation interviews are often presented graphically as cause-and-effect diagrams. These can then be compared to show similarities and differences in how people conceive the world and measure levels of consensus on a given issue. Elicited mental models can also be combined to show an aggregated representation of participants' mental models. Some studies depict only those concepts and relationships that were common to all, or a majority of, participants to provide a shared representation of a natural resource system. This type of shared representation depicting a common view of an issue can be used to facilitate dialogue and support negotiations over how a system should be managed.

Mental model elicitation techniques can be combined with processes that support participants in collectively discussing elicited mental models to create new opportunities for managing natural resources. Dray et al., (2007), for example, used a participatory modelling approach to support stakeholders in creating a shared understanding of groundwater management in the Republic of Kiribati. This shared understanding was used to create new opportunities for overcoming conflict and engage in collective management. Participants engaged in a social learning process using a computer-based role-playing game that prompted them to question their own understanding of water management issues and build new knowledge of the resource management issue together.

*Values*: Participatory interactive techniques are effective techniques to elicit values as they assist in reflection and can elicit deep emotions. First, participants think about the topic when collecting the data (e.g., taking photos, drawing, or mapping priorities). Second, they reflect when explaining the meaning of photos in an interview; third, when they hear and compare their perspectives with others in a workshop or focus group; and fourth, if they develop shared values to present the groups' views to decision-makers and the broader community (Chandler & Baldwin, 2010). Thus, they build on traditional qualitative methods such as semi-structured interviews. The focus group method might be used among people with common interests to enable them to formulate and present their agreement on values about natural resource management. A discursive workshop might be used when bringing together people who disagree, to enable them to better understand the range of views in a non-judgemental way and identify common ground. Using photos, for example, can illustrate both tangible and intangible concepts in a way that objectifies them, but engages emotions and empathy, which may assist in posing options and finding a solution to a complex problem. Quantitative methods, such as surveys, can be used to quantify the extent to which values, such as those identified through qualitative methods, exist within a larger population. In doing so, these methods can provide important empirical evidence that can be used to inform the design of natural resource management programmes and policies at a regional level and national scale.

Importantly, mental models and values research provide insight into the 'social' aspects of social-ecological systems which has been undervalued and thus under-researched (Berkes & Ross, 2013). Uncovering and sharing deep human understanding and emotions about a management challenge can contribute to reflection and deliberation about a solution that addresses everyone's needs to some extent. It can improve 'standing in another's shoes' among people with different levels of awareness of risks and lack of understanding of distribution of benefits and costs of action, combined with deep attachment to place, as well as external resistance framing. This can be a basis for building resilient communities that are required for adaptively managing complexity of rapid human and environmental change. Adaptation pathways can enable inclusion of local knowledge and values, as well as agreement on goals that allow for working through psychological, spiritual and emotional responses (Barnett et al., 2014).

As far back as the 1990s, it has been acknowledged that both natural and socio-economic science are necessary to natural resources management but there has been limited success in synthesising (Ison et al., 1997). The challenge of applying these concepts and associated methods is the level of skill needed in eliciting deeply held meaning in interviews and facilitating groups about sensitive issues. In many cases, this requires a person or a team with a multi-disciplinary environmental and social science background.

### *Future Research*

As NRM issues become more pressing because of climate change and pressures from population growth and dispersal, there will be an even greater need to use these concepts to resolve differences and develop policy and management approaches that a range of people can live with. As cognitive constructs, mental models and values exist within a person's mind and therefore cannot be examined directly. Social scientists depend on theoretical conceptualisations to enable development and application of methods and tools to examine these parts of cognition, albeit indirectly. Future research is needed to continue to develop the theory and practice of mental model and values research to deal with natural resource management challenges.

Future research will also need to explore and develop new participatory and creative methods that work together to engage and motivate people to create a commitment to change. These changes will require stakeholders to reflect on and question their own knowledge and values, as well as that of others. Methods and processes that facilitate collaborative exploration and integration of diverse knowledge sources and values, including the social context in which they are embedded, will prove crucial for achieving the transformative changes required to steer humanity through environmental crises faced at local, regional and global scales (Fazey et al., 2020).

Future research will also need to explore the extent to which natural resource management practitioners are willing to use these methods and processes, or be willing to rely on the results from mental models and values research conducted by others. Elicitation methods developed will thus need to be assessed for their usability, their usefulness in terms of results produced, as well as credibility.

## Conclusion

Ongoing research is advancing theoretical understanding of mental models and values, with progress also being made in developing and refining methods for eliciting these cognitive constructs in the context of natural resource management. Mental model research has demonstrated strong utility in shedding light on how individuals or groups perceive complex systems to function. Values research, on the other hand, has proved useful in providing insight into how people attribute importance and significance to aspects of the world, as well as insight into preferences and judgements that guide their decisions and actions. Minimal research, however, has been directed towards studying these cognitive constructs together. Doing so would provide richer insight into the cognition linking natural resources with management practices that meet a wider range of all stakeholders' needs.

This chapter presents concepts and methods that are relevant to the major themes of this book, showing participatory techniques useful for community engagement and development, necessary for effective co-management, facilitating systemic intervention and eliciting social aspects of resilience.

## References


Allahverdyan, A. E., & Galstyan, A. (2014). Opinion dynamics with confirmation bias. *PLoS ONE, 9*(7), e99557. https://doi.org/10.1371/journal.pone.0099557

Baldwin, C. (2008). *Consensus-building in water allocation and use*. University of Queensland.

Baldwin, C., Connor, S., & Perez, E. (2019). *Investigating social acceptance for the wild catch commercial fishing industry of Southeast Queensland*.

Baldwin, C., & Ross, H. (2012). Bridging troubled waters: Applying consensus-building techniques to water planning. *Society & Natural Resources, 25*(3), 217–234. https://doi.org/10.1080/08941920.2011.578120

Baldwin, C., Smith, T., & Jacobson, C. (2017). Love of the land: Social-ecological connectivity of rural landholders. *Journal of Rural Studies, 51*, 37–52. https://doi.org/10.1016/j.jrurstud.2017.01.012

Barnett, J., Graham, S., Mortreux, C., Fincher, R., Waters, E., & Hurlimann, A. (2014). A local coastal adaptation pathway. *Nature Climate Change, 4*(12), 1103–1108. https://doi.org/10.1038/nclimate2383

Bélisle, A. C., Asselin, H., LeBlanc, P., & Gauthier, S. (2018). Local knowledge in ecological modeling. *Ecology and Society, 23*(2). https://doi.org/10.5751/ES-09949-230214

Berkes, F., Folke, C., & Colding, J. (1998). *Linking social and ecological systems: Management practices and social mechanisms for building resilience*. Cambridge University Press.

Berkes, F., & Ross, H. (2013). Community resilience: Toward an integrated approach. *Society & Natural Resources, 26*(1), 5–20. https://doi.org/10.1080/08941920.2012.736605

Biedenweg, K., Williams, K., Cerveny, L., & Styers, D. (2019). Is recreation a landscape value?: Exploring underlying values in landscape values mapping. *Landscape and Urban Planning, 185*, 24–27. https://doi.org/10.1016/j.landurbplan.2018.12.005

Blewett, A., Jacobs, M., Kok, K., Jones, N. A., & Ogle, S. (2021). Stakeholder mental model analysis supports focused conservation policy and actions for Eurasian beaver (Castor fiber) reintroduction. *Journal for Nature Conservation, 64*, 126064. https://doi.org/10.1016/j.jnc.2021.126064

Brown, T. C. (1984). The concept of value in resource allocation. *Land Economics, 60*(3), 231–246. https://doi.org/10.2307/3146184

Chandler, L., & Baldwin, C. (2010). Reflections from the water's edge: Collaborative photographic narratives addressing climate change [Paper in themed section: The Visual Narrative: Alternative Photographic Exposures. Livingston, Debra (ed.).]. *Social Alternatives, 29*(4), 30–36.

Colding, J., & Barthel, S. (2019). Exploring the social-ecological systems discourse 20 years later. *Ecology and Society, 24*(1). https://doi.org/10.5751/ES-10598-240102

Cox, M. (2016). The pathology of command and control: A formal synthesis. *Ecology and Society, 21*(3). https://doi.org/10.5751/ES-08698-210333

Craik, K. J. W. (1943). *The nature of explanation*. The University Press.

Cundill, G., Cumming, G. S., Biggs, D., & Fabricus, C. (2012). Soft systems thinking and social learning for adaptive management. *Conservation Biology, 26*(1), 13–20. https://doi.org/10.1111/j.1523-1739.2011.01755.x

Dietz, T., Fitzgerald, A., & Shwom, R. (2005). Environmental values. *Annual Review of Environment and Resources, 30*(1), 335–372. https://doi.org/10.1146/annurev.energy.30.050504.144444

Dipu, M. A., Jones, N. A., & Aziz, A. A. (2022). Drivers and barriers to uptake of regenerative agriculture in southeast Queensland: A mental model study. *Agroecology and Sustainable Food Systems, 46*, 1502–1526.

Doyle, J. K., & Ford, D. N. (1998). Mental models concepts for system dynamics research. *System Dynamics Review, 14*(1), 3–29. https://doi.org/10.1002/(SICI)1099-1727(199821)14:1%3c3::AID-SDR140%3e3.0.CO2-K

Dray, A., Perez, P., Le Page, C., D'Aquino, P., & White, I. (2007). Who wants to terminate the game? The role of vested interests and metaplayers in the ATOLLGAME experience. *Simulation & Gaming, 38*(4), 494–511. https://doi.org/10.1177/1046878107300673

Engel, M. T., Vaske, J. J., & Bath, A. J. (2021). Seal hunting in Newfoundland from the perspective of local people. *Marine Policy, 128*, 104491. https://doi.org/10.1016/j.marpol.2021.104491

Failing, L., Gregory, R., & Higgins, P. (2013). Science, uncertainty, and values in ecological restoration: A case study in structured decision-making and adaptive management. *Restoration Ecology, 21*(4), 422–430. https://doi.org/10.1111/j.1526-100X.2012.00919.x

Fazey, I., Schäpke, N., Caniglia, G., Kendrick, I., Lyon, C., Page, G., et al. (2020). Transforming knowledge systems for life on Earth: Visions of future systems and how to get there. *Energy Research & Social Science, 70*, 101724. https://doi.org/10.1016/j.erss.2020.101724

Given, L. M. (2008). *The SAGE encyclopedia of qualitative research methods*. Sage.

Grant, B., Baldwin, C., Lieske, S. N., & Martin, K. (2015). Using participatory visual methods for information exchange about climate risk in canal estate communities. *Australian Journal of Maritime & Ocean Affairs, 7*(1), 23–37. https://doi.org/10.1080/18366503.2015.1014012

Guerrero, A. M., Jones, N. A., Ross, H., Virah-Sawmy, M., & Biggs, D. (2021). What influences and inhibits reduction of deforestation in the soy supply chain? A mental model perspective. *Environmental Science & Policy, 115*, 125–132. https://doi.org/10.1016/j.envsci.2020.10.016

Holling, C. S., & Meffe, G. K. (1996). Command and control and the pathology of natural resource management. *Conservation Biology, 10*(2), 328–337. http://www.jstor.org/stable/2386849

Ison, R. L., Maiteny, P. T., & Carr, S. (1997). Systems methodologies for sustainable natural resources research and development. *Agricultural Systems, 55*(2), 257–272. https://doi.org/10.1016/S0308-521X(97)00010-3

Jones, N. A. (2012). *Eliciting mental models of natural resource systems: A procedural comparison* (Unpublished doctoral dissertation). University of Queensland.

Jones, N. A., Ross, H., Lynam, T., & Perez, P. (2014). Eliciting mental models: A comparison of interview procedures in the context of natural resource management. *Ecology and Society, 19*(1). https://doi.org/10.5751/ES-06248-190113

Jones, N. A., Ross, H., Lynam, T., Perez, P., & Leitch, A. (2011). Mental models: An interdisciplinary synthesis of theory and methods. *Ecology and Society, 16*(1). http://www.jstor.org/stable/26268859

Jones, N. A., Ross, H., Shaw, S., Witt, K., Pinner, B., & Rissik, D. (2016a). Values towards waterways in south east Queensland: Why people care. *Marine Policy, 71*, 121–131. https://doi.org/10.1016/j.marpol.2016.05.027

Jones, N. A., Shaw, S., Ross, H., Witt, K., & Pinner, B. (2016b). The study of human values in understanding and managing social-ecological systems. *Ecology and Society, 21*(1). https://doi.org/10.5751/ES-07977-210115.

Kellert, S. R. (2008). *Foundations of environmental sustainability*. Oxford University Press. https://doi.org/10.1093/acprof:oso/9780195309454.003.0021

Lebel, L., Grothmann, T., & Siebenhüner, B. (2010). The role of social learning in adaptiveness: Insights from water management. *International Environmental Agreements: Politics, Law and Economics, 10*(4), 333–353. https://doi.org/10.1007/s10784-010-9142-6

Lockwood, M. (1999). Humans valuing nature: Synthesising insights from philosophy, psychology and economics. *Environmental Values, 8*(3), 381–401. https://doi.org/10.3197/096327199129341888

Nyaki, A., Gray, S. A., Lepczyk, C. A., Skibins, J. C., & Rentsch, D. (2014). Local-scale dynamics and local drivers of bushmeat trade. *Conservation Biology, 28*(5), 1403–1414. http://www.jstor.org.ezproxy.library.uq.edu.au/stable/24480390

Pahl-Wostl, C. (2009). A conceptual framework for analysing adaptive capacity and multi-level learning processes in resource governance regimes. *Global Environmental Change, 19*(3), 354–365. https://doi.org/10.1016/j.gloenvcha.2009.06.001

Pinner, B., Ross, H., Jones, N. A., Babidge, S., Shaw, S., Witt, K., et al. (2019). A custodial ethic: Indigenous values towards water in Moreton Bay and catchments. In I. R. Tibbetts, P. C. Rothlisberg, D. T. Neil, T. A. Homburg, D. T. Brewer, & A. H. Arthington (Eds.), *Moreton Bay Qandamooka and catchment*. The Moreton Bay Foundation.

Reser, J. P., & Bentrupperbäumer, J. M. (2005). What and where are environmental values? Assessing the impacts of current diversity of use of 'environmental' and 'world heritage' values. *Journal of Environmental Psychology, 25*(2), 125–146. https://doi.org/10.1016/j.jenvp.2005.03.002

Reyes-Arroyo, N., Camacho-Valdez, V., Saenz-Arroyo, A., & Infante-Mata, D. (2021). Socio-cultural analysis of ecosystem services provided by mangroves in La Encrucijada Biosphere Reserve, southeastern Mexico. *Local Environment, 26*(1), 86–109. https://doi.org/10.1080/13549839.2020.1867836

Ross, H., Witt, K., & Jones, N. A. (2018). Stephen Kellert's development and contribution of relational values in social-ecological systems. *Current Opinion in Environmental Sustainability, 35*, 46–53. https://doi.org/10.1016/j.cosust.2018.10.007

Schulz, C., Martin-Ortega, J., Ioris, A. A. R., & Glenk, K. (2017). Applying a 'value landscapes approach' to conflicts in water governance: The case of the Paraguay-Paraná waterway. *Ecological Economics, 138*, 47–55. https://doi.org/10.1016/j.ecolecon.2017.03.033

Seymour, E., Curtis, A., Pannell, D., Allan, C., & Roberts, A. (2010). Understanding the role of assigned values in natural resource management. *Australasian Journal of Environmental Management, 17*(3), 142–153. https://doi.org/10.1080/14486563.2010.9725261

Stirzaker, R. J., Roux, D. J., & Biggs, H. C. (2011). Learning to bridge the gap between adaptive management and organisational culture. *Koedoe, 53*, 28–33. http://www.scielo.org.za/scielo.php?script=sci_arttext&pid=S0075-64582011000200004&nrm=iso

Witt, K., Ross, H., Shaw, S., Jones, N. A., Rissik, D., & Pinner, B. (2019). How do local people value rural waterways? A study in the upper catchments of south east Queensland's rivers. *Society & Natural Resources, 32*(6), 638–656. https://doi.org/10.1080/08941920.2019.1578910

CHAPTER 21

# Towards Transformative Collaboration Between Agri-Food Supply Chain and Rural Community: Insights from a Case Study in Wuhan, China

*Li Liu*

## Introduction

Agri-food supply chains and rural communities, in the broad agri-food systems, are closely connected to solve various challenges, including poverty reduction in farming dependent households (Bolwig et al., 2010; Minh & Osei-Amponsah, 2021), environmental protection in rural communities (Ponte, 2020) and operational efficiency in agri-food supply chains. For instance, poor and uneducated farmers in rural communities are likely to overuse chemicals to increase productivity for their survival, while depleting natural resources may constrain the quality of products

L. Liu (✉)
Fudan Development Institute, Fudan University, Shanghai, China
e-mail: liuli@fudan.edu.cn



C. Baldwin and S. van Bommel (eds.), *Rural Development for Sustainable Social-ecological Systems*,
Palgrave Studies in Environmental Sustainability,
https://doi.org/10.1007/978-3-031-34225-7_21

to gain a higher market price for downstream retailers. Likewise, the exploitation of price by wholesalers or retailers may decrease farmers' profits and constrain their further investments in farming. Researchers have noted that any challenges or problems relevant to the agri-food systems cannot be solved alone by an individual chain actor or community member. Instead, a transformative way is needed for these actors to work together to achieve diverse economic, environmental and social benefits, i.e. sustainability goals (Liu et al., 2020).

However, little is known about the process, mechanisms and policy interventions to encourage the transformative collaboration between agri-food supply chains and rural communities for sustainability goals. Transformative collaboration points to the importance of engaging and consulting farmers, other chain actors and rural communities for shared benefits (Majchrzak et al., 2015). Such a process transforms mechanisms of collaborative actions in the whole system by three principles: engagement of the poor, the dissolution of the power structure and creation of systemic changes (Sankaran & Presier, 2021). This study illustrates and addresses how and why transformative collaboration between agri-food supply chains and rural communities can be achievable in delivering agri-food system-wide changes. This chapter bridges the knowledge gap to link individual efforts to systemic changes at the methodological level, and contributes significant implications for transformation at the system level to provide a range of social, economic and environmental benefits (Gray et al., 2015).

This chapter aims at explaining and facilitating deeper systemic interventions (Hasle & Vang, 2021) for the agri-food systems, which highlights the most insightful implications drawing from my PhD thesis under supervision of Professor Helen Ross. I argue that systemic interventions in agri-food systems need joint adjustments of rural communities and agri-food supply chains to overcome challenges and embrace opportunities emerging from their interactions. The transformative collaboration between chains and rural communities, at the epistemological level, suffers the difficulties to cater the contextual situations. To tackle this shortcoming, this study holistically examines the interactions between supply chains and communities to inform systematic interventions for changes of agri-food systems in Wuhan, China. I apply 'walking-the-chain' approach to purposively choose three types of chains and two connected villages to fit into a framework of representation and voices of all actors of agri-food systems. To understand the chains and communities, in-depth interviews, household interviews and focus group discussions about chain capabilities, and connected community mapping were adopted, respectively. The collective examination of the villages and chains (sub-systems) helps to understand dynamics, drivers of changes and interactions in the whole food system. Interventions for systemic changes need to include collaboration or partnership in the supply chain-community nexus.

The chapter has five sections. Following the introduction, the second section develops the theoretical framework. The third section introduces the methodology. The fourth section presents the results. The last section discusses and concludes the pathways to transformative collaboration in the agri-food systems.

## Literature Review

### *Function of Collaboration*

Collaboration, a process to combine and expand individual's capabilities, emerges as a new way to solve shared challenges and respond to opportunities, as collaborative initiatives reshape and rearrange resources and relationships to create outcomes that could not be achieved by one alone (Elia et al., 2020). In this context, collaboration is characterised as 'strategically interdependent action' (Head et al., 2016; Keast et al., 2004; Ross et al., 2019), as it involves shared ideas, resources, mission and reciprocity, transcends individual organisational interests. Cooperation and coordination are very close to collaboration, but the prior two types of relationships are related to shorter time frames and less investment of time or resources compared with collaboration. Collaboration suggests interdependence, continuous contact and regular exchange information.

The process of collaboration has distinct functions in resources, innovation and empowerment (Head et al., 2016; Ross et al., 2019). At the resource level, collaboration enables organisations and individuals to expand their pool of resources and skills and develop distinctive capabilities, which promotes the synchronisation of efforts and resources (Bryson et al., 2006). At the innovation level, collaboration is significant for social learning and innovation, as it engenders synergistic knowledge creation. The collaboration encourages collective agency of individuals, and social learning from contingent or unorganised processes and outcomes (Vestergaard et al., 2021). The collaboration-based social learning processes encourage collective interpretation and actions, which share similarities with a dialogue approach to create new knowledge or practices. At the empowerment level, collaboration requires a process of capability-building among the disadvantaged actors with less resources (Salignac et al., 2018), despite the constant challenges and power asymmetry. Collaboration among a network of wide relationships empowers the participants to move beyond possible constraints in resources and innovation.

### *Transformative Collaboration*

Transformative collaboration suggests a process of collaboration for transformative outcomes, or collaboration for changes at the system level (Salignac et al., 2018). In this context, systemic intervention encourages collective agency of individuals to learn from contingent or unorganised processes and interactions for system changes. The new concept of transformative collaboration extends the function of collaboration from individual, organisational levels to the system level (Salignac et al., 2018; Vestergaard et al., 2021), which is aligned with the principles of equality, diversity and inclusiveness. Equality suggests a network of interconnected actors who are equal in access to opportunities. Inclusiveness and diversity suggest an enabling environment to involve actors with diverse characteristics, such as age, gender, social status, etc. The three principles assemble to put individuals in an enabling environment that is being created by real or prospective collaboration for system changes.

Transformative collaboration is a more powerful area of intervention that turns strategic arrangements in practices to structural empowerment for participants, not just resource building among actors. It is known that the superficial form of collaboration network offers compositional capabilities for collaborating actors (Hardy et al., 2003). Hardy et al. (2003) suggest that inter-organisational ties extend beyond specific individual relationships, shifting patterns of influence and power. The high level of equality, diversity and inclusiveness in the processes of collaboration contribute to transformative forms of collaboration structure or network, which encourages coordinated efforts from multiple stakeholders working towards a common overarching goal, which is related to the collective impact or a system view to social change (Salignac et al., 2018).

Transformative collaboration seeks to create systemic changes or exert systemic intervention through greater involvement. The systemic concept concerns multiple interaction between parts (elements, or sub-systems) which make up a whole. The system is an adaptive whole that can survive through time by adapting to changes in its environment. Being systemic is not about being comprehensive; it is about dealing with the interactions between parts of a whole. The whole of system approach considers the complexity of social situations (Duboz et al., 2018), including the paradox with the 'contradictory yet interrelated elements that exist simultaneously and persist over time'. The whole of system is kin to the Chinese idea of Yin-Yang as two contrasting aspects that stimulate imagination and creation for changes (Li, 2016). Transformative collaboration can derive from the co-existence of competitors and cooperators, the power and the powerless, successful and failed experiences that help to create an enabling holistic environment to address interlinked challenges or add implications for innovation. Meanwhile, transformative collaboration also links with a further process of moving towards the prospective systemic changes that are emerging from the innovative interactions between the various practising parts. This perspective of transformative collaboration is the focus of case study in this chapter.

The whole of system approach is an interactive process that engages holistically by incorporating views of stakeholders for systemic intervention. This is because it facilitates boundary crossing (e.g. cross sector/cross-scale/cross-domain/cross-property) and coordination across scales necessary to support transitions. Boundary crossing is enabling for social actors, which plays a critical role in forging multi-level linkages. The system, rebuilt by cross-boundary principles and practices, enables actors to regain control in its transformation (Hassanein, 2003), which helps to envision the ways in which individuals can have a positive impact on society and the broader environment. This helps us to theorise what the agri-food systems might be or might become (Bartunek, 2020; King et al., 2010; Marti & Gond, 2018).

The well-developed as well as the less-developed chains and communities suggest the diversity of the China's agri-food systems, and each actor is considered equally relevant for agri-food system development. Functional agri-food systems arise from and reinforce the interactions between agri-food supply chains and rural communities. This sense of mutual dependence between chains and communities tempers the pursuit of organisation-specific gains at the expense of other partners. It encourages collaboration between chains and communities to promote the transformation of agri-food systems. This wider conceptualisation of the whole of agri-food system through diverse supply chain-community interactions leads to reality construction, interpretation and action (Gray et al., 2015). Firstly, it overcomes the cultural bias and powers embedded in policy practices to amplify into widely shared and more enduring policy priorities (Bartunek, 2020; Purdy et al., 2019). This helps to capture meanings at the individual level equally (Purdy et al., 2019). Secondly, the bidirectional, recursive supply chain-community interactions unfold a process of ongoing change, adaptation and learning among actors, which offers a socially constructed view of collective impacts (Hoey et al., 2016). Thirdly, the mobilisation of cooperation and the contextually appropriate collective action in a network of actors is critical to the achievement of system-wide changes (Murphy et al., 2021), which acts as a tool for understanding systemic institutional building (Cornelissen & Werner, 2014). Hence, the purposefully designed network of multiple interacting chains and villages with diversity, inclusion and interconnections may drive a process of social learning and capture innovation. The mutually beneficial relationships between chains and communities will encourage individual's learning to change minds and practices, which further informs transformative collaboration among individuals for systemic changes.

## Methodology

The whole of system approach incorporates multiple perspectives and worldviews relevant to the system. Thus pragmatism that stimulates integrative research while at the same time maintaining a plurality of perspectives is the research paradigm that underpins this study. Rather than subordinating "theory" to "epistemology", the pragmatist tradition places continuous dialogue at the centre of inquiry, developing a systematic approach to understand the social fabric and diversity. Pragmatism utilises the strengths of various theories to effectively address research problems.

### *Case Study Approach*

A case study approach is valuable to understand the complex processes and mechanisms for agri-food system changes. Wuhan in central China is chosen as the study area (Fig. 21.1). It is purposively selected based on

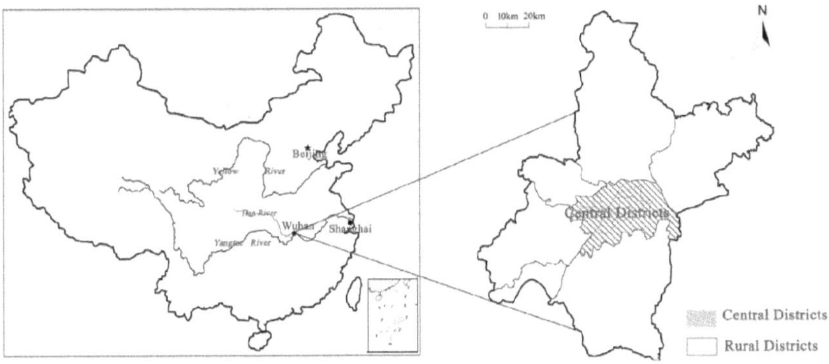


**Fig. 21.1** Location of the study area—Wuhan, Central China

three criteria: (1) it is a key agricultural area with ongoing potential for improved agri-food systems; (2) it has a policy environment that addresses sustainability challenges regarding economic development, social equity and issues on agro-environment and (3) the public and the private sectors are both involved in agri-food system transformation.

Wuhan is located in "the bowl of rice and fish", so the rice and vegetable systems are chosen as two types of production cases to understand the interactions between agri-food supply chains and rural communities. For the convenience of data collection, the study chose rice and vegetable cases from one of the districts in a rural area—JX District. In the JX district, agri-food systems have transitioned from the traditional to the modern, and there are farmer-led food chains, cooperative-led (co-op-led) food chains and small and middle enterprise-led (SME-led) food chains in the industry sector. According to the literature and policy documents, less-developed as well as well developed villages are typical in rural communities. While the rice and vegetable systems operate within a shared landscape environment, the rice and vegetable systems have differences in market value, government interventions and community structure. Thus, there are varying capitals and capabilities in the community and chains. Systemic intervention in the agri-food system is a process of holistic observation of the diverse chains and villages, and then building connections between these actors collectively to inform changes of agri-food systems to be more sustainable.

**Table 21.1** The in-depth interviews and household interviews in the rice and vegetable cases

| *Types of data* | | *In-depth interviews* | | *Household interviews* | |
|---|---|---|---|---|---|
| | | *Rice case* | *Vegetable case* | *Rice case* | *Vegetable case* |
| Studies on agri-food supply chains | SME-led | 8 | 17 | – | – |
| | Co-op-led | 3 | 4 | – | – |
| | Farmer-led | 5 | 4 | – | – |
| Studies on rural supporters or stakeholders | Government officers, extension service providers, input suppliers | 10 | 10 | – | – |
| Studies on rural communities | Village 1(well-developed) | 6 (QF village) | 6 (CJ village) | 25 | 27 |
| | Village 2 (less-developed) | 3 (XH village) | 7 (DL village) | 27 | 23 |
| Total numbers | | 35 | 48 | 52 | 50 |

### *Data Collection in Supply Chains*

The walking-the-chain observation and interview approach considers the way in which food is produced, processed, delivered and consumed. By examining activities and various actors, it helps to diagnose and identify challenges and opportunities for chain-level improvement.

Specifically, in-depth interviews were used with informed government officers, chain actors (see details on Table 21.1) to obtain information on (1) policies and strategies for sustainable development in rice and vegetable sectors, (2) the structure of various agri-food supply chains, such as product flow, information flow, financial flows and relationships and (3) capabilities of agri-food supply chains. Secondary data, including government policies, reports, was collected both before and after the fieldwork to provide complementary information on the agribusiness sectors in rice and vegetables.

### *Data Collection in Rural Communities*

For studies of villages, the community capitals' frameworks helped to examine seven capitals in the villages to understand the resources, strengths and dynamics of communities. Household interviews were used to examine village-level development processes, such as village history, the connections with the government and internal dynamics. The information was collected from village leaders and group members from the village committee. The questions included three aspects: (1) village development progress including economic, social, environmental and cultural changes; (2) village-level rice/vegetable industries' development and (3) rice/vegetable industries' impacts on villages' capital assets and village development visions. Meanwhile, four focus group discussions were added to further examine the interactions between agri-food supply chains and the connected villages.

### *Case Analysis and Cross-Case Synthesis*

This research used QSR NVivo 10 software to assist the data analysis. It was used to code, organise and sort information according to my research purposes. In both rice and vegetable cases, emerging themes from the supply chains and rural communities were organised to understand community capitals (including economic capital, environmental capital, physical capital, financial capital, social capital, cultural capital, political capital) and chain-level capabilities (including economic capabilities, social capabilities and environmental capabilities, i.e. sustainability capabilities) respectively. Based on the initial analysis of rural communities and agri-food supply chains, I adjusted the coding and analysis to reflect the interactions between community capital assets and chain-level sustainability capabilities.

One goal of this multiple-case study is to build a general explanation that fits each individual case, even though the cases vary in their details. Based on the identified thematic framework and results in the rice and vegetable cases, I made a cross-case comparison and synthesis (see Fig. 21.2). Synthesis of the interactions between supply chains and communities in two cases is a process to integrate knowledge from multiple systems and across multiple scales and levels. The holistic examination of the network of chains and communities offers opportunities to explore ways for transformative collaboration between chains and the connected villages for systemic changes.

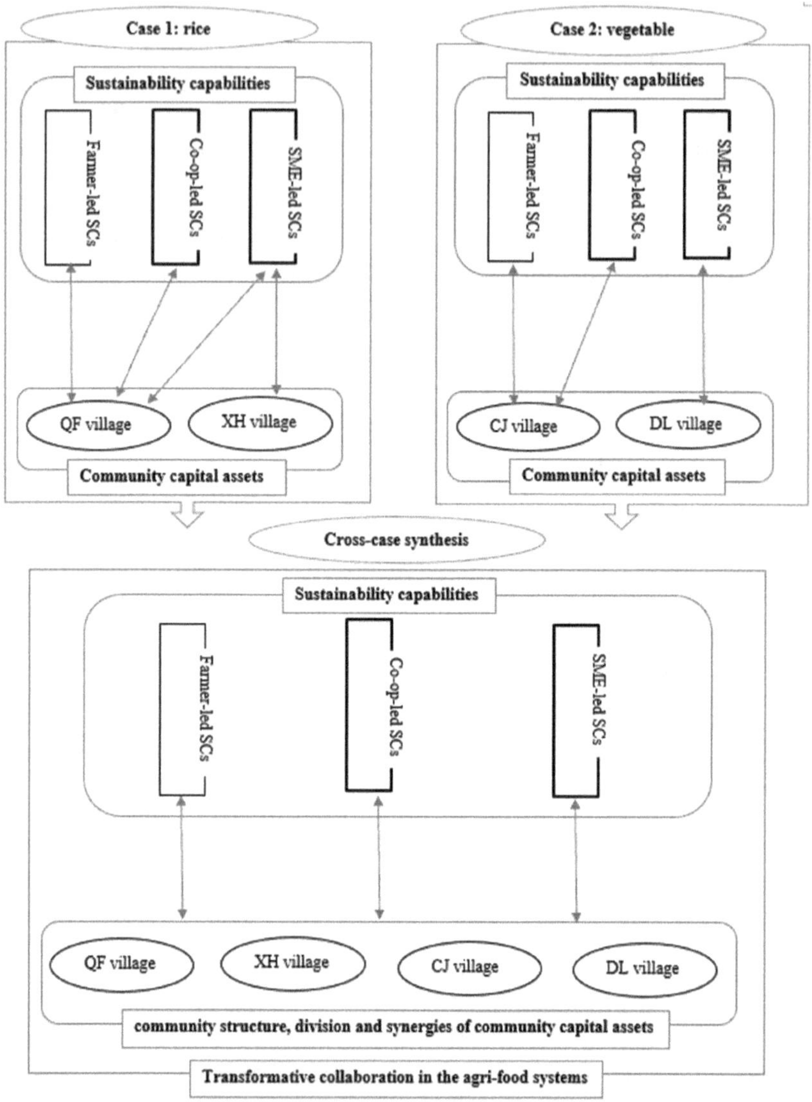


**Fig. 21.2** Summary of research design

## Results: Knowing About the Agri-Food Systems

### *Community Capitals in Rural Villages*

In the rice and vegetable cases, the less-developed and well-developed villages have different community capitals, including tangible financial capital, environmental capital and physical capital, and non-tangible social, cultural and political capitals. Although the villages had experienced different village community change processes, the synergies of capital assets across them show a similar pattern. It indicates that two competing rural community development initiatives decreased collective effectiveness. On one hand, the government invested in and mobilised community capital assets that are connected with agricultural modernisation tasks, while on the other, rural households concentrated on community assets to improve individual household well-being.

The government intervention gave attention to community "deficiencies" when building a modernised and economically viable rural place. In order to respond to the gaps, the government promoted a series of planned efforts in agro-environment improvement (environmental capital), infrastructure building (physical capital), investments in new agribusiness (economic capital and social capital), technical knowledge improvement (human capital), eco-agri-food brand building and business diversification (cultural capital) and leadership development to implement a market-oriented land tenure system as well as other institutional and policy reforms (political capital). These "purposive" government efforts sought to integrate the villages' assets with economic opportunities in agricultural modernisation.

In contrast to the task-based view of villages by the government, rural households treated the villages as a place to achieve their cultural identities, i.e. the household-based lifestyle in agro-environment protection (environmental capital), infrastructure building (physical capital) and household income optimisation (economic capital). The rural households mobilised personal social networks to improve the individual quality of their lives (social capital). As an independent economic unit, the rural household had limited commitments to community-level action. The household traditions and cultures remained to play a role in community cultural aspect practices and political affairs. For instance, the rural elites, such as village leaders, also had strong incentives to keep the economic benefits and development opportunities within the boundary of personal social networks (especially family members or kin) rather than the broader community for personal benefit, so subsidies for community development projects were not very visible (political capital). While there was a high level of collaboration within households, collaboration at the village level was eroding. The village committees played a role in village-level affairs by securing and managing funding from various government projects, while rural households lacked incentives to pursue shared interests.

### *Capabilities in Agri-Food Supply Chains*

Various types of agri-food supply chains have different sustainability capabilities which influence village-level changes and development. These capabilities suggest that the various chains have diverse connections with the four villages. Farmer-led agri-food supply chains contributed more to household-level socio-economic well-being. They considered the personal social networks of farmers in order to diversify household incomes. Such networks were created by product quality, competitive prices, hardwork and mutually dependent relationships. Besides the internal chain-level attributes, farmer-led agri-food supply chains linked their survival and prosperity with the external institutional environment and government supports. However, the system did not provide the resources to support them in increasing their competence. The deficiencies in facilities, institutional limitations and the lack of service investments restricted economic viability and environmental performance improvement, as well as social benefits in many of farmer-led agri-food supply chains. Farmer-led agri-food supply chains gain limited support from the government, but they reported that genuine government support is required in response to chain-wide problems regarding food quality and efficiency.

The co-op-led agri-food supply chains are purposively directed by the government to encourage entrepreneurial commitments in villages. Both agribusinesses and village committees played a dominant role in the formation of the co-op-led agri-food supply chains. The government investments in co-op-led agri-food supply chains increased the economic capital of some rural households, and their specialised knowledge, extensive interaction and commitments helped to overcome economic constraints for community welfare. Both co-op-led rice and vegetable supply chains stressed the significance of social networks, but the co-op-led agri-food supply chains had dual social identities. From an economic perspective, the quality of social networks promotes coordination, information sharing and efficiency, which increases the co-op's level of economic performance. The social network means social inclusion, information sharing and connections to the co-op members. From the social perspective, the organisational structure of the co-op-led agri-food supply chains was manipulated to attract government support, and the social networks of the co-op-led agri-food supply chains represent the narrow interest of co-op members. As the village elites did not show any interest in the well-being of the community, the social exclusion of other members damaged internal solidarity for economic initiatives. The transparent unfairness in the village decreased mutual trust and collective agency.

SME-led rice and vegetable supply chains sought to create value for the rural place and serve a broader market. The SME-led agri-food supply chains initiated mutual connections between community development and chain-level capabilities. The interconnections were manifested in three ways. Firstly, the employment well-being was connected with personal working conditions and social conditions of the workers' community. As employment well-being and workers' personal social networks were incorporated into chain-level economic benefits and economic risks, the lively rural community conditions increased employment welfare and chain-level value creation. This suggested that community members' benefits or community development provided alternative ways to improve chain-level competence. Additionally, the production capacities of SME-led rice and vegetable supply chains, through industry upgrading and innovation, not only collectively contributed to local development, but also profited from the interactions with tangible and intangible resources in each place. As the business competitiveness was embedded with agro-environment conditions, trust and material support in rural communities, there were different integration patterns. In the SME-led rice supply chains, monitoring and improvements of the production environment were integrated with food quality, such as genetically modified rice crop experiments (technological improvements), and agro-environment degradation (heavy metal pollution of rice). SME-led vegetable supply chains exercised tremendous initiatives to reshape rural agro-environments, as organic vegetable production and agro-tourism programmes which provided payments for such environmental protection practices. SME-led vegetable supply chains created a diverse and innovative economy to spur landscape changes in villages.

Thirdly, the community social development was interactive with the business climate, which included trust, coordination or collaborations, market knowledge and human resources support. Both SME-led rice and vegetable supply chains linked chain-level value creation capacities with the community institutional context. SME-led rice and vegetable supply chains were informal economies but with inadequate social security systems. As the labourers in the agribusinesses were mainly ageing farmers without formal social insurance, the organisational structure increased employment risks for both workers and the agribusinesses. Without formal institutions to regulate the field, the intangible community norms or cultural-cognitive institutions can be both assets and liabilities for economic advances in SME-led rice and vegetable supply chains.

### *Interactions Between Rural Communities and Agri-Food Supply Chains*

The formation of rice and vegetable supply chain networks has a structural influence on three aspects of rural communities—new social institutions, cross-boundary interactions and a revised place vision. Firstly, the rice and vegetable supply chain network contributes to new social institutions and new production patterns, such as labour, land rent income, rural restructuring and new technologies in the community. The interaction of social factors is aligned with market revolution in deregulation of rural development. Under the influence of a free-market ideology, these processes link farmers' land assets to the development goals of agribusinesses and local food providers, which also respond to changes in rural livelihoods and rural landscapes.

Secondly, the new social institutions including private agribusinesses, the commodification of farmland assets, modern farming facilities, private food certifications (like green farming, organic farming), traceability system and integrated supply chain channels, engage a network of collaborative and complementary actors to transform the vision of rural place. This is a dynamic and unpredictable process at local level, as the social institutions may either generate or obstruct innovative ways to mobilise various capital assets at the place. For instance, the emergence and evolution of private governance mechanisms lead to strategies that favour the agribusinesses' interests (like reducing transaction costs) better than small farmers' capabilities. Local farmers increasingly drop out of the process, while capable migrant farmers are involved increasingly. Farmers who lose land in the village may be forced to be net buyers of food rather than producers. Hence, most small farmers have difficulty in accessing the development benefits, as the institutional arrangements tend to overlook social demands in the rural context.

Thirdly, the agri-food supply chains have triggered economic restructuring in the rural communities, as chain-level development highlights the interdependence, interaction and synergy among various actors, like government, agribusinesses, migrant rural households and consumers. The inter-organisational collaboration is significant for economic development and innovations, while it also causes some risks and uncertainties for rural households. The improving quality of life for most rural households requires an inclusive framework to mobilise community strengths or solve community problems.

Meanwhile, the village communities, connecting with rice and vegetable supply chain network contributed to a shared context for the various actions of the supply chains. The crucial aspects of community context—capital assets, institutions and interactional patterns—impact situational capacity for chain-level sustainability in different ways. This sub-section draws conclusions about the community's influence on sustainability capabilities in vegetable supply chain networks. Firstly, community capital assets, such as initiatives in labour, skills and organisational patterns impact sustainability capabilities in various vegetable supply chains. To some extent, capability deficiencies of the vegetable supply chains derive from misallocations of capital assets in rural communities. Government promotes improvement of the agro-environment through technological innovation and improvement of farming facilities through a small number of agribusinesses, while the agency of farmers tends to be ignored in promoting sustainability capabilities of vegetable supply chains.

Secondly, community traditions have an enduring influence on the behaviour of vegetable supply chains. In the vegetable supply chain networks, the rural households are widely embedded in farmer-led, co-op-led and SME-led vegetable supply chains, showing that the household farming traditions have connected with modern agribusiness practices in different ways. The household farming is integrated into the "corporate farming". The production processes in the SME-led supply chains show that the resistance to capitalist farming is socially constructed. The passive response to employment opportunities is linked with low local autonomy. Many local farmers felt alienated and incapacitated because they find it difficult to control their lives and affect changes at the community level. Without incorporating into the social fabric of the communities, the SME-led vegetable supply chain is struggling to fulfil the holistic aspiration of sustainability in a dynamic environment.

Thirdly, and most importantly, the interactions between the vegetable supply chain networks and local communities tend to destabilise existing institutional arrangements and shape new conventions. The changing community institutions provide both value creation potential and chain-wide challenges. For example, SME-led vegetable supply chains suffer from overinvestment in farming facilities and lack of producers to make sufficient use of the consolidated farmland. Meanwhile, farmer-led vegetable supply chains lack investments in farming facilities. The imbalanced farming facilities investments are likely to be a source of threat in communities, as it breaks the community solidarity rather than building the capability of local farmers. It is important to increase community organisation to create an enabling environment and leverage farmers' capabilities, which, to some extent, could be translated into the sustainability capacities of the vegetable supply chains.

## Discussion and Conclusion: Practising Towards Transformative Collaboration

While there are many studies on supply chain collaboration (Matopoulos et al., 2007) and government-community cooperation (Ross et al., 2019), the collaboration between chains and communities is a new topic. In most cases, the trade-off and conflicts between supply chains and communities have overshadowed their potential for synergy and collaboration for changes at the agri-food system level (Selwyn, 2016; Vicol et al., 2018). Through bringing diverse chains and communities together in collective forums, this study informs understanding transformative collaboration in the network of chains and communities. The systemic changes need a new network mindset that emphasises interrelationships and interdependence, and the interactions of actors enabling structural empowerment. The network of actors widens the ways to allocate and distribute resources, information, capabilities and institutions, which also emerges as a new strategy for systemic intervention in agri-food systems.

Transformative collaboration is a creative way to combine and restructure the resources, capabilities and institutions among a network of actors for systemic effects. The creative pathways derive from an integrated process of knowing about whole system. The detailed examination of the chains, communities and interactions between chains and communities in this study helped to gain a holistic observation and inquiry of the agri-food system. As suggested by Midgley (2003), the process of observation and inquiry is a type of intervention, which helps to overcome the biases and seek information for meaning-making. By incorporating existing knowledge from chains and communities, this study suggests the importance of co-designing systemic interventions (Minh & Osei-Amponsah, 2021) in the rice and vegetable agri-food systems to achieve social, economic and environmental (i.e. sustainable) benefits in Wuhan, China. The case study suggests that a collaboration between different types of chains and supply chain-rural community nexus will promote the transformation of agri-food systems to be more sustainable.

In summary, knowing the whole system by observation and inquiry offers opportunities for intervention at the system level. A holistic understanding of chains, communities and interactions in the network enables learning from the network stocks of resources, development processes and strategies, which also suggests the opportunities and constraints of mobilising their capabilities, intervening in the processes and reshaping strategies. The knowledge and practice of the network of chains and villages foster an awareness and appreciation of the benefits of the supply chain-rural community nexus, and transcend the constraints of the separate actions for systemic changes. As this study exerts systemic intervention on the local agri-food system by just observing and inquiring, further theoretical and empirical studies are needed to unearth the processes of transformative collaboration in a broader network of actors.

**Acknowledgements** This chapter derives from the author's PhD study, conducted under the guidance of Prof. Jim Cavaye then Prof. Helen Ross, with Dr. Anoma Ariyawardana.

## References

Bartunek, J. M. (2020). Accomplishing impact by performing our theories: It can be done, though not easily. *Journal of Applied Behavioral Science, 56*(1), 11–31. https://doi.org/10.1177/0021886319885451

Bolwig, S., Ponte, S., du Toit, A., Riisgaard, L., & Halberg, N. (2010). Integrating poverty and environmental concerns into value-chain analysis: A conceptual framework. *Development Policy Review, 28*(2), 173–194. https://doi.org/10.1111/j.1467-7679.2010.00480.x

Bryson, J. M., Crosby, B. C., & Stone, M. M. (2006). The design and implementation of cross-sector collaborations: Propositions from the literature. *Public Administration Review, 66*, 44–55. https://doi.org/10.1111/j.1540-6210.2006.00665.x

Cornelissen, J. P., & Werner, M. D. (2014). Putting framing in perspective: A review of framing and frame analysis across the management and organizational literature. *Academy of Management Annals, 8*(1), 181–235. https://doi.org/10.1080/19416520.2014.875669

Duboz, R., Echaubard, P., Promburom, P., Kilvington, M., Ross, H., Allen, W., et al. (2018). Systems thinking in practice: Participatory modeling as a foundation for integrated approaches to health. *Frontiers in Veterinary Science, 5*. https://doi.org/10.3389/fvets.2018.00303

Elia, G., Margherita, A., & Petti, C. (2020). Building responses to sustainable development challenges: A multistakeholder collaboration framework and application to climate change. *Business Strategy and the Environment, 29*(6), 2465–2478. https://doi.org/10.1002/bse.2514

Gray, B., Purdy, J. M., & Ansari, S. (2015). From interactions to institutions: Microprocesses of framing and mechanisms for the structuring of institutional fields. *Academy of Management Review, 40*(1), 115–143. https://doi.org/10.5465/amr.2013.0299

Hardy, C., Phillips, N., & Lawrence, T. B. (2003). Resources, knowledge and influence: The organizational effects of interorganizational collaboration. *Journal of Management Studies, 40*(2), 321–347. https://doi.org/10.1111/1467-6486.00342

Hasle, P., & Vang, J. (2021). Designing better interventions: Insights from research on decent work. *Journal of Supply Chain Management, 57*(2), 58–70. https://doi.org/10.1111/jscm.12261

Hassanein, N. (2003). Practicing food democracy: A pragmatic politics of transformation. *Journal of Rural Studies, 19*(1), 77–86. https://doi.org/10.1016/s0743-0167(02)00041-4

Head, B. W., Ross, H., & Bellamy, J. (2016). Managing wicked natural resource problems: The collaborative challenge at regional scales in Australia. *Landscape and Urban Planning, 154*, 81–92. https://doi.org/10.1016/j.landurbplan.2016.03.019

Hoey, L., Colasanti, K., Pirog, R., & Shapiro, L. F. (2016). Implementing Collective Impact for food systems change: Reflections and adaptations from Michigan. *Journal of Agriculture Food Systems and Community Development, 7*(2), 101–115. https://doi.org/10.5304/jafscd.2017.072.014

Keast, R., Mandell, M. R., Brown, K., & Woolcock, G. (2004). Network structures: Working differently and changing expectations. *Public Administration Review, 64*(3), 363–371. https://doi.org/10.1111/j.1540-6210.2004.00380.x

King, B. G., Felin, T., & Whetten, D. A. (2010). Finding the organization in organizational theory: A meta-theory of the organization as a social actor. *Organization Science, 21*(1), 290–305. https://doi.org/10.1287/orsc.1090.0443

Li, P. P. (2016). Global implications of the indigenous epistemological system from the east: How to apply Yin-Yang balancing to paradox management. *Cross Cultural & Strategic Management, 23*(1), 42–77. https://doi.org/10.1108/ccsm-10-2015-0137

Liu, L., Ross, H., & Ariyawardana, A. (2020). Community development through supply chain responsibility: A case study of rice supply chains and connected rural communities in Central China. *Sustainability, 12*(3). https://doi.org/10.3390/su12030927

Majchrzak, A., Jarvenpaa, S. L., & Bagherzadeh, M. (2015). A review of interorganizational collaboration dynamics. *Journal of Management, 41*(5), 1338–1360. https://doi.org/10.1177/0149206314563399

Marti, E., & Gond, J. P. (2018). When do theories become self-fulfilling? Exploring the boundary conditions of performativity. *Academy of Management Review, 43*(3), 487–508. https://doi.org/10.5465/amr.2016.0071

Matopoulos, A., Vlachopoulou, M., Manthou, V., & Manos, B. (2007). A conceptual framework for supply chain collaboration: Empirical evidence from the agri-food industry. *Supply Chain Management—An International Journal, 12*(3), 177–186. https://doi.org/10.1108/13598540710742491

Midgley, G. (2003). Science as systemic intervention: Some implications of systems thinking and complexity for the philosophy of science. *Systemic Practice and Action Research, 16*(2), 77–97. https://doi.org/10.1023/a:1022833409353

Minh, T. T., & Osei-Amponsah, C. (2021). Towards poor-centred value chain for sustainable development: A conceptual framework. *Sustainable Development*. https://doi.org/10.1002/sd.2220

Murphy, J., Denyer, D., & Pettigrew, A. (2021). The role of framing mechanisms in explaining system-wide change: The case of the Northern Ireland conflict and peace process. *British Journal of Management, 32*(2), 322–341. https://doi.org/10.1111/1467-8551.12393

Ponte, S. (2020). The hidden costs of environmental upgrading in global value chains. *Review of International Political Economy*. https://doi.org/10.1080/09692290.2020.1816199

Purdy, J., Ansari, S., & Gray, B. (2019). Are logics enough? Framing as an alternative tool for understanding institutional meaning making. *Journal of Management Inquiry, 28*(4), 409–419. https://doi.org/10.1177/1056492617724233

Ross, H., Adhuri, D. S., Abdurrahim, A. Y., & Phelan, A. (2019). Opportunities in community-government cooperation to maintain marine ecosystem services in the Asia-Pacific and Oceania. *Ecosystem Services, 38*. https://doi.org/10.1016/j.ecoser.2019.100969

Salignac, F., Wilcox, T., Marjolin, A., & Adams, S. (2018). Understanding collective impact in Australia: A new approach to interorganizational collaboration. *Australian Journal of Management, 43*(1), 91–110. https://doi.org/10.1177/0312896217705178

Sankaran, S., & Presier, R. (2021). Systemic change towards sustainable development: Innovative and integrative approaches. *Systems Research and Behavioral Science, 38*(5), 579–582. https://doi.org/10.1002/sres.2820

Selwyn, B. (2016). Global value chains and human development: A class-relational framework. *Third World Quarterly, 37*(10), 1768–1786. https://doi.org/10.1080/01436597.2016.1156484

Vestergaard, A., Langevang, T., Morsing, M., & Murphy, L. (2021). Partnerships for development. Assessing the impact potential of cross-sector partnerships. *World Development, 143*. https://doi.org/10.1016/j.worlddev.2021.105447

Vicol, M., Neilson, J., Hartatri, D. F. S., & Cooper, P. (2018). Upgrading for whom? Relationship coffee, value chain interventions and rural development in Indonesia. *World Development, 110*, 26–37. https://doi.org/10.1016/j.worlddev.2018.05.020

CHAPTER 22

# The Roles of Boundary Spanning Organizations in the Governance of the Galápagos Islands

*Arturo Izurieta Valery, Melanie Zurba, and Anastasia Papadopoulos*

## Introduction

The Galápagos Islands belong to Ecuador and lay one thousand kilometers from the mainland in the Pacific Ocean (Fig. 22.1). Galápagos is famously known for its unique ecosystem and high levels of endemism

A. Izurieta Valery · A. Papadopoulos
Research Institute for the Environment and Livelihoods (RIEL), Charles Darwin University, Casuarina, NT, Australia
e-mail: anastasia.sp@dal.ca

A. Izurieta Valery (✉)
Coordinadora Ecuatoriana para la Defensa de la Naturaleza y el Medioambiente (CEDENMA), Quito, Ecuador
e-mail: arturo_izurieta@hotmail.com



C. Baldwin and S. van Bommel (eds.), *Rural Development for Sustainable Social-ecological Systems*, Palgrave Studies in Environmental Sustainability,
https://doi.org/10.1007/978-3-031-34225-7_22

(Toral-Granda et al., 2017; Tye et al., 2002). In 1978, the Galápagos Islands were inscribed by UNESCO and in 1998 the Galápagos Marine Reserve of 138,000 square kilometers was created and extended as a World Natural Heritage Site in 2001, significantly expanding the protected areas of the Galapagos. The unique features of the Galápagos Islands have resulted in keen interest from nature conservation and scientific communities the establishment of research centers and organizational hubs (Tye et al., 2006). In the 1960s, the Galápagos Islands became a major destination for ecotourism, which is now the primary driver of the local economy (Powell & Ham, 2008).

As discussed in the next section, the governance of the Galápagos Islands has evolved alongside the history of colonization and waves of national and global interest in its unique ecological characteristics (Jones, 2013). We use Kooiman's framework to better understand the Galapagos governance. This framework views governance as complex interactions of actors working at different levels and scales toward "solving societal problems, or creating societal opportunities" (Kooiman, 2003, p. 3). Kooiman's framework creates the potential to understand governance as a dynamic and evolving system with three main stages, or orders. The first order involves problem solving and the finding of opportunities. The second order involves maintaining and adapting the characteristics of institutions that are formed in the first order, and the third order involves "meta-governance" which is driven by the rules, norms, and values that have become embedded in the governance system. It is at this level where governance actors' abilities to engage substantively in decision-making occurs (Kooiman & Jentoft, 2009).

Historically, actors in the governance of the Galápagos such as the Galapagos National Park Directorate as a State institution and the Charles

---

*Present Address:*
M. Zurba
School for Resource and Environmental Studies (SRES), Dalhousie University, Halifax, Canada

Theme for Governance, Equity and Rights (TGER), International Union for the Conservation of Nature (IUCN), Gland, Switzerland

M. Zurba
e-mail: melanie.zurba@dal.ca

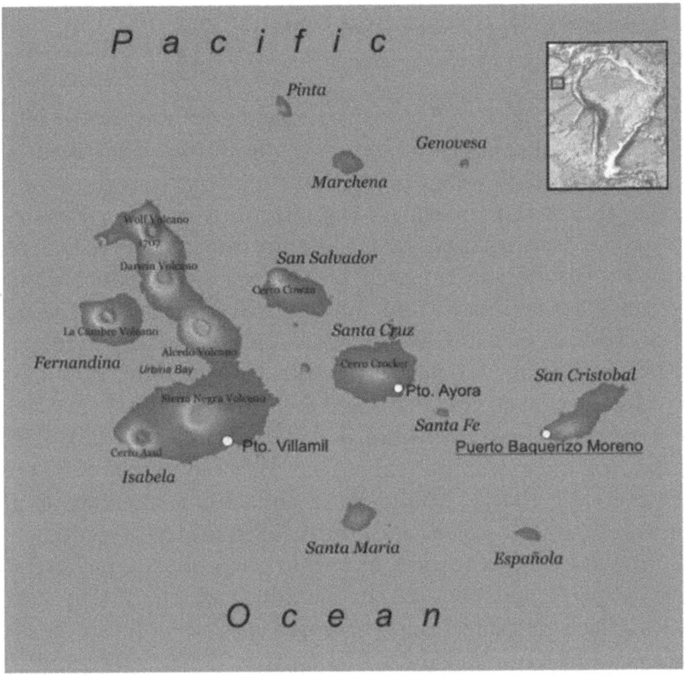


**Fig. 22.1** Location of the Galapagos Islands (*Source* Daniel Feher freeworld maps.net)

Darwin Foundation as a strong scientific and conservation NGO have almost solely focused on the preservation of the ecosystem and ecosystem services associated with resource industries such as fishing and tourism (Quiroga, 2014). However, a shift is currently occurring within the governance system toward acknowledging the interconnection and dynamics of social and economic systems with the ecological one. This is what is known as the Galapagos social-ecological system (González et al., 2008) and thus, the roles that local communities play in the governance of the Galápagos. This shift is a result of social demands, based on the fact that basic public services such as potable water, transportation, health, and education have been neglected in comparison with the attention given to conservation of the natural capital of the islands.

Different types of governmental and non-organizations have been playing different roles in creating connections among governance actors within the Galápagos social-ecological system. The creation of such connections, along with enhanced understanding and action relating to social issues, will be essential for securing the future of the islands. Organizations planning such roles can be understood as boundary or boundary spanning organizations. Boundary organizations were initially defined as formal organizations focused on facilitating interactions between policymakers and scientific and research communities to balance perspectives and to ensure that legitimacy and credibility in decision-making processes (Crona & Parker, 2012; Gustafsson & Lidskog, 2018; Hoppe et al., 2013; Kirchhoff et al., 2015; Miller, 2001). Over time, boundary organizations have evolved to fulfill an even greater role by facilitating multilateral and cross-organizational collaboration that link knowledge and action together and participate in multiple activities (Lee et al., 2014; Pitt et al., 2018). Boundary organizations, operate in different fields, across multiple scales, and with different actors—often evolving and adapting their roles to best fit within the context they are working on (Gustafsson & Lidskog, 2018; Kirchhoff et al., 2015; Pitt et al., 2018). Boundary organizations are noted in the literature as aiding in the processes of identifying the social environments, roles, and practices that can facilitate productive collaboration between governance actors and their associated knowledge and value systems and by doing so, support the establishment of stronger relationships between scientists, policymakers, resource users, and practitioners (Crona & Parker, 2012; Lee et al., 2014; Pitt et al., 2018). This chapter explores the roles of boundary organizations in connecting the various actors and facilitating social and policy transition for the Galápagos governance system as a tool to facilitate the functionality of the Galapagos socio-ecological system.

## Emergence of the Galápagos Islands' Community and Its Governance System

The colonization of the Galápagos Islands differs with many other islands in the Pacific as the settlement of people was on land that was not previously occupied by any other human and or ancient culture (Machuca, 2006). Understanding its history and how its relatively new colonization took place, combined with the fragile and unique ecosystems that later became the focus of scientific relevance—starting with Charles Dawin in

1835, have shaped the way in which the governance of the islands has evolved to guide their development.

The Galápagos Islands were discovered by Bishop Tomás de Berlanga in 1535 while sailing from Panama to Perú (Machuca, 2006). This was the first time a person set foot on the islands and opened the avenue for future visits and colonization. The island´s recent geological formation, arid lowlands, poor fresh water sources, and barren terrain framed the islands as a hide-out for pirates and buccaneers during the sixteenth and seventeenth centuries. The eighteenth and nineteenth centuries were marked by the presence of whales and fur seal hunters. Captain James Colnett drew the first map of the Galápagos in 1793 which was used by most of the posterior visitors until the arrival of Charles Darwin on board the HMS Beagle in 1835. It was not until 1832 that the recently formed Republic of Ecuador took an interest in colonizing the Galapagos and took possession as Ecuadorian territory (Jackson, 1993; Machuca, 2006).

The visit of Charles Darwin to the Galápagos confirmed the establishment of a penal colony on the island of Floreana. By 1838, this colony reached a population of 300 men, out of which half were political prisoners. This colony came to an end and left the island empty of inhabitants by 1852. However, another intent of permanency of Ecuadorian people in Galápagos took place from 1866 to 1904 on the island of San Cristóbal. This attempt gave way to the permanent presence of human settlements on the Galapagos on four main islands: San Cristóbal, Floreana, Isabela, and Santa Cruz. The political domain during those times were given to individuals who were sent by the government to act as "gobernadores" and rule the islands in the name of Ecuador (Latorre, 1991).

Although the early of occupation of the Galápagos was linked to penal colonies, all these colonization attempts were accompanied by big agricultural businesses led by entrepreneurial people from mainland Ecuador who used prisoners of different kinds, including political prisoners who were educated (Idrovo, 2013). On Floreana Island, the exploitation of a species of liverwort for dyes was big but not too lucrative; while on San Cristóbal a very prosperous sugarcane export farm led and owned by Manuel Julian Cobos lasted for over thirty years (Latorre, 1991). Idrovo's (2013) analysis of the perception by the current residents on San Cristóbal Island of the big enterprise of sugarcane and sugar export established back then on San Cristóbal Island indicates that the identity of the current local residents has been in crisis due to their historical link to a penal background. Idrovo (2013) raises a positive aspect that, although the current identity is associated with former penal ascendants, the shaping of the identity should also feature the tireless work of their ancestors in assuring the provision of sugar to mainland Ecuador as well as recognizing the spirit of endurance and presence on such a hostile environment.

After the abolishment of the sugarcane farm on San Cristóbal in 1904, many of the prisoners stayed on the Galápagos; some stayed on San Cristóbal and some others moved to Santa Cruz and very few to Isabela and or Floreana (Jackson, 1993). An incipient Ecuadorian colony had already set roots on the islands and the limitation of resources forced them to look for inventive methods of survival and simple livelihoods. At the same time, in Europe the end of the First World War and the political changes in Germany at the end of the 1920s gave way to the migration of many people, some of whom have heard about these isolated islands with strange creatures in the middle of the Pacific Ocean (Treherne, 1983).

Following the industrial revolution and the survival of the First World War, the European settlers who came to the Galápagos brought a different way of thinking to the way in which humans could survive in such a hostile but nevertheless pristine environment (Jackson, 1993). The Norwegians intended to develop a tuna fish canned industry on Floreana Island and on Santiago Island and the establishment of the first Biological Station (on Floreana), but it did not last but less than five years (Grant & Estes, 2016). The Germans followed in bigger numbers than the Swiss, Belgium, and some Americans in establishing on various inhabited islands (Jackson, 1993). They shared their knowledge of how to be more efficient with the available resources with the local Ecuadorians who knew very well the environment. Still, both depended from a cargo ship that would deliver goods from mainland Ecuador every four to six months.

The Second World War marks another landmark in the colonization and community living of the Galapagos Islands. The islands were lent by Ecuador to the United States of America to guard the Panama Canal against enemy forces (Idrovo, 2013). The first airport was built on the small island of Baltra, close to one of the inhabited islands, Santa Cruz. The temporary occupancy of the US troops on the Galapagos that ended in 1946 forced the government of the time to reinforce the sovereignty of Ecuador over the islands by establishing again another penal colony, this time on the island of Isabela (Rodas & Vivanco, 2012).

By the mid-1950s, the interest on the uniqueness of the wildlife of the Galápagos grew bigger and the 1957 UNESCO expedition recommended the Government of Ecuador the establishment of a Biological Station to guide the conservation and protection of its unique biota. In 1959, the Government declared 97% of the archipelago as the Galápagos National Park which, with the help of an NGO specifically created to help the Ecuadorian Government, the Charles Darwin Foundation for the Galapagos Islands, by 1964 the Charles Darwin Biological Station (CDBS) was inaugurated (Hennessy, 2018). In 1973, the Galápagos Islands were declared as the 20th province of Ecuador. This new landmark initiated the treacherous road of building the future of the community in four inhabited islands of the archipelago: San Cristóbal, Santa Cruz, Isabela, and Floreana.

The first Management Plan for the Galápagos National Park was elaborated in 1974 by the Charles Darwin Foundation (CDF) with the support of few Ecuadorian conservation officers assigned to this task and to start the presence of Ecuadorian staff on the islands. This Plan did not consider the newly geopolitical arrangement as a province, thus, concentrating only on the National Park area, leaving out the inhabited area and the role of the local communities. The CDF was delegated by the Government of Ecuador as the manager of the newly created National Park. The CDBS was the headquarters of the administration of the protected area, which slowly and with the growing capacity of the Ecuadorian conservation staff; the responsibility to administer the National Park was passed on to what is now known as the Galápagos National Park Directorate (GNPD) (Hennessy, 2018).

The first elections for dignitaries of the Galápagos took place in 1978. This included three mayors, 21 city council members and 10 provincial counselors, plus, the addition of all executive bureaucrats including the Governor and other lower rank authorities. By 1978, the population in the Galápagos was estimated at 6,800 inhabitants: 3,500 on Santa Cruz, 1,700 on San Cristóbal, 800 on Isabela and less than 30 on Floreana (a rural colony geopolitically under San Cristóbal). The majors initiated the building of the future of each of the islands they were elected to guide and lead. Each would write a "cantonal plan" that will last for at least the four years in power. The connection between these development plans and the Plan of the Galapagos National Park was very weak if not at all. With no doubt, the new political setting established for the Galapagos will set the way in which the populated areas of the islands develop. By 1979, the Instituto Nacional Galápagos was created to articulate all efforts to properly develop the Galápagos. This Institute brought central government representatives along with the three majors, the GNPD and the

CDF to plan for the future of the Islands. By 2009, the INGALA was replaced by the Galapagos Government Council, presided by a representative of the President of Ecuador, the Minister of Environment, Minister of Tourism, Minister of Agriculture and Fisheries, National Secretary of Planning, the three Majors of Galapagos, and one representative of the Rural Decentralized Governments of Galapagos.

## Current Issues Facing Communities of the Galápagos

The efforts to conserve the islands and its unique biodiversity were recognized when the Galápagos were listed as one of the first world natural heritage sites under UNESCO in September 1978 and as a Biosphere Reserve in 1986 as part of UNESCO´s Man and the Biosphere (MAB) Program (Pazmiño et al., 2018). These commitments boosted further the world interest to visit the Galapagos Islands. By the end of the 80 s, the industry of tourism was solidly growing at an exponential rate. The mix of a fantastic propaganda of scientific discoveries, the boom of ecotourism, and the growing tourism economy on the Galápagos brought the attention of Asian markets to exploit the rich marine resources.

By the beginning of the 1990s, for the first time the multimillion dollar fisheries of sea-cucumbers was established around the Galápagos Islands, engaging local fishermen in this illegal activity and generating a new wave of migration of artisanal fishermen from the mainland (Pazmiño et al., 2018). This booming economic activity was extractive, was being introduced at this time, and created conflict with the conservation and tourism activities that for almost three decades dominated the livelihoods of the local people.

Providing basic services to a growing and expanding population, many of which came from the mainland to help with such provisions, demanded the creation of special regulations for the development and conservation of the Galapagos Islands (Ruiz-Ballesteros & Brondizio, 2013). By the end of 1990s, the Government of Ecuador advised by the CDF and other international organizations such as UNESCO, made a historic change to its Constitution allowing the creation of a Special Law for the Galápagos Islands (LOREG), which will guide its future conservation and development (Jones, 2013). Restrictions to who will live and stay on the islands and funding for sustainable development, conservation of nature, the creation of the Galapagos Marine Reserve, and surveillance were key elements in this special law. This Special Law, which was passed in 1998, took five years of intense participation from the local authorities and community leaders.

By 2007, the Galápagos had been put on the list of World Heritage Sites in Danger by UNESCO, due to the lack of attention from the government to manage the fast growing of tourism, weak measures to control and avoid introduction of new species, high immigration, and illegal fishing in the marine reserve (UNESCO, 2007). By 2014, a comprehensive study of six months carried out by the Ministry of Environment with the support of academia and NGOs analyzing social, economic, and ecological data of over thirty years, showed that the Galapagos were at the "point of no return", meaning that if no immediate actions were taken to control the influx of tourists, slow down the immigration, reduce the potential arrival of introduced species, improve the energy use and properly manage basic needs of the local population such as health and education, the Galapagos could lose its natural universal values and be no longer the special and unique place to the world (Pizzitutti et al., 2014). By 2019, the number of visitors to the islands had reached over 270,000 tourists, with an increase of over 9% to the previous five years. Measures were needed to control the flow of tourists which in turn demanded goods and services, more labor and thus immigration, more energy consumption, pollution, and increased chances of new introduction of exotic species to the islands.

The arrival of COVID-19 in mid-March 2020 created a hard break to the influx of tourists. All institutions and organizations were obliged by the situation to learn and share resources and coordinate to deal with the pandemic of COVID-19 by sitting permanently as an Emergency Committee and planning ahead so less impact was created to the socio-ecological system of the Galapagos. Conservation NGOs such as the CDF went beyond their scope and donated medical equipment and instruments to help combat the pandemic. A planning process led by the Galapagos Government Council (GGC) resulted in the Galapagos Plan 2030, articulating efforts of all institutions and civil organizations coordinate toward a new post-COVID-19 era and avoiding the Galapagos from reaching critical development limits (Consejo de Gobierno del Regimen Especial de Galápagos, 2020).

## Boundary Spanners as Agents of Stability and Transformation

Planning for conservation and protected areas often happens in settings where tension, conflict, contestation, and issues of power and control exist and the Galapagos Islands are no exception (Caine, 2016; Hennessy, 2018; Pazmiño et al., 2018). In order to effectively address such complexity of these environmental governance systems, a significant amount of information and collaboration is needed (Caine, 2016). Initiatives focused on environmental governance increasingly recognize that governance structures and processes must transcend and operate across boundaries between actors (e.g. local communities, non-government organizations, and governments) to effectively address governance issues (Caine, 2016; Fudge & Hiruy, 2019; Lee et al., 2014). As such, within conservation and environmental governance, boundary organizations are increasingly identified as playing an important role in assisting with governance decision-making processes (Caine, 2016; Kirchhoff et al., 2015). Boundary organizations are noted as having the potential to advance collaborative governance approaches by facilitating open, productive dialogue, and bringing different actors together (Gustafsson & Lidskog, 2018). Some of the activities that boundary organizations conduct include convening and facilitating dialogue, analyzing policy, conducting and supporting research, advocating for different perspectives, providing technical, and capacity-building support, as well as supporting the co-production of knowledge (Kirchhoff et al., 2015; Lee et al., 2014; Pitt et al., 2018). Through these activities, they behave as connectors between different actors who must work together to effectively navigate complex environmental governance spaces.

Key to the work of boundary organizations is providing a legitimizing space for different actors to work together strategically and systematically through the use of boundary objects, fostering mediation between the science-policy domain, and offering accountability to both—or all, if more than two—actors working on different sides of a "boundary" (Kirchhoff et al., 2015; Pitt et al., 2018). Despite the fact that boundary organizations may need to incrementally address specific problems to mobilize the development of stability in governance systems, boundary objects are used to contribute to transformative change in governance systems (Le Heron et al., 2020). Boundary objects are defined by Kirchoff and colleagues (2015) as "mechanisms, processes, and material things that transcend the knowledge divides between different actors and provide space for collaboration". Boundary objects are described as the tools that transcend the boundaries of key actors working together, being understandable and relevant to all actors involved (Kirchhoff et al., 2015). Boundary objects can be abstract or concrete but in order for them to be effective; they must be malleable enough so that they retain their relevance to and agreed upon identity to all actors (Kirchhoff et al., 2015). Some examples of boundary objects include geographical spaces, funding and resources, new concepts or frameworks for operating, and specific projects or initiatives that bring different actors together.

Boundary organizations that work within conservation can be philanthropic, academic, government, or non-government organizations but often they are characterized by their multidisciplinary approaches and dedication to merging multiple social worlds as a means for achieving legitimacy (Caine, 2016; Lee et al., 2014). Some examples of boundary organizations that have been identified at the global scale include the Intergovernmental Science-Policy Platform on Biodiversity and Ecosystem Services and the Intergovernmental Panel on Climate Change (Gustafsson & Lidskog, 2018). These organizations are characterized as having the skills to communicate, translate knowledge, and negotiate priorities between different governance and decision-making actors and through such activities are able to establish stable relationships and interactions between different actors (Maclean et al., 2022; Pitt et al., 2018). The need for these organizations to therefore have relationships with and accountability to all actors involved is essential as they serve to mediate and establish spaces for engagement, interaction, and learning which requires that relevant actors place a certain level of trust on these organizations to adequately translate their perspectives and facilitate discussion (Fudge & Hiruy, 2019; Leith et al., 2016). The capacity for boundary organizations to reduce the instability of relationships is contingent on the willingness of the actors that they work with to address and solve the problems that they face (Leith et al., 2016). It also requires that these organizations recognize the complexity of the governance issues that they are mediating and that they acknowledge that addressing such issues requires incremental change that stabilizes conflict by making long-term strategies clear to actors (Leith et al., 2016).

## Boundary Organizations Working Within the Galápagos

In the context of the Galápagos Islands, there are several organizations that have emerged over time as boundary organizations who are increasingly behaving in ways that are complementary and embedded. The Galápagos National Park Directorate (GNPD) and the Charles Darwin Foundation (CDF), for example, are two of the most influential governance actors in the Galapagos Islands that are tied both to Ecuadorian national and Galapagos regional governments but who also engage on-the-ground with local communities and resource sectors (e.g. fishers) on the islands (Burbano et al., 2020; Hennessy, 2018; Ruiz-Ballesteros & Brondizio, 2013). The GNPD, a public and government institution and CDF, an international NGO with a long-term agreement to operate in the Galapagos, were erected to ensure the conservation of the Galapagos archipelago's flora and fauna and have played an important role in managing the use and access to the islands over time (Durham, 2021; Mathis & Rose, 2016). This has led to contestation with local populations around access and use of local resources, local displacement in the name of conservation, and unequal power relations between local populations and these conservation actors (Mathis & Rose, 2016). Due to increased conflicts on-the-ground (e.g. local fisher protests and tortoise killings), the GNPD and CDF have begun to recognize and advocate the need for a more community-centered approach to protecting the Galápagos Islands with a greater focus on local livelihoods and community-based tourism (Burbano et al., 2020; Hennessy, 2018). In order to achieve sustainable community-based conservation and tourism efforts, it is necessary for local and regional actors, state agencies, non-government organizations, and private actors to work together (Pazmiño et al., 2018; Ruiz-Ballesteros & Brondizio, 2013).

Some ways that the GNPD and CDF have characteristics of boundary organizations include their work on-the-ground to facilitate the conservation and protection management of the Galápagos Islands which involves interaction with local populations while also serving as advisory bodies to policy-makers at local, regional, and national scales (Benitez-Capistros et al., 2016; Burbano et al., 2020; Jones, 2013; Ruiz-Ballesteros & Brondizio, 2013). Although achieved through years of conflict, these organizations have also developed an understanding of the local livelihood concerns of the residents of the Galapagos Islands while also recognize the national objectives and ecological importance of the area (Burbano et al., 2020; Pazmiño et al., 2018). The history of these organizations with the Galapagos Islands has placed them in a position to be working at the boundaries between policy-makers and local residents because of their in situ activities on the Galápagos Islands, the research and monitoring that they conduct, as well as the recognition that protecting the Galápagos Islands will require collaboration between the different actors (Pazmiño et al., 2018; Ruiz-Ballesteros & Brondizio, 2013). These organizations must balance the socio-ecological and conservation values that their organizations are rooted within, the values of the local peoples, as well as the priorities of governments. In addition, the GNPD and CDF necessarily must hold themselves accountable to both local populations and government entities because of the history of conflict with local peoples that they wish to avoid and the responsibilities that the organizations have to government entities in Ecuador. They also have been advocating for a more legitimized space for collaboration between governance actors in the Galapagos through advocacy for community-centered tourism and a reconfiguration of governance approaches to the islands that address local concerns, perspectives, and livelihood issues.

Through collaboration with other governance actors such as national government organizations and international non-governmental organizations, the GNPD and CDF have been facilitating and participating in boundary work with these groups. Important government actors in this context include the Galápagos Government Council (GGC), Decentralized and Autonomous Governments (the local Galápagos municipalities and parishes), as well as the Galápagos Biosecurity Agency (ABG). Some notable international NGOs in this context include World Wildlife Fund Ecuador (WWF-Ecuador), Conservation International (CI), and Island Conservation (IC). In addition, FUNDAR and Fundación Un Cambio por La Vida which are two non-government organizations who also work with the GNPD and CDF. These actors have worked together in numerous ways through the establishment of local inter institutional and organizational agreements toward sustainable socio-economic programs (e.g. buying local produce, visiting tourist areas, getting to know more about where people live, etc.). The most relevant boundary organizations in the Galapagos are presented in Tables 22.1 and 22.2.

A useful framework for understanding the activities that define boundary organizations and the effectiveness of the activities that they conduct emerged from the work of Cash and colleagues (2003) and

**Table 22.1** SCL framework applied to key government organizations who conduct boundary work in the Galapagos

| *Government boundary organizations* | | | | | | |
|---|---|---|---|---|---|---|
| *Boundary organization* | *Origin* | *Scale* | *Boundary functions* | *Salience* | *Credibility* | *Legitimacy* |
| Galapagos National Park Directorate (GNPD) | Ecuadorian Presidential Decree in 1959 | Regional (Galapagos Islands) | • Administration of the Galapagos National Park and Marine Reserve<br>• Controls and regulates the environmental issues related to the municipalities (towns)<br>• High convening capacity and hires neutral facilitators to favor discussions with stakeholders to avoid unnecessary conflicts | • Emphasizes importance of conservation of unique biodiversity of Galapagos Islands to economic activities<br>• Advocates for conservation of resources to support and sustain economic activities on the Islands and to preserve way of life<br>• Protection of the Galapagos Islands contributes to conservation of biodiversity unique to Galapagos | • Relied on for things like monitoring, maintenance, and enforcement of conservation/technical activities in the GNP<br>• Despite also being a controlling agency of all activities within the protected areas of Galapagos, the credibility is relatively high | • Has evolved over time to respond to the significant institutional changes in leadership over the last 10 years which has influenced the increased inclusion of local stakeholders over time<br>• Social pressure weakens decisions of the GND often making the GNPD overlook technical and scientific information for decision making |

| *Government boundary organizations* | | | | | | |
|---|---|---|---|---|---|---|
| *Boundary organization* | *Origin* | *Scale* | *Boundary functions* | *Salience* | *Credibility* | *Legitimacy* |
| Galapagos Government Council | Created by Law in 1982 and reformed in 2008 | Regional (Galapagos Islands) with concentration within the 3% of the terrestrial, "unprotected" part of the Archipelago | • Coordinates and promotes the approval of regional policies in support of local livelihoods based on the unique conditions of the Galapagos environment<br>• Controls migration of people to the islands and vehicle import to the islands | • Since its creation, it has achieved a better articulated work from all its members (4 ministers; 3 city majors; 1 representative from the rural decentralized government) | • Credibility has increased over time as the needs of local peoples have been increasingly addressed throughout time (health, roads, education, capacity building) | • Many questionable decisions particularly on migration permits and vehicle intake to the islands |
| Galapagos Agency for Prevention, Control and Quarantine (ABG) | Created by Presidential Decree in 2012 | Regional (the 3% of the Galapagos archipelago that are inhabited by residents) | • Regulates and control all possible introduction of foreign biota to the Galapagos Surveillance and control of diseases including those in agriculture and livestock areas | • Creation of well-structured control and quarantine measures to prevent and control new invasions to the Galapagos | • Credibility has grown high as on-the-ground actions showed positive results on the control of introduced species | • Despite the strengthening of the Agency, local management of pets in the towns are still to be handled |

(continued)

**Table 22.1** (continued)

*Government boundary organizations*

| *Boundary organization* | *Origin* | *Scale* | *Boundary functions* | *Salience* | *Credibility* | *Legitimacy* |
|---|---|---|---|---|---|---|
| Municipalities of Galápagos also called Decentralized and Autonomous Governments | Created in 1974 by Presidential Decree | Regional (Galapagos Islands); share political jurisdiction of sectors of Galapagos but administrative jurisdiction is limited to each inhabited island | • Leads the administration, planning and human growth in the Galapagos<br>• As an autonomous and decentralized government, they lead the development of urban and rural areas and influence the economic growth of the towns | The urban and rural planning has been weak and disorganized. Poor basic infrastructure of potable water and sewage treatment. Some recycling of waste has been implemented | • Low credibility. The municipalities' leaders are elected by the community and in many cases with not much professional preparation and/or experience | • Infrastructure for recreation and street and rural road maintenance has been seen as major achievements. Lack of transparency in many contracts makes the municipality administration weak in legitimacy |

**Table 22.2** SCL framework applied to key non-government organizations who participate and facilitate boundary work in the Galapagos

*Non-government boundary organizations*

| *Boundary organization* | *Origin* | *Scale* | *Boundary functions* | *Salience* | *Credibility* | *Legitimacy* |
|---|---|---|---|---|---|---|
| Charles Darwin Foundation (CDF) | International NGO created for Galapagos in 1959 under the auspices of UNESCO and IUCN | Regional, the Galapagos Islands | • Advisor to the Government of Ecuador in conservation and sustainability of the Galapagos based on scientific information<br>• Main partner is the GNPD | • Key in advising and supporting the Government of Ecuador with proposals (e.g. for declaration of Galapagos as World Heritage Site, Biosphere Reserve and Ramsar Site to UNESCO)<br>• Contributes to research and technical advise related to protected areas on the Galapagos (e.g. World Heritage Site, Biosphere Reserve, etc.) | • Research organization with a long trajectory in Galapagos<br>• Well respected locally, nationally and internationally | • Local community has complained the NGO has not put much attention to the human dimensions of Galapagos |

(continued)

**Table 22.2** (continued)

*Non-government boundary organizations*

| *Boundary organization* | *Origin* | *Scale* | *Boundary functions* | *Salience* | *Credibility* | *Legitimacy* |
|---|---|---|---|---|---|---|
| WWF-Ecuador | International NGO which had initial action in Galapagos in the 90 s and now has action in whole of Ecuador | Its presence and action in Galapagos has been dependent on funding. The WWF-Galapagos was reduced to a WWF office in Galapagos as part of a WWF-Ecuador program. Office in Santa Cruz island | • Have played key role initially financially supporting the CDF and lately strengthening the planning capacity of the GNPD<br>• Mobilizes eco-tourism as a niche focus area to generate information that supports decision making about tourism within the national park and marine reserve | • Supports GNPD in participatory planning processes<br>• Works with CDF with local participatory planning processes and information-sharing | • Well respected locally, nationally and internationally<br>• The WWF Galapagos program concentrated its efforts to support the conservation efforts of the GNPD, including capacity building of the staff of GNDP and strengthening of internal administrative processes | • The support provided to the GNPD included funding and technical assistance for planning. This planning included the participation of key stakeholders of the Galapagos community |

*Non-government boundary organizations*

| *Boundary organization* | *Origin* | *Scale* | *Boundary functions* | *Salience* | *Credibility* | *Legitimacy* |
|---|---|---|---|---|---|---|
| Conservation International-Ecuador | International NGO with a sub-organization that works at the national Ecuadorian level | Its presence in Galapagos has remained constant as part of the organization's country program. Office on Santa Cruz Island | • Came to Galapagos to support conservation actions directed to artisanal fisheries and lately to sustainable agriculture | • Active in providing technical assistance to improve the management of fisheries and sustainable agriculture | • Consistently focuses its technical support in areas of use of resources of economic interest (e.g. fisheries and agriculture)<br>• Well respected locally, nationally and internationally | • Has facilitated consistent participation of local interested groups, particularly fishermen and farmers |
| WildAid | International NGO with presence in Ecuador and with a program for Galapagos | Work primarily focused on the Galapagos Islands with some learning lessons for the mainland Ecuador | • Has concentrated its work supporting GNPD and ABG's surveillance and capacity building | • The surveillance infrastructure of both the GNPD and ABG have been reinforced, including capacity building of rangers and biosecurity agents | • The support in areas of surveillance has been strategic so local communities understand the need to comply with already established norms<br>• Well respected locally | • Its support to the GNPD and ABG has been concentrated on prevention and education of local communities regarding laws, regulations, surveillance and legal actions if rules are broken |

(continued)

**Table 22.2** (continued)

| *Non-government boundary organizations* | | | | | | |
|---|---|---|---|---|---|---|
| *Boundary organization* | *Origin* | *Scale* | *Boundary functions* | *Salience* | *Credibility* | *Legitimacy* |
| Island Conservation | International NGO with presence only in Galapagos | Focus on restoration of degraded islands of the Galapagos with learning lessons for the world. Office on Santa Cruz Island | • Focuses on restoration of degraded islands of the National Park. Lately included the inhabited island of Floreana | • It has been crucial in the successful eradication of rodents from various islands. This has highlighted the efforts from Ecuador to restore natural ecosystems | • Since 2012 the proposed project to restore the first inhabited island of Galapagos has given this NGO recognition for presenting well documented evidence that these restoration efforts can be achieved with the support of the local community<br>• Well respected locally and nationally | • The participation of the GNPD and the local community of Floreana Island has been permanent. The project is well recognised as per the transparency in participation of key stakeholders |

| *Non-government boundary organizations* | | | | | | |
|---|---|---|---|---|---|---|
| *Boundary organization* | *Origin* | *Scale* | *Boundary functions* | *Salience* | *Credibility* | *Legitimacy* |
| FUNDAR | Local NGO created for the Galapagos with ample radius of action | Regional focus on the Galapagos Islands. Based on Santa Cruz Island | • It has executed various projects on conservation, environmental education and technical assistance with socio-economic information | • Projects executed have made effort to have participation of many local professionals and local community members | • Being a local NGO generated expectations and hope from local people. The lack of experience sent the NGO to a halt as financial problems hit the organization | • It has managed to involve many locals in projects and to link actions with government institutions as well as with other NGOs |
| Fundación Un Cambio por La Vida | Local NGO created less than 10 years ago to tackle social problems | Galapagos based in Santa Cruz Island | • It has concentrated on tackling social problems such as better attention to elders, family violence and kids with disabilities | • Recently it played a role in helping the delivery of services and goods to people during COVID-19 and facilitated initial efforts to insert the concept of Sustainable Development Goals in the Galapagos | • The NGO was not involved in providing information on the status of COVID-19 or on the efforts to reach people in need, but it was well seen as locals supporting locals. Gained recognition as facilitator of informing the community on the SDGs | • The projects executed required close communication and interaction with local communities |

applies three key characteristics as a framework for unpacking boundary organization functions that includes: *salience, credibility,* and *legitimacy* (SCL). This framework (hereafter "the SCL framework") has been expanded and applied by Fudge and Hiruy (2019) as well as Leith and colleagues (2016). The literature describes how effective and adaptive governance can occur when work is done at "boundary" interfaces that bring together relevant actors to establish "credible, salient, and legitimate" knowledge, values, and practices that are co-created by the relevant actors (Fudge & Hiruy, 2019; Leith et al., 2016). Organizations that work at such boundaries can make information and decision-making processes *salient* (through articulating and translating knowledge such that they are relevant to the actors involved), *credible* (by adequately handling evidence and information), and *legitimate* (by openly addressing fairness, organizational and actor biases, and creating respectful spaces for dialogue) (Leith et al., 2016; Pitt et al., 2018). Although it is challenging to assess the success of boundary organizations since they often play an ever-evolving role in governance settings, the literature notes that boundary organization success can be linked to mobilizing and maintaining the stability of interactions between governance actors, supporting learning and reflection, and contributing to change that can align assumptions, values, and context actors (Leith et al., 2016; Pitt et al., 2018).

In some cases, boundary organizations create links and partnerships with other organizations to respond to multi-faceted; cross-scalar issues that transcend jurisdictions to enhance their capacity to address governance issues (Fudge & Hiruy, 2019; Kirchhoff et al., 2015; Leith et al., 2016). These links between boundary organizations, known in the literature as "boundary chains" require that the organizations involved possess a level of complementarity and embeddedness to work well together (Fudge & Hiruy, 2019). In these contexts, *salience* is described as an organization's ability to make information relevant to actors involved (Cash et al., 2003; Fudge & Hiruy, 2019; Leith et al., 2016). *Credibility* is described as organizations being viewed as having the capacity and reliability to handle information and evidence (Cash et al., 2003; Fudge & Hiruy, 2019). Organizations are described as having *legitimacy* when their activities, advocacy, or suggestions about how an issue or initiative can be addressed respect the different views of the actors involved and address biases and fairness to articulate how different—and possibly divergent—perspectives are balanced (Cash et al., 2003; Fudge &

Hiruy, 2019; Leith et al., 2016). By applying the SCL framework to the context of the Galápagos Islands, insight into the role of key boundary organizations for mobilizing a more collaborative and locally-informed governance system will be highlighted in this chapter. The SCL framework was applied in this context to the GNPD and CDF—along with the several other organizations working "at the boundaries" on the Galapagos Islands to advance collaborative local governance systems to investigate the ways that boundary work is being done in this context and how relationships are being established to connect these boundary activities.

## Discussion and Conclusions

The boundary organizations that are playing a prominent role in the Galapagos are primarily both governmental and non-governmental organizations. Our analysis has shown that the boundary organizations of the Galápagos governance system have different boundary functions and varying degrees of salience, credibility, and legitimacy. Government boundary organizations follow established procedures and regulations that in the end are translated into actions that regulate public and private activities. However, in this relatively small and faraway socio-ecosystem, these government boundary organizations have been naturally forced to interact with non-government organizations, local and non-local, in order to succeed and or move forward by generating policies to maintain as much as possible a balance between the natural universal values of the Galapagos and the development of a fast growing local population. Collaboration between government and non-governmental organizations is necessary to ensure that salience, credibility, and legitimacy are retained in support of local peoples. As population growth continues in the Galápagos, the presence of other institutions and organizations seems to be a natural process as also the need for them to interact with one another so the needs of the people are in balance with the needs of nature.

The Charles Darwin Foundation (CDF), as the leading non-governmental boundary organization, and the Galapagos National Park Directorate (GNPD) as the leading governmental boundary organization both have salience, credibility, and legitimacy to mobilize boundary work together, forming a boundary chain (i.e. they are both complementary to one another and their activities are embedded). These boundary organizations have worked effectively together for six decades. Other organizations have worked as complementary actors in specific empty niches (i.e. sustainable agriculture and fisheries; environmental education). Since government organizations are vulnerable to shifts in power and leadership, they often change their way of action and interaction with other boundary organizations depending on the inclination of the government in power. Partially due to changes in government, the relationship between the CDF and GNPD has undergone ups and downs throughout time. However, the sole goal of working for the conservation of nature has made it possible for this relationship to last, even if this goal has often marginalized communities.

Boundary organizations have the potential to evolve over time in response to local needs—in times of biodiversity and human health crisis and recognition of human rights/local livelihoods in relation to environmental protection, such organizations are essential for facilitating positive change that does not overlook the role of "people in nature". Furthermore, the value of boundary organizations in fostering resilience and adaptive capacity-building is of great importance in these contexts as we learn more and advocate for more socially-just approaches to conservation that are centered around people in nature ideals rather than people without nature make advancements. The evolution of boundary organizations, particularly governmental organizations, has been influenced by the "injection" of global thinking coming from those of international origin, bringing global perspectives into the local development and thus, improving the pool of thinking so sustainable livelihoods are reached in such isolated island systems. Boundary organizations are important for bringing the shifting global norms and values (i.e. the recognition that community rights and wellbeing are essential to nature conservation) to local spaces and making them relevant to localized contexts. Even though the global scale can sometimes be quite broad and make it difficult to visualize local applications/implications, boundary organizations may be able to translate the global scale to localized contexts given their capacity to transcend scales and their engagement with local actors.

Boundary organizations have played a strong role in influencing the social dimension of the local population and have taken an interesting protagonist role in the last ten years. Social and economic aspects have been a priority versus those related to conservation of nature during the COVID-19 pandemic. Future research should investigate the long-term effects on communities, boundary organizations and how this affects the governance system and conservation efforts in the Galápagos.

## References

Benitez-Capistros, F., Hugé, J., Dahdouh-Guebas, F., & Koedam, N. (2016). Exploring conservation discourses in the Galapagos Islands: A case study of the Galapagos giant tortoises. *Ambio, 45*(6), 706–724. https://doi.org/10.1007/s13280-016-0774-9

Burbano, D. V., Meredith, T. C., & Mulrennan, M. E. (2020). Exclusionary decision-making processes in marine governance: The rezoning plan for the protected areas of the 'iconic' Galapagos Islands, Ecuador. *Ocean & Coastal Management, 185*, 105066. https://doi.org/10.1016/j.ocecoaman.2019.105066

Caine, K. J. (2016). Blurring the boundaries of environmentalism: The role of Canadian parks and wilderness society as a boundary organization in Northern conservation planning. *Rural Sociology, 81*(2), 194–223. https://doi.org/10.1111/ruso.12094

Cash, D. W., Clark, W. C., Alcock, F., Dickson, N. M., Eckley, N., Guston, D. H., Jäger, J., & Mitchell, R. B. (2003). Knowledge systems for sustainable development. *Proceedings of the National Academy of Sciences, 100*(14), 8086–8091.

Consejo de Gobierno del Regimen Especial de Galápagos. (2020). *Plan Galapagos 2030: Plan de Desarrollo Sustentable y Ordenamiento Territorial del Régimen Especial de Galápagos 2020–2030*. https://unidosporgalapagos.files.wordpress.com/2021/05/plan-galacc81pagos-2030-2.pdf Accessed on 13 February 2022.

Crona, B. I., & Parker, J. N. (2012). Learning in support of governance: Theories, methods, and a framework to assess how bridging organizations contribute to adaptive resource governance. *Ecology and Society, 17*(1), 32–32. https://doi.org/10.5751/ES-04534-170132

Durham, W. H. (2021). *Exuberant life: An evolutionary approach to conservation in Galápagos*. Oxford University Press.

Fudge, M., & Hiruy, K. (2019). Linked boundary functions: Examining the role of 'research for development' organizations in integrating levels of resource governance. *Society & Natural Resources, 32*(3), 255–274. https://doi.org/10.1080/08941920.2018.1517914

González, J. A., Montes, C., Rodríguez, J., & Tapia, W. (2008). Rethinking the Galapagos islands as a complex social-ecological system: Implications for conservation and management. *Ecology and Society, 13*(2), 13. https://doi.org/10.5751/ES-02557-130213

Grant, K. T., & Estes, G. B. (2016). Alf Wollebæk and the Galapagos archipelago's first biological station. *Galapagos Research, 68*, 33–42. http://hdl.handle.net/1834/36362

Gustafsson, K. M., & Lidskog, R. (2018). Boundary organizations and environmental governance: Performance, institutional design, and conceptual development. *Climate Risk Management, 19*, 1–11. https://doi.org/10.1016/j.crm.2017.11.001

Hennessy, E. (2018). The politics of a natural laboratory: Claiming territory and governing life in the Galápagos Islands. *Social Studies of Science, 48*(4), 483–506. https://doi.org/10.1177/0306312718788179

Hoppe, R., Wesselink, A., & Cairns, R. (2013). Lost in the problem: The role of boundary organisations in the governance of climate change. *Wires Climate Change, 4*(4), 283–300. https://doi.org/10.1002/wcc.225

Idrovo, H. (2013). *Baltra-Base Beta: Galápagos y la Segunda Guerra Mundial*. Fondo Editorial, Ministerio de Cultura del Ecuador.

Jackson, M. (1993). *Galapagos: A natural history* (Rev. and Expanded ed.). University of Calgary Press.

Jones, P. J. S. (2013). A governance analysis of the Galápagos Marine Reserve. *Marine Policy, 41*, 65–71. https://doi.org/10.1016/j.marpol.2012.12.019

Kirchhoff, C. J., Lemos, M. C., & Kalafatis, S. (2015). Creating synergy with boundary chains: Can they improve usability of climate information? *Climate Risk Management, 9*(C), 77–85. https://doi.org/10.1016/j.crm.2015.05.002.

Kooiman, J. (2003). *Governing as governance* (III, pp. 115–117). Sage.

Kooiman, J., & Jentoft, S. (2009). Meta-governance: Values, norms and principles, and the making of hard choices. *Public Administration, 87*(4), 818–836.

Latorre, O. (1991). *Manuel J. Cobos, emperador de Galápagos*. Fundación Charles Darwin para las Islas Galápagos.

Le Heron, E., Le Heron, R., Taylor, L., Lundquist, C. J., & Greenaway, A. (2020). Remaking ocean governance in Aotearoa New Zealand through boundary-crossing narratives about ecosystem-based management. *Marine Policy, 122*https://doi.org/10.1016/j.marpol.2020.104222

Lee, E., Su Jung, C., & Lee, M.-K. (2014). The potential role of boundary organizations in the climate regime. *Environmental Science & Policy, 36*, 24–36. https://doi.org/10.1016/j.envsci.2013.07.008

Leith, P., Haward, M., Rees, C., & Ogier, E. (2016). Success and evolution of a boundary organization. *Science, Technology, & Human Values, 41*(3), 375–401. https://doi.org/10.1177/0162243915601900

Machuca, J. E. (2006). *Historical chronology of Galapagos 1535–2000*. Gráficas PATO.

Maclean, K., Robinson, C., Bock, E., & Rist, P. (2022). Reconciling risk and responsibility on Indigenous country: Bridging the boundaries to guide knowledge sharing for cross-cultural biosecurity risk management in northern Australia. *Journal of Cultural Geography, 39*(1), 32–54. https://doi.org/10.1080/08873631.2021.1911078

Mathis, A., & Rose, J. (2016). Balancing tourism, conservation, and development: A political ecology of ecotourism on the Galapagos Islands. *Journal of Ecotourism, 15*(1), 64–77. https://doi.org/10.1080/14724049.2015.1131283

Miller, C. (2001). Hybrid management: Boundary organizations, science policy, and environmental governance in the climate regime. *Science, Technology, & Human Values, 26*(4), 478–500. https://doi.org/10.1177/016224390102600405

Pazmiño, A., Serrao-Neumann, S., & Low Choy, D. (2018). Towards comprehensive policy integration for the sustainability of small islands: A landscape-scale planning approach for the Galápagos Islands. *Sustainability (Basel, Switzerland), 10*(4), 1228. https://doi.org/10.3390/su10041228

Pitt, R., Wyborn, C., Page, G., Hutton, J., Sawmy, M. V., Ryan, M.,& Gallagher, L. (2018). Wrestling with the complexity of evaluation for organizations at the boundary of science, policy, and practice. *Conservation Biology, 32*(5), 998–1006. https://doi.org/10.1111/cobi.13118

Pizzitutti, F., Mena, C. F., & Walsh, S. J. (2014). Modelling tourism in the Galapagos Islands: An agent-based model approach. *Journal of Artificial Societies and Social Simulation, 17*(1), 1–21. https://doi.org/10.18564/jasss.2389

Powell, R. B., & Ham, S. H. (2008). Can ecotourism interpretation really lead to pro-conservation knowledge, attitudes and behaviour? Evidence from the Galapagos Islands. *Journal of Sustainable Tourism, 16*(4), 467–489. https://doi.org/10.1080/09669580802154223

Quiroga, D. (2014). Ecotourism in the Galapagos: Management of a dynamic emergent system. *The George Wright Forum, 31*(3), 280–289.

Rodas, F. P., & Vivanco, A. K. (2012). *Galapagos: Prisión de Basalto Terror y Lágrimas en la Isla Isabela*. Ministerio de Cultura del Ecuador.

Ruiz-Ballesteros, E., & Brondizio, E. (2013). Building negotiated agreement: The emergence of community-based tourism in Floreana (Galápagos Islands). *Human Organization, 72*(4), 323–335. https://doi.org/10.17730/humo.72.4.4767536442q23q31

Toral Granda, M. V., Causton, C. E., Jäger, H., Trueman, M., Izurieta, J. C., Araujo, E., Cruz, M., Zander, K. K., Izurieta, A., & Garnett, S. T. (2017). Alien species pathways to the Galapagos Islands, Ecuador. *PLoS One, 12*(9), e0184379–e0184379.https://doi.org/10.1371/journal.pone.0184379

Treherne, J. (1983). *The Galapagos affair*. Random House.

Tye, A., Atkinson, R., & Carrión, V. (2006). Increase in the number of introduced plant species in Galapagos. *DPNG, GCREG, FCD, GC. Galapagos Report, 2007*, 133–135.

Tye, A., Snell, H. L., Peck, S. B., & Andersen, H. (2002). Outstanding terrestrial features of the Galapagos archipelago. In R. Bensted-Smith (Ed.), *A biodiversity vision for the Galapagos Islands* (pp. 12–23). Charles Darwin Foundation and World Wildlife Fund.

United Nations Educational, S. a. C. O. U. (2007, June 23–July 2). Decisions adopted at the 31st Session of the World Heritage Committee. *Convention concerning the protection of the world cultural and natural heritage: World Heritage Committee*. Christchurch, New Zealand. Available at: WHC-07/31.COM/24. unesco.org

PART IV

# Synthesis

CHAPTER 23

# Realising Just and Sustainable Social-Ecological Systems

*Séverine van Bommel, Claudia Baldwin, Katrina Brown, Jim Cavaye, and Helen Ross*

## Introduction

This book set out to celebrate trends and progress in research and practice on the role of communities in the sustainable management of social-ecological systems. It features contributions from scholars from different parts of the world, invited due to their connection with Emeritus Professor Helen Ross. The authors originate from eleven different

S. van Bommel (✉) · H. Ross
School of Agriculture and Food Sciences, The University of Queensland, Brisbane, QLD, Australia
e-mail: s.vanbommel@uq.edu.au

C. Baldwin
Sustainability Research Centre, School of Law and Society, University of the Sunshine Coast, Sippy Downs, QLD, Australia
e-mail: CBaldwin@usc.edu.au; claudiab@westnet.com.au



C. Baldwin and S. van Bommel (eds.), *Rural Development for Sustainable Social-ecological Systems*,
Palgrave Studies in Environmental Sustainability,
https://doi.org/10.1007/978-3-031-34225-7_23

countries and case studies are from three continents: Australia, Asia and Latin America. The book chapters address a diversity of contexts, scales, landscapes and cultures. The first chapter described how the field has evolved over time. It mapped the major changes that have occurred in the domain of community-environment interactions in the intersecting areas of rural community development, community-based natural resource management and social aspects of resilience. As Berkes pointed out in the preface, these changes have led to three paradigm shifts, namely in relation to (1) participation in rural development; (2) environmental governance; and (3) the adoption of a systems view. While authors and cases come from different parts of the world, throughout the book, the role of communities in the sustainable management of social-ecological systems has remained a central theme, and the authors have engaged with the paradigm shifts in various ways. In this chapter, we revisit the three paradigm shifts and discuss some of the insights that can be gleaned from the chapters. Next, we draw out some patterns that emerged throughout the chapters and that resonate with a path set by Helen Ross on putting communities first in the management of social-ecological systems. Her inspiration has guided much of this work. We end this chapter by discussing future directions of the field.

## Three Paradigm Changes

### *Participation in Rural Development*

Section 1 provides examples of Rural Community Development programmes such as Australia's Landcare and international agricultural extension as they metamorphosed with reduced government funding, necessitating greater ownership and agency by on-ground participants. Community development approaches aim to build leadership, identify locally-agreed priorities and promote collective action on issues important to communities, building on their strengths, relying on partnership with or support of government agencies. The importance of appropriate

K. Brown
Environment and Sustainability Institute, University of Exeter, Exeter, UK

J. Cavaye
University of Tasmania, Hobart, TAS, Australia

community engagement in rural development is highlighted in two of the international cases in this part, one revealing the real issues for dairy farmers in Brazil and the other for female recipients of aid. Community development values local knowledge and expertise, and if carried out well, creates stronger and more connected communities. However without some level of resources and attention by governments, volunteer and engagement fatigue as well as cynicism can set in. Communities may equally lead their own development, with or without some enabling governance and funding. This is illustrated in a case study from Indonesia, where women in small businesses self-organise to empower themselves and their community members.

The more recent concept of community resilience, as promoted by Prof Ross, has drawn on community development practice highlighting the importance of building social networks and social capital, social learning and multi-level and multi-sector collaboration (discussed in the next part), with progress accelerated by the need to deal with rapid change and disasters.

### *Environmental Governance*

Section 2 on community-based natural resource management, conservation and co-management extends the community development concept by emphasising the interests of multiple communities collaborating together to understand each other and build capacity. It identifies that the form of collaboration must match the interests and needs of the participants, with support to the range of voices where required. Community-based management recognises that instead of competing interests, there is increased value when we all care for shared resources (salute to Ostrom). Of importance is ensuring inclusive processes where interests of those not actually present—those in the past and future, as well as the environment—are also represented, acknowledging that interests can shift over time. Ultimately, we see the fundamental respect of diverse views (multiple voices) promoted by Prof Helen Ross and shared among her collaborators.

Much of the international innovation in environmental governance, particularly community-based natural resource management and co-management, has been driven by need for governments officially tasked with managing conservation areas, fish and wildlife to cooperate, often collaborate, with Indigenous peoples, local resource users or both. This has arisen from a combination of recognition of rights, efficacy of local institutions and the cost-effectiveness of committed local management and practicality (for instance migratory species which do not stay in single jurisdictions). Part of Helen Ross' legacy has been to understand and help create space for these processes, and to work closely with some communities and regional organisations in developing innovations in co-management.

### *The Adoption of a Systems View*

Section 3, on the theme of social aspects of resilience, builds on the participatory concepts and agency-building discussed in previous chapters. The chapters by Witt, Cheung and Simpson, and by Maclean and Witt, rigorously examine the influence of Prof. Helen Ross's work in the area of social-ecological resilience, providing a foundation for the rest of this Section. It not only features building knowledge, skills, networks and agency within the community but involves understanding the synergistic relationship between people and environment. The interdisciplinary nature of the work requires collaboration not only among researchers from different disciplines, but also community members with different backgrounds. Recognising and valuing different forms of knowledge are key to understanding and representing the diverse perceptions of SES and to identifying how change happens, its impacts and most importantly the leverage points and means of effecting sustainable and just management strategies. Inherent in these change processes are trade-offs. Identifying and evaluating these across space, time and different domains and individuals is a major challenge for the emerging science of resilience. The analyses presented in this Section bring insights and demonstrate how community-centred and social science-learning approaches are critical to not only understand the complex dynamics of SES, but also to co-identify and co-develop strategies for sustainable management. Various creative and participatory techniques, including mental models, participatory modelling, narratives and scenarios, can assist in building a common understanding of social-ecological resilience. Learning from many fields across rural development, community participation and international development, the cases here highlight different applications of these and other techniques. A challenge, of course, is how this knowledge is applied, the extent to which meaningful empowerment results, and whether deliberative and democratic mechanisms and structures exist to influence governance. As part of Helen Ross's legacy, colleagues continue to advance the theory, providing a basis for further applied studies.

## Putting Communities First in the Sustainable Management of Social-Ecological Systems

The three paradigm changes in the intersecting areas of rural community development, community-based natural resource management and social aspects of resilience point to elements of a new approach to community-environment interactions that puts communities first in the management of social-ecological systems. In this section, we identify three elements in the scientific practice of the chapter authors that characterise this approach, namely (1) challenging the dominant technocratic, neoliberal rationality; (2) developing alternative scientific narratives that support concerted or connective action; and (3) a commitment to an ethics of care for communities and the environment. The purpose of this section is to open up dialogue and further develop the contours of this rapidly advancing approach.

First, all chapters in one way or another challenge the technocratic neoliberal rationality, which is still the status quo in scientific literature on community-environment relations. This dominant technocratic neoliberal rationality assumes that environmental problems can be studied objectively by an outside observer who can produce universal 'truths'. Much scientific environmental knowledge 'speaks' for the environment as if it was produced in a value-free way by someone who does not have any interests or stakes. Biophysical scientific knowledge separates people from the environment and subsequently maps environmental phenomena—such as environmental degradation—into indicators and criteria which can then be used in evidence-based environmental management strategies. The strategic narratives of economics increasingly dominate scientific environmental discourse in the evidence-based environmental management strategies (Apostolopoulou et al., 2021; Baldwin et al., 2019; Turnhout et al., 2014). Market driven institutions are created and global networks of competing actors are designed to deliver the greatest 'good' to the greatest number. The chapters in this book challenge this dominant technocratic neoliberal rationality by (1) drawing attention to the way in which social aspects and ecological aspects of the system are intricately intertwined; and (2) showing that sustainable management is an emergent property of social interaction and thus an internality of the social-ecological system.

In regard to the first point about neoliberal rationality, the authors show that the biophysical environment cannot be understood in isolation without understanding communities, their lived experiences and their relationships with that environment. Peoples' practices are embedded in local situations which cannot be understood without taking into account situated perceptions, power relations, institutions (including markets) as well as the environmental contexts. Chapter 11 by Bellamy and Head, for example, describes in detail how water quality in South East Queensland is intricately intertwined with partnerships, flooding and drought, urbanisation, policy change and conflicts. Similarly Chapter 22 by Izurieta, Valery et al. illustrate how the environmental protection of the Galapagos Islands cannot be understood without understanding COVID-19, tourism, globalisation, local governance, human rights and boundary spanning organisations. Thereby the authors draw attention to the way in which people—including market institutions—and their environments mutually influence each other.

Regarding the second point about the neoliberal paradigm, the authors show that social interactions among stakeholders are essential to sustainable management. Chapter 15 by Phelan and Setianto shows how ocean plastic pollution in Indonesia is an emergent property of wider social interactions which cannot be tackled at the local level alone. Similarly in Chapter 7, Liao and Liu discuss how effective innovation support of smallholder farmers in China needs a whole of system approach that takes into account technologies (including social technologies) to mediate the relationship between smallholder farmers and their environment. If the sustainable management of social-ecological systems is dependent on social interactions (such as sensemaking, conflict management, power dynamics, learning) among multiple, interdependent stakeholders then sustainability becomes an inherent part of the social-ecological system. Similarly, ecosystem degradation and related social problems can be summarised as a "pathology" of the system (Ravetz, 2006, p. 275). Indeed Chapter 8 by Palaniappan, for example, illustrates three natural resource dilemmas with common pool elements in which the dominant technocratic neoliberal rationality is part of the problem as it creates conditions in which the rational choice of one group of stakeholders is at the expense of other groups of stakeholders. From a social-ecological systems perspective, this strategic behaviour is not an external effect but rather an internality of the social-ecological system itself. This does not mean that biophysical science and markets cannot be part of the solution. It does imply that something is currently missing from the mix.

Second, the authors develop an alternative scientific narrative that supports concerted or connective action which has the potential to play an enabling role in the management of sustainable social-ecological systems. Research accounts do not 'just' represent reality but actively construct it. Choices need to be made in terms of what to include and what to emphasise as well as what to exclude and what to de-emphasise. This is important because the way in which we perceive a situation is closely related to the actions that we take (van Bommel & van der Zouwen, 2013). Already in 1999, Maarleveld and Roling noticed that "Social science can be as powerful as natural science because it can equally affect people's sense making. It is not the power of its predictions that give social – or any – science its influence, but the extent to which its perspectives or narratives take hold of people's imagination and enthusiasm, and especially the extent to which that sense making begins to justify policies and shape enduring practices, institutional design, and the use of natural resources and ecological services" (Röling & Maarleveld, 1999, p. 297). A social-ecological systems perspective is a framing choice which allows us to draw the boundaries of our system of interest in such a way that it allows us to see interdependencies and envision solutions that often remain invisible in other approaches.

The authors in this book all tell a story of contexts that are characterised by uncertainty, a diversity of ambitions and values, sometimes in the context of competing claims on natural resources. In such cases, the representation of problems and solutions can be highly contested. The chapters share an acceptance and appreciation of these diverse perspectives, including those of the communities that experience the lived effects of environmental change. What we learn from this book is that diversity can go hand in hand with reducing power struggles and marginalisation, especially when supportive conditions are created. Chapter 9 by Keith, for example, discusses how groups of farmers are jointly working on water management issues in a Landcare group. This illustrates that when there is a strong sense of interdependence among stakeholders; they can engage in effective concerted or connective action that supports sustainable management of social-ecological systems, even reducing their own use in order to conform to an agreed approach. Key to this is the presence of enabling institutions that create conditions for trust. Accordingly, the authors share a search for, and examination of, ways in which people can be supported in their efforts to take concerted action or connective action that fosters the greater good. Thereby they reconstruct the dominant technocratic; neoliberal narrative into an alternative that enables people to recognise themselves as being part of a larger social-ecological system and thereby encourage the creation of institutions in a society that supports deliberation and negotiation instead of competition. By opening up the dominant, neoliberal narratives in science, the chapters in this book make them debateable and changeable and allow us to imagine and pursue alternative futures.

Third, underlying the paradigms in this book is an ethic of respect and care, for communities and the environment. The foundation is respect for people and acknowledgement of their right to help shape their relationships with others, with institutions and with the environment, based on shared knowledge and social learning. As mentioned, the chapters also show that these practices are not universal but are specific to the local context. The authors of the chapters attend deeply, listen, feel, respect and practise research methodologies that start from a position of 'passionate humility' (Yanow, 2009, p. 582) and honour, care for and support those that they are working with. But the chapters also show that a methodology and ethics of care and respect for community and country is more than 'just' listening deeply. They are all committed to making a difference in one way or another to improve the situation for communities and their environment by supporting the coming into being of new types of interdependencies, supportive institutions and concerted or connective action that improve well-being. This is similar to 'the relational ethics of care' discussed by Alam and Houston (2020). It also aligns with the work of Fisher and Tronto (1990) who define care as: "a species activity that includes everything that we do to maintain, continue, and repair our world so that we can live in it as well as possible. That world includes our bodies, ourselves, and our environment, all of which we seek to interweave in a complex, life-sustaining web" (p. 40). The ethics of care and respect as enacted by the authors of the chapters in this book is attentive to the needs of communities and country and promotes "social relationships of mutuality and trust rather than *one-way* dependence" (Lawson, 2007, p. 3). This is very different to conventional ethics in research on people-environment relationships which expresses an ethics of control by rhetorically using imperialist values such as manipulation and domination as a moral organisational force (Brannelly & Boulton, 2017). In contrast, the chapters in this book share the vision of engaging communities genuinely to meet a variety of perceived needs, for people and for the environment. It leads us to ask ourselves as researchers "how to care in each situation?" (after de la Bellacasa, 2011, p. 100), thereby bringing into being more caring worlds and ways of being.

## Future Directions

Overall, this book advocates continuing convergence across the three themes of:

- fostering and finding creative and locally sensitive means towards participation in rural development;
- continual improvement to contextualised forms of environmental governance, connecting with institutions that nurture rural development; and
- consideration within a systems, and particularly a social-ecological systems, view.

Many directions are possible from here. We suggest some broad ones for particular attention. In all of these, we advocate reciprocity between theory and practice: learning from practice in problem solving and participatory methodologies to evolve theory and use theory to inform continual improvement in practices. In this, we believe firmly in the power of reframing issues in order to tackle them.

We envisage opportunities for new participatory and community-led approaches being developed in diverse local areas then adapted by others. Following the interactions of theory and practice promoted by the contributors to this book, scholars and practitioners can collaborate with local actors to make space for new approaches, to support and sometimes participate in them, and document and disseminate promising social and environmental experiments.

### *Enrich the Paradigm of Social-Ecological Systems*

As systems views and the paradigm of social-ecological systems are so central to our thinking, a logical next step is to continue to enrich—and indeed to challenge—our understandings of these systems. While applying systems thinking can be flexible, a philosophical leaning towards embracing Indigenous perspectives makes it important to reconsider the notion of social-ecological systems as "coupled". As a step forward in western scientific thinking, this does not accord with the ancient wisdom of many Indigenous cultures, which view these as a single system, humans within nature, not coupled with it. Belief systems and institutional arrangements in many Indigenous societies present humans as having reciprocal responsibilities for nature, not rights of domination or control. Whether or not our modest set of fields can provide a niche that leads eventually to transformation in acceptance of Indigenous ways of thinking, theorists and practitioners in our fields could well take a lead here. Incorporating Indigenous worldviews sincerely has many consequences and can be challenging for western practitioners since it entails quite different ways of thinking. It is not sufficient to weigh up knowledge, but to comprehend how each particular aspect of knowledge sits within another's worldview, and in the social systems involved in rights to attain and disseminate knowledge.

The paradigm of social-ecological systems has been strong in its foundations in complex adaptive systems. As social scientists become more involved, it is important to attend to all aspects of humans in "nature": the cognitive and emotive such as mental models and values, identity and sense of place, as well as the cultural, behavioural, and social arrangements (such as those reflected in power and gendered relationships) and behavioural. While institutions continue to be well explored in this field, with excellent contributions from commons theory and practices and recognition of polycentric governance, we should recognise and continue to explore the ways in which interdependencies play out using and changing nature, relating to nature cognitively, and behaving within nature. Some interesting dimensions worthy of more exploration are multi-species relationships (e.g. multi-species ethnography), and specific problem domains such the emerging field of "One Health" (Ross, 2014), an interdisciplinary field which joins human, animal and ecosystem health towards understanding their interactions and building better institutional arrangements and practices for the safety of all three. Further, we need to recognise and build on the ideas that social-ecological systems are not necessarily direct interactions of humans with "nature", although we all ultimately depend on it. They involve built environments (e.g. housing, cities, transport infrastructure), modified environments (e.g. farmland, engineered waterways) and technology (e.g. vehicles, electricity, internet).

We need much more focus on how people co-create these settings, and the ripple effects and implications throughout SES of doing so. Decades ago Amos Rapoport, a mentor of Helen Ross, began to explore the idea of the "house-settlement system" (Rapoport, 1969, 1977), and also the meaning of the built environment (Rapoport, 1990). These and many more ideas about how people relate to their environments, and how and why they construct them as they do, deserve a place in social-ecological systems thinking.

### *Understanding Resilience and Transformation*

A western tendency to dichotomise has become—in our view—dysfunctional, by intellectually separating processes of resilience from transformation, where resilience is about strengthening the capacity to cope and adapt after disturbances, often in a continual learning process. These are about constructive and complex change processes, regardless of how one chooses to label the outcome. The outcome also can take a long time to detect, e.g. many incremental demonstrations of resilience could eventually look like a transformation—it is not an "either-or" relationship. The dichotomy is particularly damaging when resilience is portrayed—erroneously—as "bouncing back". There is no "bounce", indeed the idea can create false expectations of quick recovery, and change to somewhat new states, e.g. "building back better" is more desirable than return to the old. Much more importantly, we need to learn how to foster resilience. What attributes and processes are fundamental in different societies and circumstances? How can community development and participatory approaches be evolved to work with communities to re-shape resilience across multiple systems and spaces, and to co-occurring multiple hazards?

### *Learn from Problem Solving*

Turning to our specific themes, new directions in both rural development and innovations in environmental governance tend to arise from problem solving (for instance Carlsson & Berkes, 2005 observation that co-management is a continuous problem solving process rather than a "fixed state" framed by formal structures). This suggests focusing on problem solving as a route to new strategies, sometimes enabling new forms of governance in the process. Multi-stakeholder collaborative processes, in which parties create new "framings" of problems and look for opportunities to solve the problems together, are a particularly fruitful opportunity for learning.

### *Joining Separated Fields*

We do not wish to restrict evolution in our general domain to the three themes explored in this book. One of Helen Ross's contributions mentioned above has been social science leadership within One Health. As well as being a highly important new problem solving field (remember COVID-19), this field is also one of many opportunities to take a community development approach to solving land use (ecosystem damage), animal (livestock and wildlife) and human well-being together, under a systems perspective. One Health inevitably entails governance and could perhaps learn from the co-management and community-led approaches that have evolved within environmental management. Community development takes a problem-centric approach, so is an excellent starting point for One Health, and many other conjunctions of fields that could benefit environment and society.

### *Conjoined Community and Environmental Development*

Within this sketch of opportunities in systems thinking, social-ecological systems and joining new fields, we see opportunities for conjoined community development that is more explicitly environmental and social, and that goes beyond classic goals of confidence building, social capital and material improvement to focus on developing sustainability and resilience (Cavaye & Ross, 2019). We already see institutional innovations in environmental governance as creating enabling spaces rather than prescribing outcomes, many reflected in this book. Such institutions are more easily adaptive, a necessity in the rapidly changing conditions of our environment and societies, especially under the 'terrible twins' of climate change and biodiversity loss (Nursey-Bray, chapter 12), as well as the rapid exceedance of the 'planetary boundaries' (Rockström et al., 2009; Steffen et al., 2015).

It is especially important in these global to local challenges to build strategic alliances and different forms of social capital for collective action to effect change at multiple levels, scales and across systems. This in turn is necessary in order to develop effective communication and learning and as basis for collaboration.

When working in the evolving spaces of community development, community-based natural resource management, conservation and co-management and enabling institutions, it is important to emphasise that communities are not just place-based or interest-based, but are multi-dimensional, overlapping and dynamic. Meanwhile technology is rapidly changing the nature of communities, people-environmental relationships and ways of communicating. It is changing opportunities for connection among people, vastly expanding 'communities of interest' as well as providing a virtual dimension to local communications and belonging. In conjoined approaches to rural community and environmental development, internet technology has already eased the distance burden and reduced carbon footprint in communications, though with compromises in the rounded experience of face to face interaction. More secure social infrastructure, particularly for remote areas and those less privileged in access to internet, is of course needed to facilitate open interaction, collaboration and social learning among a wider range of people and communities. Meanwhile facilitation of online interaction is a new skill for community developers and others fostering collaboration. The implications of its future development, adaptation and application appear limitless and not yet widely understood.

The need to navigate across multiple levels in conjoined social and environmental rural development suggests further new approaches and development of joint 'visions'. Traditionally, the focus of community development has been on discrete geographic towns or districts. While this local focus aligns with the identity and connection that most people have, there are also major opportunities in applying development processes across multiple communities to foster multi-community collaboration towards new regional potentials. Much can be learnt here from the many collaborative experiments in environmental governance, from their structures to their processes.

Our promotion of a conjoined approach also entails broadening of the goals in rural development from typically piecemeal to much more holistic. Much of community planning and government supported community development has focused on specific goals and actions that are implemented and evaluated. Yet, this can lead to quite prescriptive actions limiting co-creation and empowerment. Community development processes are looking to better orchestrate existing community-based actions and motivation, rather than 'deliver' improvements, albeit identified by citizens. This potential requires much more use of 'untied' grants. Whether from government or the philanthropic sector, funding for community improvement has traditionally been structured to deliver specific outcomes. This 'line of sight' from funds to outcomes is quite understandable, yet restrictive. Communities, and their regions, would be permitted far greater scope for constructive change under flexible development funding that enables them to pursue their priorities more independently, and to take systems views to catalysing change. This now partially applies to every village in Indonesia, under a community empowerment programme (see Dwiastuti this volume).

## Towards Socially-Just Decision-Making

The rapid, profound and irreversible changes occurring in our social-ecological systems and our global biosphere require new knowledge, new approaches and robust decision-making, based on the ethics of care, transparency and inclusion. Social justice is fundamental to achieve sustainability and vital now more than ever. Working through—and joining up—community networks to embrace social learning and build community capacity and agency will be critical to achieve more just and sustainable societies and environments. There is a continual need to develop and share how to do this, in ways that are sensitive to different cultures and contexts. It is hard to contemplate our changed physical environment and society in 20 years' time, but it is essential to prepare our students, our communities and our institutions to be proactive and resilient. Fundamentally, a value underlying the transformation is that *we put communities first*.

## References

Alam, A., & Houston, D. (2020). Rethinking care as alternate infrastructure. *Cities, 100*, 102662.

Apostolopoulou, E., Chatzimentor, A., Maestre-Andrés, S., Requena-i-Mora, M., Pizarro, A., & Bormpoudakis, D. (2021). Reviewing 15 years of research on neoliberal conservation: Towards a decolonial, interdisciplinary, intersectional and community-engaged research agenda. *Geoforum, 124*, 236–256.

Baldwin, C., Marshall, G., Ross, H., Cavaye, J., Stephenson, J., Carter, L., Freeman, C., Curtis, A., & Syme, G. (2019). Hybrid neoliberalism: Implications for sustainable development. *Society and Natural Resources, 32*(5), 566–587. https://doi.org/10.1080/08941920.2018.1556758

Brannelly, T., & Boulton, A. (2017). The ethics of care and transformational research practices in Aotearoa New Zealand. *Qualitative Research, 17*(3), 340–350.

Carlsson, L., & Berkes, F. (2005). Co-management: Concepts and methodological implications. *Journal of Environmental Management, 75*, 65–76. https://doi.org/10.1016/j.jenvman.2004.11.008

Cavaye, J., & Ross, H. (2019). Community resilience and community development: What mutual opportunities arise from interactions between the two concepts? *Community Development, 50*(2), 181–200. https://doi.org/10.1080/15575330.2019.1572634

de La Bellacasa, M. P. (2011). Matters of care in technoscience: Assembling neglected things. *Social Studies of Science, 41*(1), 85–106.

Fisher, B., & Tronto, J. (1990). Toward a feminist theory of caring. In E. Abel & M. Nelson (Eds.), *Circles of care: Work and identity in women's lives* (pp. 35–62). Sunny Press.

Lawson, V. (2007). Geographies of care and responsibility. *Annals of the Association of American Geographers, 97*(1), 1–11.

Rapoport, A. (1969). *House form and culture*. Prentice-Hall.

Rapoport, A. (1977). *Human aspects of urban form: Towards a man-environment approach to urban form and design*. Pergamon Press.

Rapoport, A. (1990). *The meaning of the built environment: A nonverbal communication approach*. University of Arizona Press.

Ravetz, J. R. (2006). Post-normal Science and the complexity of transition towards sustainability. *Ecological Complexity, 3*(4), 275–284.

Rockström, J., Steffen, W., Noone, K., Persson, Å., Chapin, F. S. III, Lambin, E., Lenton, T. M., Scheffer, M., Folke, C., Schellnhuber, H., Nykvist, B., De Wit, C. A., Hughes, T., van der Leeuw, S., Rodhe, H., Sörlin, S., Snyder, P. K., Costanza, R., Svedin, U., Falkenmark, M., Karlberg, L., ... Foley J. (2009). Planetary boundaries: Exploring the safe operating space for humanity. *Ecology and Society, 14*(2), 32. http://www.ecologyandsociety.org/vol14/iss2/art32/

Röling, N., & Maarleveld, M. (1999). Facing strategic narratives: In which we argue interactive effectiveness. *Agriculture and Human Values, 16*(3), 295–308.

Ross, H. (2014). One health from a social-ecological systems perspective: Enriching social and cultural dimension. In J. S. Mackenzie, M. Jeggo, P. S. Daszak, & J. A. Richt (Eds.), *One Health: The human-animal-environment interfaces in emerging infectious diseases* (pp. 217–229). Springer.

Steffen, W., Richardson, K., Rockström, J., Cornell, S. E., Fetzer, I., Bennett, E. M., Biggs, R., Carpenter, S. R., de Vries, W., de Wit, C. A., Folke, C., Gerten, D., Heinke, J., Mace, G. M., Persson, L. M., Ramanathan, V., Reyers, B., & Sörlin, S. (2015). Planetary boundaries: Guiding human development on a changing planet. *Science, 347*(6223). https://doi.org/10.1126/science.1259855

Turnhout, E., Neves, K., & de Lijster, E. (2014). 'Measurementality' in biodiversity governance: Knowledge, transparency, and the intergovernmental science-policy platform on biodiversity and ecosystem services (Ipbes). *Environment and Planning A: Economy and Space, 46*(3), 581–597.

van Bommel, S., & van der Zouwen, M. (2013). Creating scientific narratives: Experiences in constructing and interweaving empirical and theoretical plots. In B. Arts, J. Behagel, S. van Bommel, J. de Koning, & E. Turnhout (Eds.), *Forest and nature governance: A practice-based approach* (pp. 217–239). Springer.

Yanow, D. (2009). Ways of knowing: Passionate humility and reflective practice in research and management. *American Review of Public Administration, 39*(6), 579–601.

# Index






C. Baldwin and S. van Bommel (eds.), *Rural Development for Sustainable Social-ecological Systems*, Palgrave Studies in Environmental Sustainability, https://doi.org/10.1007/978-3-031-34225-7

MIX
Papier | Fördert gute Waldnutzung
FSC® C083411

Kolibri 360 GmbH
Ferdinand-Jühlke-Straße 7
99095 Erfurt, Deutschland
produktsicherheit@kolibri360.de